THE MAN WHO WOULD BE PRESIDENT

As summertime bathed downtown Boston in warm sunshine, Obama led a gaggle of reporters, aides, and a couple of friends—a group occasionally two dozen deep—around a maze of chain-link security fences guarding the large-scale FleetCenter indoor arena. A former high school basketball player who, at forty-two, still relished a pickup game, the rail-thin Obama was carrying his upper body as if he were heading for the free throw line for the game-winning shot, a shot he believed he was destined to sink. His shoulders were pitched backwards. His head was held erect. His blue-suited torso swayed in a side-by-side motion with every pace forward. His enormous confidence appeared at an all-time peak. And for good reason: Hours later, the Illinois state lawmaker and law school lecturer would take his first steps onto the national stage to deliver his now famous 2004 Keynote Address to the Democratic National Convention.

Indeed, Obama's time in the bright sunshine had finally arrived. And though this moment itself had come upon him rather quickly, unexpectedly, and somewhat weirdly, with only two weeks of notice, his opportunity to prove to the world that he could play in this most elite league was at hand. Finally.

OBAMA
From Promise *to* Power

DAVID MENDELL

HARPER

An Imprint of HarperCollinsPublishers

This book was originally published in hardcover August 2007 and trade paperback April 2008 by Amistad, an Imprint of HarperCollins Publishers, and special mass market printing June 2008 by Harper.

HARPER

An Imprint of HarperCollins*Publishers*
10 East 53rd Street
New York, New York 10022-5299

Copyright © 2007 by David Mendell
Front cover photo by Ted Soqui/Corbis; back cover photo by AFP/Getty Images
ISBN 978-0-06-173666-7

First Harper mass market printing: October 2008
First Harper mass market special printing: June 2008
First Amistad trade paperback printing: April 2008
First Amistad hardcover printing: August 2007

HarperCollins® and Harper® are registered trademarks of HarperCollins Publishers.

Printed in the United States of America

Visit Harper paperbacks on the World Wide Web at www.harpercollins.com

10 9 8 7 6 5 4 3 2 1

For my father,
whose enduring example of moral courage
has always shown the way

CONTENTS

ACKNOWLEDGMENTS

As much of America knows by now, Barack Obama has written two books that detail, from his perspective, his first forty-five years of life. During the course of an interview for this book, an Obama friend once asked me, "He hasn't left much for you to write, has he?" Indeed, Obama's own personal story is exhaustively well told—more than eight hundred pages of type.

I strove with this book to add another perspective—various outside perspectives, really—to what Obama has filtered through his personal lens. I also endeavored to fill in the gaps that Obama left open, and to fill them both with my own observations and with astute reflections from some of the many people who make significant appearances throughout Obama's restless existence.

I tried to assess the senator's entire political career up to the point of his presidential candidacy announcement and to show how his formative years, coupled with the currents of the country's political tide, have carried Obama to the crest of this enormous wave. More than anything, however, I wanted to provide a straightforward, honest account of how a talented, exceedingly driven man ascended so quickly from relative anonymity to political superstardom. I tried not to muck up the narrative with too many deep thoughts of my own but to simply carry readers along on this incredible journey that Obama has taken and, to some extent, the wild ride that I unwittingly hopped onto for several years.

There are many people to thank for helping me maintain my balance along the way. My literary agent, Jim Hornfischer, was among the first to see the promise of this project; his steady hand from the first proposal through the final manuscript was invaluable. Stacey Barney, the editor who purchased this book for Amistad, deserves my gratitude for her enthusiasm and foresight. Dawn Davis steered me through some difficult moments as this story unfolded and I am grateful to her. Laura Klynstra designed a simply wonderful book jacket. And Rakesh Satyal, this book's final editor, struck just the right touch of firmness and compassion, but I especially owe him for his patience with a first-time author.

I thank all of those in Obama's universe who allowed me to harvest their thoughts and kept me in the loop. Robert Gibbs, David Axelrod, Dan Shomon, Julian Green, Jim Cauley, Pete Giangreco, Maya Soetoro-Ng, Michelle Obama and, especially, Barack Obama, have my utmost appreciation. Many journalists who have written about Obama enhanced my observations about him, with Jeff Zeleny, Ben Wallace-Wells, Laurie Abraham and Lynn Sweet at the top of that list.

At the *Chicago Tribune,* the number of my colleagues deserving thanks is innumerable. The short list: Hanke Gratteau, Ann Marie Lipinski, George de Lama, Bob Secter, Jim Webb, Mike Tackett, John Chase, Liam Ford, John McCormick, Ray Long, Rick Pearson, Pete Souza, Rick Kogan and Flynn McRoberts and the Cypriot brotherhood. Jim O'Shea, now at the *Los Angeles Times,* gave his blessing to this project, and I thank him for that. Darnell Little deserves special thanks for listening to me drone on, day after day, as I collected my scattered thoughts into something that resembled coherent prose.

Many journalists and others also deserve mention for helping me climb the mountain to this point. Again, the short list: Steve Rohs, John Cole, Steve Bennish, Susan Vinella, Abdon Pallasch, Regina Waldroup, Armelia Jefferson and Jim Bebbington. My good friends Mark Adams, Greg DeSalvo and Dave Doran helped keep me sane; and without Glenn Gamboa, this book would not have happened. Shawn Taylor provided early inspiration and encouragement, and I cannot thank her enough. Finally, the happy innocence of my son, Nathan, lifted me from the abyss at countless desperate moments.

The Ascent

1

I'm LeBron, baby.
—BARACK OBAMA

For those who know Barack Obama well, this might sound close to impossible, but the swagger in his step appeared even cockier than usual on the afternoon of July 27, 2004.

As summertime bathed downtown Boston in warm sunshine, Obama led a gaggle of reporters, aides and a couple of friends—a group occasionally two dozen deep—around a maze of chain-link security fences guarding the large-scale FleetCenter indoor arena. A former high school basketball player who, at forty-two, still relished a pickup game, the rail-thin Obama was carrying his upper body as if he were heading to the free throw line for the game-winning shot, a shot he believed was destined to sink. His shoulders were pitched backward. His head was held erect. His blue-suited torso swayed in a side-by-side motion with every pace forward. His enormous confidence appeared at an all-time peak. And for good reason: hours later, the Illinois state lawmaker and law school lecturer would take his first steps onto the national stage to deliver his now famous 2004 keynote address to the Democratic National Convention.

Indeed, Obama's time in the bright sunshine had arrived. And though this moment had come upon him rather quickly, un-expectedly and somewhat weirdly, with only weeks of notice, his opportunity to prove to the world that he could play in this most elite league was at hand. Finally.

Having covered Obama for the *Chicago Tribune* since the early days of his U.S. Senate candidacy more than nine months before, I had already established a rapport with the state senator, and I was mostly trying to stay out of the way and watch the day unfold, watch the story of Barack Obama unfurl. Still, as a skeptical newspaper reporter, I was not completely convinced that, by day's end, all would come out well. I was still trying to gauge if this strut was something of an act, whether his winning free throw would clang on the rim and bounce away or whether he was on the verge of hitting nothing but net and making a national name for himself.

After Obama and I slipped through a security checkpoint and he momentarily broke free from the entourage, I sidled up to him and told him that he seemed to be impressing many people of influence in this rarefied atmosphere.

Obama, his gaze fixed directly ahead, never broke his stride.

"I'm LeBron, baby," he replied, referring to LeBron James, the phenomenally talented teenager who at the time was shooting the lights out in the National Basketball Association. "I can play on this level. I got some game."

I wasn't so sure. I fell back amid the marching gaggle of the Obama entourage and chatted with one of his closest friends, Marty Nesbitt, who had flown in from Chicago to accompany Obama during the convention week. I asked Nesbitt how he thought his friend would perform that night, given all the media attention and political pressure. "He sat down with Ted Koppel earlier this week and he hit the cover off the ball, didn't he?" Nesbitt asked. "Barack reminds me of a player on my high school basketball team back in Ohio. He could elevate his game to almost any situation. And when we needed a shot, he always hit it. Always."

That evening, Obama introduced himself to America. He delivered a keynote address of historic proportions, so inspiring that even some conservative commentators would concede they were moved by it. His rich baritone voice resolute and clear, he hearkened back to his beloved mother's philosophy of a common humanity, a philosophy that had been ingrained in him throughout his childhood. He declared that America is a land of good-hearted people, a nation of citizens who have more unifying traits than dividing traits, a country of individuals bound by the common purpose of freedom and opportunity for all. "There's not a liberal America and a conservative America—there's the *United States* of America. There's not a black America and white America and Latino America and Asian America—there's the *United States* of America. . . . We are one people. . . ."

Across the arena, many Democrats from various states, various walks of life, various races, had tears in their eyes. And as the woman seated next to me in an upper level of the FleetCenter joyously shrieked—"Oh my god! Oh my god! This is history! This is history!"—I looked around at the energized and emotional crowd and heard myself speak aloud to no one in particular.

"Yes, indeed. Tonight, Barack, you are LeBron, baby."

THROUGHOUT 2004, THE POLITICAL AND CULTURAL MOOD IN Obama's home state of Illinois—and much of the country—was sharply polarized. A bevy of Democratic presidential aspirants had vied to challenge President George W. Bush, who had led the country into war in Afghanistan, and then Iraq, in the wake of the September 11, 2001, terrorist attacks in New York City and Washington. Chagrined from being in the minority in both chambers of the Congress, Democrats desperately craved a strong candidate who could defeat Bush in the November election. Among those Democrats, Senator John Kerry of Massachusetts had won the party's nomination, but despairing Democrats were having difficulty warming up to him. They hungered for something more than Kerry could offer—a political savior, an inspirational figure who could lead them out of one of the darkest periods in their party's history. Kerry surely seemed electable, but his reserved nature and plodding public style made him far from a savior who could stir the souls of the masses.

At this point the nation was evenly divided on the Iraq war, but the Democrats were not. In the eyes of many moderate Democrats who had initially supported the war, the nationalistic fever that had washed over America in the wake of the 9/11 terrorist strike was beginning to wear off. For most left-leaning party members, the war had been nothing short of a colossal mistake from the very beginning. Thousands gathered for anti-war rallies in Chicago, San Francisco and other urban centers. In some cities, pro-war demonstrations attracted partisans on the other end of the spectrum. In Chicago, several anti-war events were largely well-behaved. They were arranged and attended not solely by young radicals but also by established members of the city's lakefront liberal crowd. A rally in the city's Federal Plaza seventeen months earlier, in October 2002, was assembled by an aide to former U.S. senator Paul Simon and a mainstream liberal public relations expert. The rally featured a pointed anti-war speech from Obama, then a fairly anonymous state lawmaker, who deemed the impending Iraq engagement "a dumb war." The event drew old and young alike, as 1960s protest veterans mixed with suburbanites and college students and journalists. Another war protest was more demonstrative, spilling out of the city's downtown Loop area one evening and choking off the main transportation artery of Lake Shore Drive. Protestors marched aimlessly to a destination unknown, before being cornered on the drive and arrested en masse by Chicago police attired in riot gear.

In spring 2003, the *Washington Post,* the daily bulletin board of the D.C. elite, published a story illuminating the nation's growing political and cultural divide. The article explored a burgeoning theory that political scientists informally labeled as the red state–blue state phenomenon. The theory was gaining prominence in both political and intellectual circles. Red states were core Republican Party strongholds, populated primarily by culturally conservative Christian whites. The entire South and most of the Midwest fell into the red category. Blue states were Democratic Party territories, home to larger percentages of gays, minorities and college-educated intellectuals than the nation as a whole. West Coast and Northeastern states generally voted blue. The red-blue notion resonated with so many Americans that it soon took hold as something of a truism.

This tug-of-war was reflected in the nation's capital by a bitterly partisan battle between minority Democrats and majority Republicans. After the fall elections of 2003, the Pew Research Center released a study summarizing eighty thousand interviews over the previous three years that found an evenly split and polarized nation, with sharp differences along political, cultural and religious lines. The country's most religious states were in the South, and their residents tended to be the most socially traditional and the most hawkish on national security issues. People living in New England and on the Pacific Coast, meanwhile, were less religious, more dovish and less socially traditional.

On the night of that Democratic National Convention in Boston, a handsome, caramel-skinned man who seemed to embody all of these disparate parts appeared on television screens throughout the country and called for unity. Barack Obama's mother was a white, middle-class Kansan with a naive, wandering spirit that she would pass along to her son. His father was a poor, black Kenyan whose pioneering trek to the United States for Western schooling represented the immigrant experience. Obama had been raised mostly among whites but also among Polynesians and Asians. He was a devout Christian who had married into a black family from Chicago's South Side. He had been a community organizer in the poorest African-American neighborhoods of Chicago. He had gone to the country's finest law school and excelled. He was then serving in the state legislature in Middle America. And he seemed to be offering himself as the very vision of what America should be—a place where race, class and cultural differences mix together to make a republic whole, not to divide it.

Later that year, his maternal grandmother, Madelyn Dunham, would tell me, "When he was a young man, I asked him what he wanted to do with his life. He said, 'I want to leave the world a better place than when I came in.' And I believe that has been his guiding light."

Obama, without argument, is imbued with an abiding sense of social and economic justice. He is an earnest, thoughtful, occasionally naive man who has a strong sense of moral purpose, a trait driven into him by his ardently progressive mother. But Obama is far more complex than just a crusading dreamer aiming to "give voice to the voiceless and power to the powerless," in his own oft-spoken words. He is an exceptionally gifted politician who, throughout his life, has been able to make people of wildly divergent vantage points see in him exactly what they want to see. "He definitely has this yin and yang quality to him," said Robert Gibbs, one of Obama's top political aides. He is not of the same specific ancestry as most blacks in the United States, nor has he lived the typical black experience in America. Yet he is accepted by most as a brother, in large part because his physical appearance is decidedly African and his wife and children are African American. He was raised by a white family and educated in elite white institutions, giving him nonthreatening appeal and instant credibility with the white cognoscenti. His parents and grandparents were of modest means, and for most of his life he has not had excessive material wealth, making him aware of the daily concerns of the middle class mostly because they have been his concerns.

But it is his easygoing public temperament and ingenious lack of specificity that perhaps have most abetted his career in politics. Whatever setting Obama steps into—a black church, the Senate floor, a rural farmhouse—he blends comfortably into the atmosphere, as if he has spent a lifetime there. With his relentlessly reasonable tone and a studied thoughtfulness, he can turn even the most jaded journalist into a mild fan. He emanates supreme confidence at almost every moment. But he also has a self-deprecating sense of humor and can express humility to an almost unnecessary degree, at least in public. While talking or writing about a deeply controversial subject, he considers all points of view before cautiously giving his often risk-averse assessment, an opinion that often appears so universal that people of various viewpoints would consider it their own. In settings with everyday voters, he dons his college lecturer hat and explains arcane policy matters in easy-to-understand terms that invite unanimity, not argument.

What the public has yet to see clearly is his hidden side: his imperious, mercurial, self-righteous and sometimes prickly nature, each quality exacerbated by the enormous career pressures that he has inflicted upon himself. He can be cold and short with reporters who he believes have given him unfair coverage. He is an extraordinarily ambitious, competitive man with persuasive charm and a career reach that seems to have no bounds. He is, in fact, a man of raw ambition so powerful that even he is still coming to terms with its full force. This drive is rooted in an effort to atone for his absent father's tragic failures, both as a Kenyan politician and as a family man. "He's always wanted to be president," Valerie Jarrett, a close friend of Obama's, would confide shortly after his Boston speech. "And I'm not sure that he's even still fully admitted it to himself that he does, but I know he does. I know he does."

The journey toward that admission, finally arriving while he vacationed in his native Hawaii in December 2006, would be unlike any other journey by an American politician. Indeed, for all of Obama's intellectual heft, for all of his genuine sense of mission, for all of his aching desire for personal success, no man could be fully prepared for what lay ahead of him after that heady star turn in Boston. And no man could be left unaffected by it. In just a couple of years, he would rise from obscure state lawmaker to national celebrity pursued by paparazzi on his family vacation. He would struggle through a self-described "painful year" of just three or four hours of sleep per night in order to write a best-selling book that would assure his family's financial security for his lifetime and nurture his burgeoning political career. He would spend many weekends and even more weeknights away from his devoted wife and two young daughters in helping the national Democratic Party raise millions of dollars to retake control of Congress. He would be discussed endlessly in the mainstream and alternative media as potentially the first African American to hold the Oval Office.

From Chicago's South Side, an entirely new and unique politician had emerged on the American scene. Obama and his mother's hopeful message of inclusion and brotherly compassion would land squarely in the middle of a massive political void. In a nation that perpetually seeks to anoint the next New Thing, Obama would bask for months in a remarkably intense media spotlight.

Over those two years, this new brand of politician would do the unthinkable: he would transcend race as he embodied it; he would throw an outstretched hand to conservatives as he enchanted progressives; he would beguile hard-bitten national political reporters (at least some of them, at least for a while) with his newness, intellect and affability; he would become a prideful and iconic symbol for millions of black Americans; and he would secure his role as a major national voice for Democrats. Obama would take the optimistic vision instilled in him by his mother—that, at our essence, all humans are connected by compassion and generosity of spirit—and mold it into a compelling political theme. He would offer this theme of reconciliation to a politically and culturally splintered republic as a form of national salvation.

Galloping onto the national stage, Obama would become a source of hope and optimism for disillusioned Democrats from California to New Hampshire. Obama's come-from-nowhere ascent would make it starkly evident just how passionately many Americans

yearned for an inspirational leader who could mend the various divisions within the country—racial, political, cultural, spiritual. "Hours
before he gave the speech, Democrats were excited," *Chicago Tribune*
columnist Clarence Page said. "You know why? Because they finally
got a black face for the party who's not Jesse Jackson or Al Sharpton.
Let's be frank. That's how this thing got launched. It went beyond
party, because the whole country right now is looking for that kind
of a come-together kind of feeling." Movements to draft him to run
for the presidency in 2008 would take hold on the Internet, among
Hollywood celebrities and on college campuses across the country.
As a faraway war claimed thousands of American lives and descended
into bloody chaos, as an inordinate number of power and finance
scandals engulfed Congress, as Americans grew more disenchanted
with the policies and leadership of Republican president George
W. Bush, America seemed ready for a shepherd to lead the country
in an entirely new direction.

Not since the days of Jack and Bobby Kennedy, and their lu-
minous political Camelot, had a politician captured so quickly the
imagination of such a broad array of Americans, especially the significant voting bloc of black Americans. And even the Kennedy
comparison would not characterize Obama's fame properly. Not
since Ronald Reagan had a politician been so adept at sharing his
own unwavering optimism with a disheartened electorate. Cunningly using the broad power of the modern media as his launching
pad, Obama and his small team of skilled advisers would plot a
course that catapulted him from little-known state lawmaker to
best-selling author to U.S. senator to national celebrity. A mixture
of idealist and pragmatist, Obama would move almost overnight
from a critic of the established political system inside the Beltway to
a player within that system. He would represent both outsider and
insider.

Wherever he went, Obama would draw impassioned crowds in
the thousands. Throughout 2006, Obama would grace the covers of

national magazines and the front pages of national newspapers. Like Obama's mother, who reinforced his self-esteem constantly, nearly all these journalists saw something special in him. "It's like nothing I've ever seen before," said Julian Green, a former aide. "We actually have fans among the media. I've never run across that for any other politician." "Dreaming of Obama" and "Great Expectations" and "Why Barack Obama Could Be the Next President"—these headlines would feed his growing legend. Among politicians, Obama would be in the highest demand for television's endless talk circuit, exchanging serious discourse with Charlie Rose, cracking wise with Jay Leno and opening his heart (relatively) to Oprah Winfrey. He received both an NAACP Image Award and a Grammy, the latter for a voice recording of his first memoir. In a society that worships celebrity, no political figure in the country could come close to projecting such megastar wattage. "We originally scheduled the Rolling Stones for this party," the New Hampshire governor, John Lynch, told fifteen hundred Democrats who had shelled out twenty-five dollars apiece to see Obama in December 2006. "But we canceled them when we realized Senator Obama would sell more tickets."

Internally and certainly politically, Obama would embrace this adulation. But he would also struggle mightily with its deleterious effects. As his fame spread worldwide and his public life roared ahead at a ferocious pace, his familiar world collapsed on him and he became more confined to a low-oxygen celebrity bubble. "I can't, for example, walk down the street by myself and watch people go by anymore, and that's a very difficult thing to accept," he lamented. He became more cautious of his public comments and his public image. He carefully restricted access to reporters. A highly active legislator in Illinois who sponsored and passed an assortment of bills, he now steered clear of any single controversial issue. He grew ever more reliant on key aides, family and others close to him—and these people would grow more protective of him. "For us, his family, he hasn't changed," his Kenyan-born half sister, Auma Obama, said.

"But the people around him have changed. I feel the vulnerability in him and I see him being more guarded than he ever used to be." His small cadre of dedicated advisers, who had attached their own careers to Obama's soaring rocket ship, would feel this intense pressure as well. Said David Axelrod, his chief media strategist: "It's like you are carrying this priceless porcelain vase through a crowd of people and you don't want to be the guy who drops it and breaks it."

Intimate aides like Axelrod would question privately whether Obama's warp-speed ascension had gone too fast, whether the earnest, occasionally thin-skinned Obama was prepared for the unyielding rigors of a presidential contest, even as that contest already seemed to begin with urgency inside the media and among activists. "David always worries about a meltdown," a confidante of Axelrod's told me. And yet every time that private worry had been raised previously, Obama would confidently step onto the court and hit that key shot. He did that again in early February 2007, when his final conclusion about running for the presidency was rendered with rich political pageantry and yet another speech that electrified supporters and earned praise from analysts.

After months of agonizing indecision, on a bone-chilling winter's morning in Springfield, Illinois, Obama spoke directly to America's yearning for new leadership and offered himself for that role. Before a crowd of nearly seventeen thousand shivering people gathered in the shadow of Illinois's Old State Capitol, Obama announced his candidacy for president. He drew on the powerful historic symbolism of where he stood—outside the building where Abraham Lincoln served as an Illinois lawmaker, the place where Lincoln delivered his famous "House Divided" antislavery speech in 1858. He reasserted that America was ready for a new generation of leadership, ready to withdraw from the war in Iraq, ready to be united. He merged the themes of Lincoln's long-ago call for a unified citizenry and his own mother's most endearing characteristic— seeing the best in people, rather than the worst. "Let us transform this nation," he said through a confident voice and a clenched jaw. "It was here we learned to disagree without being disagreeable— that it's possible to compromise so long as you know those principles that can never be compromised; and that so long as we're willing to listen to each other, we can assume the best in people instead of the worst. . . . The life of a tall, gangly, self-made Springfield lawyer tells us that a different future is possible. He tells us that there is power in words. He tells us that there is power in conviction, that beneath all the differences of race and region, faith and station, we are one people. He tells us that there is power in hope."

For Obama's growing legion of followers, this was the moment they had been waiting to experience. For them, it didn't seem to matter that since the aggressively liberal state lawmaker had gone to Washington he had taken a dramatic turn toward calculation and caution, or that he had yet to propose anything philosophically new, or that Obama was, in his own words, "a blank screen on which people of vastly different political stripes project their own views," or that the higher he soared, the more this politician spoke in well-worn platitudes and the more he offered warm, feel-good sentiments lacking a precise framework. It also didn't seem to matter that in his two years in the minority party in the U.S. Senate, he had the clout to pass only one substantial piece of legislation or that he avoided conflict at all costs, spending none of his heavily amassed political capital on even a single controversial issue he believed in. Indeed, through his first year in the Senate, he had to argue with his cautious political advisers to speak out, however carefully, on a topic dear to him—the impact of Hurricane Katrina and its racial and economic ramifications.

And yet his irreducible confidence, his undeniable intellect and, not least, his compelling biracial life story had transformed Obama into America's most alluring new political face. To voters of various political persuasions, perhaps, just perhaps, here was the antidote to what ailed their fractured national leadership. Here was a pragmatic politician who also possessed a capacity to dream, who spoke convincingly of a better tomorrow at a time when Americans were profoundly worried about that tomorrow. Here, at last, seemed to be the magnetic leader who conveyed the perfect blend of confidence, character and, in his own words, hope. "People don't come to Obama for what he's done," said Bruce Reed, president of the Democratic Leadership Council, a group devoted to centrist policies. "They come because of what they hope he can be."

Whether by design or destiny, ambition or purpose, Barack Obama had climbed aboard the ride of his life. And he seemed determined to take America with him.

BUT AS OBAMA STRUCK OUT FOR THE DEMOCRATIC PARTY'S 2008 nomination for the U.S. presidency, large questions lingered in the public domain: Exactly how had Obama moved this far, this fast— and was it too fast? Was he a man of substance or of media hyperbole? Did he have the experience and toughness to inhabit the White House? Was his mixed racial ancestry a political hindrance, a political asset, or, in a country still confused about race, was it both? Our media can invest people with power in the blink of an eye, but was this particular investment wise? Could this young senator with an idealistic message of inclusiveness survive the boiling cauldron of presidential politics, or would Obama fall victim to his own burning ambition? Would the mercurial media turn on the mercurial Obama?

And most of all, even though a solid section of the voting public was enthralled with Obama, would the rest of America embrace his message and entrust this newcomer with the world's most important leadership role?

CHAPTER

2

Dreams from His Mother

I know that she was the kindest, most generous spirit I have ever known, and that what is best in me I owe to her.
—BARACK OBAMA ABOUT HIS MOTHER, ANN DUNHAM

Barack Obama prepared for elective office by moving to Chicago to work as a community organizer in his mid-twenties, obtaining a Harvard Law School degree in his late twenties and authoring an exhaustive personal memoir in his early thirties. His book, originally published in 1995 by a division of Random House, was called *Dreams from My Father: A Story of Race and Inheritance.* As the title suggests, the book chronicled Obama's life in relation to his East African father, emphasizing Obama's search to find his own racial and spiritual identity amid America's divisive racial spectrum. He finally reached a comfortable place after undertaking a thorough review of his mother's and her parents' journey from the midwestern United States to Hawaii, and then deeply exploring the life and ancestry of his father, who left Obama and his mother when he was two years old.

This wasn't the book Obama originally sold to his literary agent and publisher. He had pitched them a work about his experiences as the first African-American president of the prestigious *Harvard Law Review.* After all, at the time, Obama was a modest thirty-three years old, and his *Law Review* presidency was his only claim to any modicum of fame. Besides, it might have seemed a bit presumptuous to try to sell a life's memoir at the age of thirty-three. Nevertheless, when Obama began writing, an autobiographical memoir poured forth.

Upon its release in 1995, the book sold a few thousand copies, generated mostly positive reviews—although there were a few mixed ones (one critic considered it overwrought and self-indulgent)—and then it faded into obscurity. For years after its publication, the book was difficult to find for those few who tried, with copies hidden in corners of small independent or Afrocentric bookstores. By the early 2000s, with the advent of Internet websites that sold used merchandise in a worldwide flea market, a used paperback copy could be picked up for as little as four or five dollars, but it still was not a hot seller by any measure. That changed dramatically when Obama shot to national fame in 2004 after his keynote address at the Democratic National Convention. Random House quickly ran off several new printings, promoted it vigorously, and the book landed on best-seller lists, where it remained for dozens of weeks, giving Obama the first shot of financial wealth in his life.

Obama, who began reading voraciously in college, had harbored some thoughts of writing fiction as an avocation, although it's an open question whether he seriously considered fiction writing as a full-time profession. Obama himself said he never dabbled in fiction, but others dispute that. When I asked him during the course of his U.S. Senate primary campaign to name his favorite author, he cited E. L. Doctorow, the critically acclaimed novelist and outspoken political liberal. The next day, during a phone conversation on a different matter, he made it a point to say that he wanted to change his answer—to William Shakespeare. (It's probably safe to say that Mr. Doctorow would not feel slighted.) Some politicians are infamous among reporters for casually mentioning a high-minded work that is currently on their nightstand in order to give the impression of being a deep thinker. Even so, it is difficult to imagine most politicians digesting the heavy works of Shakespeare before extinguishing the bedroom light. Yet Obama's erudite nature and his own ambitious writings made that answer seem quite plausible. Jerry Kellman, the community organizer who brought Obama to Chicago to help poor blacks on the Far South Side retrain for jobs, said Obama possessed a fertile, introspective mind that wandered from scene to scene and place to place—a rare trait that often lends itself to good fiction writing. In short, Obama was a dreamer. Of Obama's fiction writing, Kellman said, "He wasn't really talking about it as a career, for that is a whole different animal. He was talking about it as a muse kind of thing—the arts and exploring emotions and that kind of stuff." Beginning in his college years in New York City, Obama began to chronicle the day-to-day events of his life on a pad of lined white paper that he toted around. The notebook would be filled with word sketches of everyday occurrences, conflicting emotions and personal observations of the various people who passed through his world. Later, he would upgrade that notebook, wrapping it in a more professional-looking leather-bound folder. It was from those early handwritten pages that he harvested *Dreams from My Father*.

To a great extent, the narrative is the story of self-discovery, of a young man in the American middle class ambiguously tethered to an unknown family tree on a faraway continent. On his quest to meet and learn about these blood relatives, Obama also seeks a sense of belonging in his home society, a country still riven by racial, cultural and economic divisions. This might be an interesting but not extraordinary tale for the many people who fit neatly into a human demographic. But Obama's mixed racial ancestry of black and white placed him in a different category, straddling two cultures and two races that, in the United States especially, often collide. The fact that he grew up mostly in Hawaii, where there were few blacks and many people of Asian ancestry, added to Obama's feelings of racial isolation. Indeed, in the book's introduction, Obama conceded that he could not deliver the story of the typical African-American experience, which is often marked by deprivation of financial, educational or other resources. This was to the chagrin of his project's promoters. Obama, again in the introduction, explained that one Manhattan publisher, who presumably rejected his book proposal, once told him, "After all, you don't come from an underprivileged background." Although his childhood was unique to his circumstances, Obama did claim a long-standing genetic link with African Americans. He followed that quote with this: "That I can embrace my black brothers and sisters, whether in this country or in Africa, and affirm a common ancestry without pretending to speak to, or for, all our various struggles—is part of what this book is all about."

The book's chief message is consistent with Obama's overriding political theme of optimism and multiculturalism—hope in the face of despair, hope in the face of centuries of struggle, harmony among people of all races and cultures and kinds of families. The first paperback cover for *Dreams* captured this optimism with a bright color photograph of a broadly smiling Obama on a visit to Kenya. The photo was taken by his half sister Auma, with whom he would ultimately form a close association. He is at his father's small farming compound and seated beside his paternal grandmother, Sarah Onyango Obama, who is attired in white head scarf and traditional Kenyan dress. She is smiling just as broadly as she lovingly caresses Obama's neatly trimmed Afro with the back of her hand.

The remarkably candid memoir is much rawer than the typical book from a politician, no doubt because Obama wrote it before firmly deciding to run for public office. It purposefully tracks the thoughts and movements of a young black American male with more frankness than a political consultant would probably advise—the teenage parties, the pursuit of young women, alcohol and drug use, anger at the white establishment, questioning of organized religion.

The book opens with an anecdote describing an abrupt telephone call Obama had received from a relative in Kenya who informed him that his father, a notoriously poor driver, has been killed in an automobile accident. At the time, Obama is twenty-one and living in a spare Manhattan apartment while he attends Columbia University. He is shocked by the call, and is utterly clueless about how to respond emotionally to this news, chiefly because his father had left Obama's family to attend Harvard University when Obama was a toddler. Obama then launches into a rich, chronological narrative of his life, beginning with his childhood in Hawaii and Indonesia and concluding shortly before he goes into politics by running for the Illinois State Senate. Obama states in the introduction that he has melded real people from his life into fictional characters and inserted imprecise dialogue in order to move along the narrative and protect the identities of certain individuals. These literary devices are not uncommon in personal memoirs, but over his political career they would become a source of scrutiny for journalists scouring Obama's past for possible exaggerations or outright mendacity.

Dreams from My Father would prove an invaluable resource for political reporters, as well as a selling point for Obama when he ran for public office. In covering his Senate campaign for the *Chicago Tribune*, I would consistently ask Obama questions about his life. When he would grow weary of these personal inquiries, which was often, he would brush them aside by referring me to his book for answers. "I wrote four hundred pages about myself," he would say. "What more could you want from me?" When I explained that my job was to obtain fresh anecdotes and quotes, as well as to make sure that the story in his book checked out, he would wave me off and say he understood. But his perpetually furrowed brow, together with his imperious manner, gave the appearance of being personally offended that I would dare question his authorial and personal integrity. Still, he would concede what some critics asserted: that the book was too long. "I probably should have trimmed it by fifty or a hundred pages," he confided.

Obama's portrayal of his childhood (when he was known as "Barry," like his father, to better fit into American culture) was rife with vivid stories of his life in Honolulu, where he was born, and poverty-ridden Jakarta, where he lived for several years, leading up to adolescence, by which time he had returned to Hawaii. He also offered rather brief but descriptive portraits of the primary caregivers who molded his character—his mother and her parents. And he recounted his struggles adapting to the African-American community of the United States, as well as his frustrations as a community organizer in an impoverished neighborhood on Chicago's Far South Side. But the book's essence is Obama's search for a heritage that was a mystery to him as a child and adolescent—the story of his once estranged and now deceased father, a gifted African politician whose personal demons prohibited him from fulfilling his great early promise. Partially because one of his parents virtually abandoned him to the other, and then that primary parent led a searching, peripatetic existence, Obama often told close friends that he grew up feeling "like an orphan." His wife, Michelle, and close friends would later speculate that his isolated childhood and parental loss had played a significant role in feeding his desire for public attention.

So it comes as no real surprise that Obama wanted to investigate his paternal heritage and gain an understanding of why his father disappeared. The book reaches its apogee when a sobbing Obama falls to the ground between the graves of his paternal grandfather and his father in rural Kenya, his yearning questions about his absent father answered, their ghosts, which had haunted Obama, finally laid to rest. "When my tears were finally spent, I felt a calmness wash over me," Obama wrote. "I felt the circle finally close. I realized that who I was, what I cared about, was no longer just a matter of intellect or obligation, no longer a construct of words. I saw that

my life in America—the black life, the white life, the sense of abandonment I'd felt as a boy, the frustration and hope I'd witnessed in
Chicago—all of it was connected with this small plot of earth an
ocean away, connected by more than the accident of a name or the
color of my skin."

WITH OBAMA'S ILLUSTRATIVE EXPLORATION INTO HIS FATHER'S
heritage an open book—literally—I had a wealth of information
about this aspect of his life. But his wife, Michelle, advised me that to
truly understand her husband, it was necessary to visit Hawaii. No
matter how much Obama had philosophized in print about his Kenyan father, she told me, that Pacific island held even more answers to
Obama's complex persona. "There's still a great deal of Hawaii in
Barack," she said. "You can't really understand Barack until you understand Hawaii." In fact, the Obamas still make an annual sojourn
to Honolulu every Christmas season, sometimes inviting close friends
to join them. The trip is cemented into Obama's schedule, and nothing has ever dislodged it, not even his hectic political campaigns.
Hearing this, I convinced my *Tribune* editors that a trip to Honolulu
was essential in order to write a newspaper profile of Obama in the
weeks leading up to the fall 2004 Senate election in Illinois.

At the time of my visit, Hawaii's population and tourist industry
had grown considerably since Obama's youth, but the essence of the
islands' mix of various Asian, Polynesian and Western cultures had
persevered, as had their tropical serenity. Obama and his campaign
staff decided to make his family available to me, most likely because
Obama was a candidate for the highest legislative body in the land
and the *Tribune* was the largest and most influential newspaper in
Illinois. Turning down the *Tribune*'s request for family interviews
would not seem a wise political decision at this point in his Senate
campaign. However, Obama's top aides must have been wary about
what I would turn up. After many discussions among those aides,

they elected to send a deputy press aide, Nora Moreno Cargie, to track my reporting and monitor the content of my interviews. No reporter would be thrilled by this idea, and I certainly wasn't. I balked mildly but realized that if I wanted access to Obama's family, I had no choice but to acquiesce to Moreno Cargie's presence.

As it turned out, she was far from obstructive, and she even became helpful in drawing out some sources. She had worked as a researcher for National Public Radio and had a habit of asking questions herself. She also brought a woman's perspective to my discoveries. And like much of Obama's campaign staff, she also had a great curiosity about her employer. So with Moreno Cargie in tow, Obama's other half sister, Maya Soetoro-Ng, a high school history teacher in her mid-thirties, provided a tour of Obama's favorite spots on the island, as well as places where seminal moments of her brother's childhood transpired. Obama's grandmother, Madelyn Dunham, then ailing with various maladies, reluctantly agreed to a forty-five-minute interview. I also toured the Punahou School, the private academy where Obama studied, and interviewed teachers, coaches and childhood friends and their parents.

Meeting his grandmother, in particular, opened my eyes to Obama's formative years in a way that I hadn't foreseen. I initially had thought that Obama gained his practical and pragmatic side from studying his father's life. But the interviews revealed that it surely came from Madelyn, who was his primary caregiver while his mother traveled the globe studying other cultures.

Madelyn Dunham was known in the family as "Toot," which is short for "Tutu," meaning "grandparent" in Hawaiian. She still lived in the same modest apartment where Obama was raised, across an inland waterway from the teeming tourist area of Waikiki Beach. Maya had moved back to Hawaii from New York City just a couple of years before, in part to look after her grandmother. She lived in an apartment a few floors below Madelyn's unit, which was on the tenth floor of the twelve-story building in an urban section of Honolulu. The structure was rather prosaic in its beachfrontlike design. It was made of white concrete and shaped like a rectangle plopped on a short side. Triangular arches, the only real decorative treatment, rose slightly above the entrance on the first floor, and balconies just large enough for a few chairs adorned every apartment, their railings the overriding architectural feature of the building. Glass sliding doors opened to each of the balconies. The interiors of the apartments were as modest as the exterior suggested, with each unit resembling a beachfront condominium. Some families visiting Hawaii undoubtedly stay in bigger oceanfront condos than this apartment where Obama grew up.

Madelyn clearly was not one for frills. The walls of her unit were bleach-white with only a couple of pieces of artwork. A short hallway to the right led to two small bedrooms, one of which was Obama's as a child. To the left, the apartment opened into a kitchen area and then to a medium-sized living room, with the glass doors leading to the adjoining small balcony at the far end. The most prominent ornament in the room sat on a bookshelf beside the television—a handsome family portrait of Obama, Michelle and their two young daughters, Sasha and Malia.

Moreno Cargie and I assumed seats at opposite ends of the living room couch as Madelyn, a small woman shrunken even more by a perpetual stoop, dropped into a chair. At nearly eighty-two, her age was surely getting the best of her physically, but her mind seemed taut and clear. Her black hair now was thinning and graying, and a yellowish streak ran through the middle of it, the result of years of cigarette smoking. She confessed that she still smoked too much. Her family said she was also still a consistent drinker.

Throughout our interview, Madelyn left no question that she was every bit as pragmatic and self-assured as she had been described. This was a woman who had no college education and began her professional life as a bank secretary, yet retired as the bank's vice president. She said she retired, in part, because her colleagues and superiors were pushing her to learn how to work with computers. "New-fangled gadgets that are running the world," she said disdainfully. I tried my best to put her at ease, but her somewhat brusque demeanor indicated that she was dubious of my intentions. She eyed my small digital voice recorder with deep suspicion. "It's one of those new-fangled gadgets," I explained to her with a smile, an expression she did not return.

While Maya was gracious and warm, Madelyn was cautious and protective. This interview was clearly a chore, and she had obviously agreed to it only at her grandson's urging. Even though Obama was now a rising star in the national Democratic Party, she told me that she wished her grandson had entered a more esteemed profession than politics after obtaining his Harvard Law School degree. "International law or something like that," she said. When I suggested that his Harvard credentials would allow him to move back easily into the legal profession if his political career grew unsatisfying or went awry—he perhaps would even be offered a federal judgeship—she replied: "The Supreme Court would be all right." Her matter-of-fact manner about suggesting the Supreme Court for her grandson made me chuckle. After all, on this day, Obama surely looked as if he was headed to the U.S. Senate, but he was still just an Illinois state lawmaker. I wasn't sure if this was an example of naïveté or moxie, and I eventually settled on a mixture of the two. After a half hour's worth of questions, Madelyn announced that she was tired, and effectively ended the interview. As we stepped toward the door, she grabbed my arm gently and gave me a motherly instruction: "Be kind to my grandson." To which I answered: "It's not really my job to show him kindness, but to be as impartial and accurate as I can." She responded with typical pragmatism and brevity. "Oh, I know," she said.

DURING THIS TRIP, AND ESPECIALLY DURING MY INTERVIEW WITH Madelyn, it grew clearer that, even if the spine of Obama's written narrative involved his absent father, it was Obama's mother who played the most vital role in shaping the crux of his character. (His mother's full given name was Stanley Ann Dunham because her father wanted a boy, but she understandably led her life as "Ann." The name Stanley was "one of Gramps's less judicious ideas," Obama wrote in *Dreams*.) Indeed, the extreme importance of his mother's part in forming Obama's character perhaps didn't even strike Obama himself until after she passed away in the mid-1990s from ovarian cancer at the age of fifty-three.

"His mother was an Adlai Stevenson liberal," said Madelyn, who is nonideological herself, although prone to voting Republican. "And he got a heavy dose of her thinking, you know, as a youngster." Obama told me that his mother's influence was ever-present in his life. And it is apparent from private and public conversations with him that he set his moral compass not only from his readings of Mahatma Gandhi, Martin Luther King Jr. and the Bible but from his mother's guidance. Said Obama: "It's always hard to talk about your mother in any kind of an objective way. I mean, she was just a very sweet person. She just loved her kids to death. And you know, [she] was one of these parents who, you know, was the opposite of remote, was always very present and would be your biggest cheerleader and your best friend and had sort of complete confidence in the fact that you were special in some fashion. And so, as a consequence, there was no shortage of self-esteem." He once told a grassroots women's group: "Everything that is good about me, I think I got from her."

In an updated preface to *Dreams,* for its 2004 edition, Obama wrote a brief tribute to his mother: "I think sometimes that had I known she would not survive her illness, I might have written a different book—less a meditation on the absent parent, more a celebration of the one who was the single constant in my life. . . . I know that she was the kindest, most generous spirit I have ever known, and that what is best in me I owe to her."

* * *

ANN DUNHAM WAS THE ONLY CHILD OF A DEPRESSION-ERA COUPLE originally from Witchita, Kansas. Her father, a lifelong wanderer perpetually unhappy in his paid occupation of selling furniture, was an unceasingly restless and curious soul, consistently moving his family throughout Ann's childhood—from Kansas to Berkeley, California; then back to Kansas; then through some small Texas towns; then to Seattle for Ann's high school years. "It was this desire of his to obliterate the past," Obama wrote, "his confidence in the possibility of making the world from whole cloth that proved to be his most lasting patrimony." After Ann graduated from high school, her father accepted a job in Hawaii and took his family to their final destination far in the Pacific, the city of Honolulu. There, Ann enrolled at the University of Hawaii at Manoa. During all these childhood moves, and especially through middle school and high school, Obama's mother sought comfort in the same place: books. She loved to read, a love that she would pass along to her only son. Academically precocious, Ann would sit quietly and immerse herself in studies of foreign cultures and deep works of philosophy. She was so gifted academically that, while still in high school, she was offered early admission to the University of Chicago, but her father forbade her to attend because he felt she was too young to be on her own. "She was extremely brilliant. She read a lot, a very great deal at a very young age," Madelyn Dunham recalled. "She had very advanced ideas. . . . She was into all of these heavy philosophers by the time she was sixteen."

By spending so much time alone with books, Ann could be socially awkward and innocent about the world's less altruistic ways. She carried those qualities well into adulthood. "I remember one time when we went out to a play or concert or something," Obama's sister Maya recalled. "And right across the street was a TGIF [restaurant] and we decided to slip in and get some drinks. And the [waitress] asks what we want and so Mom looks at the menu and she says, 'I think I'd like a lemonade.' And the waitress says, 'Would you like that virgin?' And my mother looks and says, 'Uh, no, I'd like a little sugar in that.'" Maya laughed at this memory. "It's so funny because here's this woman who traveled around the world and yet was not so worldly," she added. "I mean, it was so sweet." Obama wrote that "this running strain of innocence, an innocence that seems almost unimaginable," was the most striking quality of his family.

Ann Dunham was sweet and kind-hearted, yes. But perhaps more than anything, she was a dreamer, an idealist who refused to see the flaws in humankind even as they were strewn before her. She shared a wildly romantic streak with her father, another trait that Ann undoubtedly handed down to her son. "Her feet never touched the earth," the no-nonsense Madelyn Dunham said, in such a way that made it seem as if she had spent much time trying, and failing, to pull her daughter back to the ground. Ann had her own mind. Her intense immersion in the works of brilliant thinkers and her own intellectualism shaped her into a nonconformist. Though Obama described his grandparents as "vaguely liberal," his mother's political beliefs were precise. She was a "secular humanist" and an avowed New Deal, Peace Corps–loving liberal. In choosing the study of cultural anthropology and marrying an African student, she showed that she had a great attraction to other cultures, and not just to study them but to live among them. Maya's doll collections as a child reflected Ann's United Nations view of the world. It contained representations of various races and nationalities—a black doll, a Chinese doll with braided hair, even an Eskimo doll. "It wasn't at all politically correct," Maya said, again with a laugh.

When I first inquired about his mother, Obama slipped into a

softer and more somber voice. He seemed to choose his words
even more carefully than when talking about an important policy
position that could have significant repercussions on his career or
the national debate. Obama said his mother's extreme idealism—
her continued ability to see the good in people, even when they
failed to live up to her lofty ideals—was the quality that he most
admired. It is also the central message that he imparts in his political speeches—that all of us are bound together as one, and if we
are to prosper as a country and, indeed, as a species, that we must
focus on the good we see in others. In fact, he would invoke similar language in his presidential announcement speech. "With her
friends, with colleagues at work, even with her ex-husbands, she
was always very generous with her estimation of people and her
willingness to see the best side of them," Obama said of his mother.
"It is a value that I care deeply about because I saw not only was it
how she operated, but I also saw the good effect it had on other
people."

Ann had a penchant for surprising her parents about major life
choices that often deviated from societal norms. "She was not exactly conforming from the time she was born," Madelyn said. "You
know, most children are stubborn. Or at least the ones in my family
are. . . . I mean, you adjust to a child as she grows, and she was
sometimes startling." When I asked exactly how she was startling,
Madelyn did not hesitate: "She married Barry's father."

Ann Dunham met a Kenyan foreign exchange student named
Barack Hussein Obama in a Russian language class at the University of Hawaii. They fell in love, he just twenty-three and she just
eighteen. Sometime in late 1960, the two slipped off alone to the island of Maui and apparently married. Obama later confessed that he
never searched for the government documents on the marriage, although Madelyn insisted they were legally married. On August 4,
1961, Ann gave birth to Barack Hussein Obama Jr.

Obama's father had been living in a nondescript concrete dormitory building just inside the campus while he and Ann dated. ("I assume that's where my brother was conceived," Maya said with a chuckle, as we both looked up at the dorm building through a brilliant Hawaiian sun.) After the marriage and his son's birth, Obama's father moved his new family into a small, one-story white house situated not far down a hilly, narrow road from the university and across from a small park. When Obama was two years old, his father won a scholarship to study at Harvard but did not have the money to take his family with him. He accepted the scholarship and never returned to the family, leaving toddler Barack in Hawaii in his mother's care.

In his portrayal, Obama initially offers a romanticized view of his parents' courtship and marriage, a story of love and understanding that was recounted to him again and again by his mother and her father, who have both since died. He wrote that his grandparents, after initial wariness, accepted the union quite naturally and that his grandfather, who considered himself extremely enlightened and something of a Bohemian, viewed the interracial relationship with a sense of pride. In 1960, interracial marriage in the United States was rare, especially between blacks and whites. In fact, more than half the states still considered miscegenation a felony, even if those laws were rarely enforced. The civil rights movement was in its embryonic stages. When Obama was in his early twenties, his mother would reveal to him that her parents were livid about the marriage.

Obama's grandfather was more accepting of the union, but Madelyn said she was not pleased by it and gave me the impression that she let her disenchantment be known at the time. Like nearly all boys, Obama idolized his father, or at least the image of his father that was presented to him—brilliant, powerful, confident, successful, moral. But Madelyn viewed her daughter's young African husband with a healthy amount of skepticism. In contrast to the dreamer personalities of her daughter and husband (and grandson), Madelyn is a reliable purveyor of skepticism. She might even fall on the outer fringes of cynical. Her advanced age at the time of our interview could have accentuated this skeptical nature. But Obama said that his grandmother was often the reality check to the many pie-in-the-sky tales he would hear as a youngster, especially those about his exceptional African father. Madelyn also rarely feared that her words might offend the listener.

Madelyn did appear to hold back some in our interview, but it was easy to gather that she had great concerns about the cultural differences between Ann and her young husband. Her many life travels notwithstanding, Madelyn maintained a sense of midwestern provincialism. She said she was suspicious of some of the tales that Obama's father would put forward to his in-laws. "I am a little dubious of the things that people from foreign countries tell me," Madelyn said. When I suggested that Obama's father, in addition to storytelling skills, had a great deal of charm—that his own father was a medicine man in a Kenyan tribe—she raised her eyebrows and nodded to herself. "He was . . . ," she said with a long pause, "strange." She lingered on the *a* to emphasize "straaaaaange." She then continued: "He wasn't that handsome in a way, exceptionally dark-skinned, but he had a voice like black velvet . . . with a British accent. And he used it effectively."

"And," she added a bit hesitantly, "he was extremely brilliant."

Obama's father was the first African exchange student at the University of Hawaii. After studying in London, he arrived in the United States in 1959 in "the first large wave of Africans to be sent forth to master Western technology and bring it back to forge a new, modern Africa," Obama wrote in his memoir. Obama's father was the son of Hussein Onyango Obama, a prominent elder and farmer in Kenya's Luo tribe. As a boy, Barack Sr. herded goats on the family farm near a poor village called Kolego in the province of Nyanza near Kenya's Lake Victoria. He stood out academically in a local school established by the British colonizers and won a scholarship to attend school in Nairobi before being sponsored for study in the United States at the University of Hawaii. "He studied econometrics, worked with unsurpassed concentration, and graduated in three years at the top of his class," Obama wrote proudly of his father. But when he came to America, his father left a pregnant wife and child back in Kenya. When he returned to Africa, he took another American woman with him, eventually marrying her and having two additional children.

Although he clearly had a gifted mind and an imperious presence that immediately captured attention, Obama's father had a raft of personal issues, not the least being alcohol indulgence, that prohibited him from reaching his full potential once he returned to his native country. An atheist with an analytical mind, he worked for a petroleum company, and for a time he was a chief economist for the Kenyan government. But he maneuvered poorly in the thicket of Kenyan tribal politics; his influence waned and his finances fell apart. "What my father became was a victim of the clash between the Kenyan and Luo cultures and the Western culture and the expectations that were on him," Auma Obama told me. "As the head of a family of Luo, he felt a responsibility to take care of everyone, all these many relatives, and that was just impossible."

In a 2006 speech in Nairobi, Obama said that his father's life "ended up being filled with disappointments. His ideas about how Kenya should progress often put him at odds with the politics of tribe and patronage, and because he spoke his mind, sometimes to a fault, he ended up being fired from his job and prevented from finding work in the country for many, many years." Barack Sr.'s relations with women seemed to be in a consistent state of disarray— by most counts, he fathered nine children by four wives. Before coming to the United States, he had been married in a tribal ceremony in Kenya, and he told Ann he had divorced that wife. She later learned to her discontent that there had been no official divorce. "Because he never fully reconciled the traditions of his village with more modern conceptions of family—because he related to women as his father had, expecting them to obey him no matter what he did—his family life was unstable, and his children never knew him well," Obama said of his father. After leaving Hawaii, his relationship with his American wife was less than sterling. "A lot of grandiose plans, a lot of promises that never worked out," Madelyn said with a sigh. Ann "subsequently divorced him."

Just Call Me Barry

Every man is trying to live up to his father's expectations or make up for his mistakes. In my case, both things might be true.
—BARACK OBAMA

As a child, as an adolescent and especially as a teenager, Barack Obama showed relatively few signs that he would evolve into a nationally recognized politician in adulthood. He mostly spent his upbringing in Hawaii soaking up the serene environment of the tropical island in the Pacific—playing basketball, bodysurfing and socializing with friends. "Hawaii was heaven for a kid and consequently I was sort of a goof-off," Obama conceded, although that self-assessment might be a bit harsh. Two things, however, did set him apart from most of his peers: his skin color and his nearly five years living in Jakarta, Indonesia. "I was raised as an Indonesian child and a Hawaiian child and as a black child and as a white child," Obama said. "And so what I benefited from is a multiplicity of cultures that all fed me."

Ann Dunham remarried after divorcing Obama's father. True to her nature, she again did not take the conforming path. This time, she wed an Indonesian native named Lolo Soetoro, another foreign student at the University of Hawaii. After spending two years in Hawaii, he was forced by political upheaval in his native Indonesia to suddenly return to Jakarta. About a year later, Ann and Barry moved to be with him.

Indonesia was an exotic experience for the boy, then six years old. He encountered new food, wild animals and an entirely foreign culture. He played in rice paddies and rode water buffalo. Wrote Obama in *Dreams:* "I learned how to eat small green chili peppers raw with dinner (plenty of rice), and away from the dinner table, I was introduced to dog meat (tough), snake meat (tougher), and roasted grasshopper (crunchy)." For the first time, he also bore witness to the unpleasantness of dire poverty. Beggars would come to their door, and even his mother, who had a woman's "soft heart," according to Lolo, eventually learned to "calibrate the level of misery" before handing out money. Obama wrote that, over time, he also developed his own calculations, a result of lectures from Lolo advising him not to give all his money away. Said Obama in a 2004 radio interview: "I think [Indonesia] made me more mindful of not only my blessings as a U.S. citizen, but also the ways that fate can determine the lives of young children, so that one ends up being fabulously wealthy and another ends up being extremely poor."

Beyond his advice not to be loose with his money, Lolo imparted a store of tough-minded, masculine wisdom to young Barry. Amid the widespread privation of Third World Indonesia, Lolo had lived a hard existence, extremely different from the relatively comfortable, middle-class American experience to which Obama and his mother were accustomed. As a soldier in New Guinea, Lolo told the boy, he would dig leeches from his military boots with a hot knife. Lolo explained that he had seen a man killed "because he was weak." "Men take advantage of weakness in other men," Lolo had told him. "The strong man takes the weak man's land. He makes the weak man work in the fields. If the weak man's woman is pretty, the strong man will take her. . . . Which would you rather be?"

Obama heard such words and realized that he was slowly slipping out of the orderly and secure cocoon that his mother and grandparents had carefully woven around him. He wrote: "The world was violent, I was learning, unpredictable and often cruel."

Yet, as Lolo preached the importance of strength and courage to Barry, Ann stressed moral values to her son. She emphasized four things: honesty, fairness, straight talk and independent judgment. On the last, she gave Barry this example: Just because other children are teasing a boy for something awkward about him, such as a bad haircut, that does not mean you should do it as well.

Ann also gave him an education in African-American history, albeit it was skewed heavily toward the positive. With his father absent, she wanted him to take pride in this side of his racial heritage. So she pushed on him books about Martin Luther King Jr. and the civil rights movement, and played recordings of the soaring gospel singer Mahalia Jackson. She filled him with stories of accomplishments by African-American heroes such as U.S. Supreme Court justice Thurgood Marshall and Hollywood movie star Sidney Poitier. "To be black was to be the beneficiary of a great inheritance, a special destiny, glorious burdens that only we were strong enough to bear," Obama wrote. Despite the unfulfilled promises of her ex-husband, Ann also talked up the positive traits of Obama's biological father, building up Barry's ego by telling him that he acquired his intellect and character from his father.

But mostly, Barry was given a lesson that he would consume over and over: His unique racial ancestry made him someone who certainly was not to be ostracized or shunned. Far from it—he was a special person worthy of others' deep admiration.

During this time, Ann gave birth to Maya, her second and final child. Ann pushed Barry and Maya to assimilate to Indonesian culture as much as possible so they would not appear to be arrogant children on foreign soil, conducting themselves as spoiled Americans. But over time, her naïveté as an American raising children in Indonesia came to the fore. After several years in Jakarta, she considered the lives of Indonesians and the lives of Americans—and she began thinking about how many more opportunities would be available to her son in the United States. First, Obama could receive a superior education, an education of a quality that even private Indonesian schools could not give him. Barry had attended Muslim and Catholic schools in Indonesia, but his advancement was not up to the standards that Ann desired. So five days a week, she would roust Barry out of bed before dawn and teach him three hours of English before she went to her job instructing Indonesian businessmen in English at the American embassy.

Meanwhile, over the phone from Hawaii, Ann fielded lectures from her mother about her children's well-being amid the Third World environment of Jakarta. One day in particular, Obama recounted in *Dreams,* was instrumental in Ann's coming around to her mother's way of thinking. Obama had been out playing and didn't arrive home until well after dark. As he approached the house, his greatly worried mother noticed that he had a sock tied around his arm. Underneath the sock was a long gash. When he explained that he had cut his forearm on a barbed wire fence, his mother rushed to Lolo for parental assistance. But Lolo seemed nonplussed at her concern and suggested the boy be taken to the hospital in the morning. This evoked Ann's ire. So disregarding Lolo's nonchalance, she borrowed a neighbor's car and rushed Barry to the hospital. There, in a darkened back room, she found two men playing dominoes. When she asked where the physicians were, they informed her that *they* were the doctors. She explained her son's injury, and they told her to wait until after they finished their game. After what surely was an agonizing wait for Ann, the doctors placed twenty stitches in Obama's arm. Soon afterward, Ann reached a conclusion that she had been edging toward for some time—her only son belonged back in America.

ARRANGEMENTS WERE MADE, AND BARRY WAS SENT TO LIVE WITH his grandparents back in Hawaii, where he was enrolled in the private Punahou School (pronounced Poon-a-ho). Founded in 1841 by missionaries, Punahou had evolved into a prestigious college preparatory academy that served Hawaii's upper crust. Obama's senior class was more than 90 percent white, with just a smattering of Asians, and so it was known informally among Hawaiians as "the white school," or the school for the *haole,* the derogatory moniker for Caucasians used by island natives. Obama's grandparents maneuvered him into Punahou; his grandfather's boss, an alumnus, intervened to have Obama accepted. And Madelyn's job at the bank helped pay the steep tuition. By living in a modest apartment and sending Obama (and eventually Maya) to private school, his grandparents had sacrificed their own prosperity for the sake of Obama and his sister. "We never suffered," Madelyn answered when I asked what specific things were given up to send her grandchildren to Punahou. "As you can see, we live in an apartment instead of a house. . . . But I think we could have done the other if we had wanted. But I traveled, you know, and spent money on the kids—the kids and traveling were priorities. We're not poverty-stricken." This was a trend that would follow Obama throughout his charmed personal life and into his political career—people going out of their way to clear a path for him to succeed. At Punahou, Obama took his first step into an educational institution for elites, a rarefied world where he would remain through all of his formal education.

Punahou was situated just beyond walking distance of his grandparents' apartment in a rather densely populated section of Honolulu. The campus itself was an island on an island, a fenced-in tract of several scenic acres largely hidden from public view. Under the silvery morning sunshine, Maya gave me a tour of her alma mater. A misty rain sprinkled on us from the lone cloud in an otherwise clear blue sky, giving the day a perfect tropical feel. On his first visit to Punahou, Obama's grandfather described the school campus as "heaven," and it would be hard to disagree with that assessment. The school featured elegant theaters and stately buildings set amid shady palm trees, verdant fields and neatly sculpted walking paths.

High, tree-covered hills served as backdrop to the campus. The scene was so idyllic that it resembled a Hollywood set. That landscape, in conjunction with the surrounding fence, provided visitors and others with a feeling of safe insularity. And after a long day of hectic travel, it provided me with a sense of calm.

As we strolled across the campus, I could envision a relaxed, smiling teenage Barry Obama walking from one class to the next, stopping to chat up a girl or joke with a buddy. Absorbing the atmosphere of this campus gave me a sense of the cool, unflappable Hawaiian nature at Obama's core. The night of his Senate primary election victory, for example, reporters marveled curiously at Obama's exceptionally cool exterior as others around him exhibited jubilation. One of Obama's greatest talents is that, even in the midst of chaos, he has the ability to slow things down internally, to project serenity, a sense of emotional control. It is a quality that superb professional athletes often possess—the ability to slow the game down and see everything around them clearly. Hawaii, if not fully responsible, most certainly contributed heavily to this trait. As I recollected that election night and mulled over my observation, Maya chimed in: "Hawaii is such a generally sweet place. You can come back here from almost anywhere and refresh yourself mentally."

Besides the island charm and physical beauty, another aspect of Punahou immediately jumped out: how incredibly slowly students moved about the campus. They sauntered around in flip-flops and blue jeans, seemingly in no hurry to be anywhere but under a palm tree to gossip or study. Even students who gathered at picnic tables were not engaged in typical teenage jawing or horseplay but in peaceful chitchat. It's almost as if there were a perpetual state of serenity in the air. Racially, most of the students were still white, although a few more kids of Asian descent were mixed in compared with Obama's years here. As the tour concluded and we headed to our domestic rental cars, which were parked amid Saabs and Volvos and Lexuses, Obama's introduction to his memoir flashed to my mind. I realized that the Manhattan publisher who turned down Obama's book idea was correct on one account—Obama certainly had not come from an underprivileged background.

IN HIS FIRST YEAR AT PUNAHOU, WHEN HE WAS TEN, OBAMA received a quite unexpected Christmas gift: a visit from his father. To this day, those several weeks are the only memories of Barack Sr. that Obama holds. At the time, they left young Barry even more confused about the man who had been served up to him as a legend. He learned that his father had remarried and Barry had five half brothers and one half sister living in Kenya. To prepare him for his father's arrival, his mother had plied him with information about Kenya and its history. She told him the Luo tribe had migrated to Kenya from its first home along the banks of the world's greatest river, presumably the Nile. This was another attempt by Obama's mother to create an almost godlike figure out of Barack Sr. But Barry began discovering that his mother and grandfather tended to relay only half the story when it came to his father. When Obama went to the library himself to read about his father's tribe, he discovered that the Luo raised cattle, lived in mud huts, and their main sustenance was cornmeal. He did not know it at the time, but his father had herded goats as a child, and only through sheer intellectual ability and educational promise did he escape living an African tribal farm life. Barry left the book on the table and stomped out of the library in a huff.

This research, unfortunately, dovetailed with something he had run across a bit earlier in a story about a black man who tried to lighten his skin with a certain chemical. Barry had wondered why on earth a black man, who was a member of such an esteemed people, would want to make himself appear less than black? After such discoveries, his young inquisitive mind became suspicious of his mother's portrayal of blacks as superior beings.

Obama recalls most vividly only a few aspects of his father's visit. First was his father's commanding physical presence—he could make himself the focus of a room simply by walking in, speaking in his confident manner, moving in his debonair fashion. An act as simple as wrapping one leg over the other would conjure a feeling of elegance from a person in Barack Sr.'s presence.

One incident from the visit also stuck out: Obama recalled starkly how his father drew the wrath of his grandmother and his mother when he stepped in and forbade Obama to watch a classic holiday television cartoon, *How the Grinch Stole Christmas*. His father believed the boy had been studying too little and watching too much TV. The orders barked from Barack Sr. led to a bitter argument, in which family members accused each other of one thing or another. In his book, Obama uses the fiction of the *Grinch* story as a metaphor for the fiction of happiness and comfort that was now evaporating around him, revealing instead the harsh reality of life— his parents' failed marriage and his father's abandonment of his familial responsibilities. Like the premise of blacks belonging to a superior race, the stories he had been told of his father's greatness were becoming ever more dubious to Barry. His youthful innocence was being lost, and he was confronted with a wholly different image of his heritage.

"My father's absence in my life, it was just so complicated," Obama explained in my first extended interview with him in December 2003. "I mean, here is a guy who spanned, who sort of leapfrogged from the eighteenth century to the twentieth century in just a few years. He went from being a goat herder in a small village in Africa to getting a scholarship to the University of Hawaii to going to Harvard. He was active in high government positions, but then sort of tragically was destroyed, although *destroyed* is too strong a word. But he was somebody who never really achieved his potential because of problems of tribalism and nepotism in politics in Kenya and partly his own failures, his inability to truly reconcile his past with modern life. . . . He was a brilliant guy, but in so many ways, his life was a mess—children by different women, a political career that turned in on itself."

Obama then used a quote that he would repeat often about his father's influence on his own life. In essence, he said he derived his personal ambition, arguably the most powerful force in his life, from his father's shortcomings. He also gained his own high expectations of himself from his father's sterling image and talents. Said Obama: "Every man is trying to live up to his father's expectations or make up for his mistakes. In my case, both things might be true."

BY THE TIME BARRY REACHED HIGH SCHOOL, ANN HAD SEPARATED from Lolo and returned full-time to Hawaii with Maya, where she began studying postgraduate anthropology at the university. The family lived in a small apartment just a few blocks from Punahou. Barry's days were filled with the normal adolescent activities of a teenager in Hawaii in the mid-1970s—seeking to understand what attracts the opposite sex; attending parties in which alcohol (and marijuana) were the main courses; hanging out at the beach; body-surfing on the rolling Pacific waves; and playing sports, especially basketball—lots and lots of basketball. At the time, basketball was fast becoming *the* sport of choice for African-American youth in the United States, as a particular blend of acrobatic street game made its way into the college and professional ranks. For Obama and many other youth, Julius Erving, better known by his moniker, Doctor J, was the hero of the era. It is not hard to see a teenage Obama being drawn to Erving, a gentlemanly, smart and thought-ful African-American sports figure. Erving was also one of the key players who changed the game from a dribble-pass-jumpshot affair to the driving-and-soaring-above-the-rim highlight reel of today. Doctor J's dunks were wondrous to the eye. He seemingly could fly through the air, with a long, slender arm extending to its full-est, effortlessly pinwheeling the basketball and slamming it into the cylinder. His game was awe-inspiring, graceful and ferocious, all at once.

Most of Obama's friends were white, although in his book he describes long conversations with an older black friend who had moved from Los Angeles to Hawaii. He named the friend "Ray" and portrayed these talks as his first serious struggle to manage the complicated racial questions brewing in his mind. Ray generally saw racism around every corner, while Barry was much less cynical about these matters, often providing a calming influence on Ray's angry nature. When Ray would complain that girls refused to date him because they were racist, Barry would smile and suggest that it wasn't necessarily that these girls didn't like blacks, but maybe they were just attracted to men they'd been socialized to connect with, people who reminded them of their father or brother. Ray, whose real name is Keith Kakugawa, surfaced in April 2007 and confirmed the intense discussions between the two teenagers. Of mixed race himself, Kakugawa was homeless on the streets of Los Angeles after being recently released from prison where he had served time for a parole violation. He called Obama's presidential campaign in search of money from his old high school chum, now a famous politician. Kakugawa said that, as a teen, Obama had a troubled side that stemmed from issues of parental abandonment and his mixed racial heritage. "He wasn't this all-smiling kid," Kakugawa told *ABC News*. "He was a kid that would be going through adolescence, minus parents, feeling abandoned and, you know, inner turmoil with himself. He did have a lot of race issues, inner race issues, being both black and white."

While these teenagers dealt with their blackness, Obama's various white friends were largely unaware that Obama was grappling with this weighty matter. Bobby Titcomb, one of Obama's closest high school friends, who remains a friend today, said he never sensed his friend wrestling with identity issues. In fact, Titcomb said he thought Obama was one of the most emotionally secure teenagers on the island. "He was just a normal Hawaiian kid, a normal guy," Titcomb said.

Titcomb, however, did remember that Obama had his own style and personality. He said Obama was much larger than most of his peers. Indeed, photos of him in his high school yearbook show a much heavier boy, almost chubby. As a high school freshman, he played defensive line on the football team and Titcomb described him as a strong lineman, "a real people mover." Yet even though he had a much larger physical presence, he never lorded it over others, Titcomb said. Obama had his own mind and, like his mother, never conformed to whims of the day. "When somebody was getting teased, he kind of gave that look, almost a look of disapproval," Titcomb recalled. "So that's kind of just the way he was. He was different in a way in that he didn't buy into the normal. He didn't tease kids just because it was the cool thing to do." Indeed, his mother's Lesson Number Four was "independent judgment"—never run down another child just because others are doing so. It was a lesson that Ann obviously had ingrained in her son successfully.

Bobby Titcomb was not the type of individual I expected to find as one of Obama's lasting friends. It's not that Titcomb wasn't a likable and bright guy. But as Obama traveled through Harvard and into national politics, Titcomb would seem like the kind of childhood friend who might get left behind. When I interviewed Titcomb, he was building a house on Oahu and working as both a flight attendant and a commercial fishermen. Physically, he was short with a wind-swept and perpetually tanned appearance, and he was laid-back enough that he looked as if he would be most at home relaxing with a margarita at a Jimmy Buffet concert. He was like a good many of Obama's Punahou classmates, perfectly happy living a settled Hawaiian existence. When I asked Titcomb again about Obama's racial confusion during high school, Titcomb reiterated that he was oblivious. "You know, in Hawaii, you go to school where you don't have to wear shoes until, I think, ninth grade. So you go barefoot," Titcomb explained. "And in Hawaii, you know, you have five best friends and one's Chinese, one's Japanese, one's Hawaiian, and so on. It is kind of just a melting pot. It was cool to have a black friend, you know. So I, you know, I never saw it."

While Obama was in high school, his mother's studies called for her to do fieldwork back in Indonesia. When she suggested that Barry return with her and Maya, Barry resisted. He was happy at Punahou and with his life in Hawaii. He had a circle of friends with whom he felt comfortable and he was in no mood to leave that for a deprived existence in Indonesia, even if he would go to an international private school. So he struck an agreement with his grandparents that when his mother left, he would move in with them, and they would not get in his way so long as he gave them no reason to. Madelyn Dunham, who had nagged her daughter to bring her grandchildren back to Hawaii, encouraged this idea. "I suppose I provided stability in his life," Madelyn said, noting that her daughter's global curiosity could take her away from her son for months at a time. Even with his mother gone, Madelyn said, Barry was essentially a well-behaved teen who spent most of his time involved in sports. "He was a jock," she said.

But out of sight of his white grandparents and his white friends, Barry's struggles to understand the African-American experience became a stronger force, a lonely force, and he mostly suppressed it from others. "I was trying to raise myself to be a black man in America," he wrote, "and beyond the given of my appearance, no one around me seemed to know exactly what that meant." Various racially driven incidents in his life had piled up in his psyche, causing him anguish. A small-minded tennis coach joked about his color rubbing off, prompting Barry to quit the tennis team. While waiting for a bus, Madelyn was harassed by a black panhandler, and Barry's grandfather offered her little sympathy, thinking she overreacted to the incident because of the man's race. Barry took two white friends to a party thrown by blacks and noted angrily to himself their obvious discomfort in the all-black surroundings.

Hoping to clear up his own racial confusion, he would forgo doing homework and bury himself in the works of prodigious black authors who sought to explain or amplify the feelings of powerlessness and anger embedded in the hearts of black men: Langston Hughes, Ralph Ellison, James Baldwin, Richard Wright, W. E. B. DuBois. Of these readings, he said he most closely identified with the *Autobiography of Malcolm X.* Since becoming a politician, Obama has steered clear of quoting such a militant and revolutionary figure as Malcolm X. But in his book, he wrote that the activist's "force of will" and "repeated acts of self-creation spoke to me."

The teen years are difficult enough for any male. It is the time when boys begin to establish their adult persona—how they fit into social groups and settings, how they interact with women, how they register their masculinity with other males. At this age, even if boys tend to act as if they disregard their father, in fact they are looking at him as the blueprint for how to conduct themselves in these vital areas. In this regard, Obama had something in common with many African-African males—he lacked a father in his life to counsel him through these confusing times. Obama looked to popular culture for male role models to fill that paternal void, and he tried to mimic African Americans on television and in the movies. "Some of the problems of adolescent rebellion and hormones were compounded by the fact that I didn't have a father," Obama said. "So what I fell into were these exaggerated stereotypes of black male behavior—not focusing on my books, finding respectability, playing a lot of sports." He grew a thick Afro and donned a stylish white open-collared leisure suit with fat lapels, making himself stand out from the crowd in an urban black way. Displaying the touch of personal vanity that he still carries, Obama cherished his full Afro and could spend an inordinate amount of time picking at it in order to make it appear just right. The plastic pick would protrude from Barry's back pocket and he carried it wherever he went. Playing the role of pesky little sister, Maya enjoyed brushing a hand through Barry's hair to rile her brother, prompting an angry Obama to cop a cool street attitude and admonish her: "Hey, don't touch the 'fro!"

Barry found some solace from his deep racial questions on the basketball court, the place he said he felt most at ease. He spent hour after hour on a court located outside a grade school behind his grandparents' apartment building. He developed a swift crossover dribble and an idiosyncratic, street-influenced style of shooting, cocking the ball far behind his left ear and then shot-putting it toward the rim. His specialty was his left-hand shot from the corner, placing a bit more backspin on the ball and launching it with a low arc. "On the basketball court I could find a community of sorts, with an inner life all its own," he wrote. "It was there I would make my closest white friends, on turf where blackness couldn't be a disadvantage." For Madelyn, there was comfort in knowing the whereabouts of her grandson. She could track him by ear. "I used to know when he was coming home because I could hear the basketball bouncing all the way from over there to over here," Madelyn said in our interview in her apartment.

The teenage years mark a period of rebellion and disaffection for males, and Obama's racial turmoil only exacerbated those natural feelings. He was always a solid B student, but by his senior year, he was slacking off in his schoolwork in favor of basketball, beach time and parties. He also, as he described it later, "dabbled in drugs and alcohol." He would buy a six-pack of Heineken after school and polish off the bottles while shooting baskets. He also smoked marijuana and experimented with snorting cocaine but demurred from heroin when he said a drug supplier seemed far too eager to have him experience it. His stream-of-consciousness passages in *Dreams* about drinking and drug use, and turning down the heroin offer, are written in the tough, boastful, self-absorbed words of a teenager—prose that marries his racial angst to his drug use. "Junkie. Pothead. That's where I'd been headed: the final, fatal role of the young, would-be black man." In the next section of the book, however, Obama reveals a broader mind and displays less self-pity. He noted that white kids, Hawaiian kids and wealthy kids also turn to drugs to soothe whatever causes them pain—an astute observation that others perhaps were suffering as much as or worse than he.

His grandmother recalled that she and her husband discussed Barry's declining grades and grew concerned about his possible drug use and overall lack of direction. She said she urged her husband to give Barry a stern lecture about the pitfalls of this lifestyle. She said she worried that, because Barry was so clearly African-American in appearance, he would be used as a courier by his white friends to buy drugs and distribute them, and that being a drug supplier could land him in long-term trouble with the law. Obama, however, told me that he recalled having no such conversation with his grandfather and questioned his elderly grandmother's memory. "This was really a very transitory or very short-term period in his life," Madelyn said. "I mean it wasn't something that lasted long. . . . Barack didn't really talk about things too much, you know. He must have suffered some racial prejudice, but he didn't talk about it with us."

At Punahou, Barry was popular among his classmates and teachers, but in an understated way. Eric Kusunoki, his homeroom teacher through all four high school years, recalled that on the first day, he reviewed Obama's first name several times on the class roster and finally admitted that he could not pronounce it. Obama showed no sign of annoyance and spoke up with a smile, "Just call me Barry."

Kusunoki categorized Obama as a good student but one who failed to reach his vast potential. Despite eventually going on to Harvard Law School, Obama did not stand out academically in high school, mostly from lack of effort. "All of the teachers acknowledged that he was a sharp kid," Kusunoki said. "Sometimes he didn't challenge himself enough or he could have done better." Obama impressed some instructors as being a deep thinker, but his grades did not reflect that. Kusunoki also recalled that he and Obama chatted quite often and Barry would mention problems with individuals he believed were racist. But Kusunoki, like so many others, said Obama gave no outward impression that he felt burdened or confused by his race. "We had discussions along these lines, but he didn't really express a lot of the thoughts or feelings that I read in his book," Kusunoki said.

As much as Obama loved basketball, it did not always come easy. Like most sports, it taught him a valuable life lesson—humility. Punahou's team was excelling his senior year, thrashing its competition, and ultimately the squad won the state championship. Unfortunately for Barry, he played only a small role in the team's success, since the coach relegated him to the bench most of the year. That meant he saw little public glory, and most of his contributions came during practice by pushing the starters to be better players. In *Dreams*, Obama wrote that his friend and black confidant, Ray, blamed race for keeping Obama out of the starting five during his first couple of years in high school. Obama, however, took up for the coach, countering that the team played a "white" style of game, more of a half-court game and less of a running street game. Perhaps he didn't fit well into that makeup, Barry said. Obama also noted that the team was winning regularly without him, so he couldn't complain too loudly.

His coach, Chris McLachlin, told me that Barry accepted his lack of playing time with aplomb and was a wonderful role player for the team. The coach said he had an immensely talented squad and Barry was just shy of being a starter. "I recall his sincere eagerness to want to get better, his positive attitude despite not getting as many minutes as he probably wanted," the coach said. "He was very respectful and understood his role."

Despite all the magnanimity that Obama displayed in his book toward the coach, I discovered a different attitude when I mentioned to Obama in October 2004 that I had interviewed his high school coach. His viscerally emotional reaction, even though twenty years had passed, was one of the moments when Obama's short fuse and healthy ego slipped out from behind his cool exterior. Obama is able to emanate Hawaiian calm in times of strife, but he also has a fiery, highly competitive streak. He has little tolerance for losing or being shoved into the background. "I got into a fight with the guy and he benched me for three or four games. Just wouldn't play me. And I was furious, you know," Obama said, a twinge of unresolved bitterness in his voice, along with an implied presumption of "How dare he bench me!" Obama's former Illinois press secretary, Julian Green, said his boss has mentioned McLachlin in less than glowing terms in private conversations. "There's still something there between those two guys," Green said. (Obama kept playing basketball into adulthood, and one of the regulars in his weekly games said Obama was not blessed with inordinate talent, but he was a reliably steady player. "He would have been a very good high school player, very good at that level," said Arne Duncan, chief executive of the Chicago Public Schools, who has played professional basketball in Europe and Australia.)

Even if Barry was not an exemplary student or a standout athlete, I ran across an occasional acquaintance from his Hawaiian years who spotted something deeper than a likable but underachieving jock. Suzanne Maurer, the mother of one of Obama's close friends, Darren Maurer, said she sensed an energetic, ambitious spirit in Barry, as well as a precocious wisdom. She said her son and Barry rode the bench together on the high school basketball team, and she believes that basketball "centered" both of them. Obama's limitations as a player didn't pull his spirits too low, she said. She added that, even if he was undergoing some internal racial turmoil and disappointment on the basketball court, Barry displayed an optimistic attitude about life in general in her presence. "I recall that he was the type that if he had a dream, he would pursue it," Maurer said. "The sky seemed to be the limit, and Barry was very much a can-do type person, even with sports, even as a benchwarmer."

The fact that Obama harbored dreams extending beyond Hawaii was something that perhaps even Barry didn't realize at the time. But he would soon enough. Maurer said that when she heard in the summer of 2004 that Obama was running for the U.S. Senate and that his political career had rocketed into national stardom, she looked up his campaign website and wrote him an e-mail. A political conservative, Maurer mentioned that she didn't agree with his political stances, but she admired his integrity and devotion. To her delight, he wrote back. And while she recognized the same wise-beyond-his-years mentality from twenty years earlier, she noted that a personal evolution evidently had occurred after his departure from Hawaii. The boyish, basketball-happy Barry Obama of the Punahou School was someone in the past, someone who no longer existed. "We agreed to disagree on the issues," she said. "And even though I called him 'Barry,' in the e-mail he sent back, he signed his name 'Barack.'"

The Mainland

Do you mind if I call you "Barack"?
—A FEMALE COLLEGE FRIEND

Hawaii was a wonderful place to grow up for Barack Obama. With his grandparents looking after him, his home environment was secure and loving enough to provide him with some stability as he negotiated his racially conflicted adolescence. "I'll admit that I did give him a few kicks in the pants," his grandmother said. "Not many, but a few." When the realities of adulthood hit him in his twenties, Obama looked in the rearview mirror of his life and called his youth in Hawaii "a childhood dream."

Yet Hawaii, with all its tropical splendor and melting pot of ethnicity, could also be stifling. His half sister Maya Soetoro-Ng said that island inhabitants were keenly aware of their serene seclusion, dropped on a spot of land in the middle of an ocean. Civilizations that drove world commerce, the arts and popular culture were on continents far, far away. That peaceful solitude was both a blessing and a curse. "It could seem very insular here. And one can feel trapped here, believe it or not," Maya said. "It seemed back then that news traveled here so slowly. There was no Internet, and you couldn't find the *New York Times*." Even Obama's grandmother, who seemed satisfied working a nine-to-five bank job and providing a secure home life for her grandchildren, complained that people felt detached from their native country. For example, with the six-hour time differential between the United States' East Coast and Hawaii, presidential elections had largely been decided by the time many Hawaiians even bothered to leave for the polls. "We're disenfranchised," Madelyn Dunham complained.

Consequently, both Barack and Maya felt the same yearning desire to experience life beyond their island. Attracted by their mother's wandering spirit of adventure and her liberal sensitivities, neither sibling considered going into business or seeking financial wealth once they reached the mainland United States. But they did want to consume the globally powerful American culture—at its epicenter. In time, both would land in the Western world's most important city, New York.

On my visit to Honolulu, Maya was in her mid-thirties. She had lived in Manhattan from 1995 to 2000, and while there, she helped to launch an alternative school. But by 2000, she had reached the moment in her educational career and her personal life when she was prepared for a family. In addition, her grandmother's health was declining—and so back to Hawaii she came. After returning, she found an extremely affable and responsible husband in a Canadian, Konrad G. Ng, a film manager at the Honolulu Academy of Arts. "Just an all-around great guy," Obama said of his brother-in-law. Four months before my 2004 visit, Maya had given birth to her first child (Suhaila Kala Kai-La Ng).

Much like her older half brother, Maya emanated a refined, confident presence. A pretty woman with a roundish face and a wide smile (like her brother's), she was a mixture of European and Indonesian heritages, with the bushy dark eyebrows of her mother and Obama, and the olive skin of her father, Lolo. She wore her thick brown hair flowing over her shoulders, which softened her features and accentuated her large brown eyes. "Maya's just beautiful!" Nora Moreno Cargie gushed. At the university, Maya taught evening classes in multicultural education, education philosophy and education history. This was appropriate, I thought, given her own family's various cultures—an African-American brother, a white American mother, an Indonesian father and white midwestern grandparents. ("Our family get-togethers look like a meeting of the United Nations," Obama is fond of saying.) Maya spoke the King's English in a hybrid of East Coast accents, enunciating words in a precise way that gave her an air of sophistication and prompted me to wonder how she picked up such an elite manner of speaking. A lifetime of private schooling, I imagined. She practiced Buddhism, taught global dance as an avocation and was a connoisseur of the arts. Considering how successful her brother had become in his career, as well as her own sophistication, it was somewhat surprising to see her driving a well-worn white Ford Contour that lacked hubcaps and had been purchased at Budget Car Sales. But she and her husband subsisted on the modest salaries of an educator and an arts school manager. And true to their mother's cerebral and benevolent nature, the pursuit of expensive material possessions was not high on the list of priorities for either Maya or Barack.

On Maya's tour of the island, we kept pace behind the compact Ford by maintaining sight of the bright blue-and-white, Illinois-imported sticker emblazoned on its back bumper: "OBAMA: Democrat for Senate." The most amusing aspect of meeting Maya was hearing her pet name for her brother: "Obama." This tickled Moreno Cargie to no end: "I just love that—she calls her brother Obama!" Using her brother's last name singularly connoted several things to me. It highlighted that she and her brother had different fathers; it emphasized the mystique surrounding her brother's African surname; and it gave Obama a certain importance, like the music stars Prince or Bono. When Maya did refer to her half brother by his given first name, it was usually in formal conversation. And in doing so, she would break "Ba-rack" into two distinct syllables and roll the *r*.

Maya's tour of all-things-Barack included guiding us up a winding road through Pu'u Ualakaa State Park, a gorgeous deep-green rain forest where Obama's family often picnicked. The park, whose Hawaiian name means "rolling sweet potato hill," was set in a thousand-foot-high mountainous area that, at its peak, provided a clear and breathtaking panoramic view of Waikiki, Diamond Head and Honolulu, a sweeping canvas of almost the full length of the island of Oahu. We also visited the nearby military gravesite at the National Memorial Cemetery of Obama's grandfather, Stanley Dunham, who died from prostate cancer in 1992 at the age of seventy-five. And we drove by the massive seaside rocks near Sandy Beach, from which a despondent Obama and his sister spread the ashes of their mother when she passed away from ovarian and uterine cancer in 1995.

My interview with Maya came over dinner at her brother's favorite restaurant, the Hau Tree Lanai. It was just a couple of miles from the bustling Waikiki Beach strip, nestled amid the wooded Kapiolani Park section of lower-rise hotels and apartments along the oceanfront—a place that tourists could easily find if they wanted to take a short drive, and certainly a favorite among Oahu residents. We dined in an elegant beachfront courtyard. Several hau trees with thick trunks and long, angular branches spotted the area. The setting provided the ideal outdoor dinner atmosphere, with light ocean waves topped with white foam softly lapping against the beach, a gentle evening breeze wafting through the night air and a black cloudless sky wrapping us in comfort. The brightly lighted Waikiki strip shone in the distance to our right, looking almost lonely amid the mass of darkness in all other directions. With its menu of fresh seafood and tender steaks at slightly high prices, I could see why Obama religiously patronized this restaurant on his Hawaii visits. And overall, after a couple of days in such idyllic settings, I completely understood why almost everyone seems to return from Hawaii to the continental United States with their blood pressure a few points lower and speaking of their excursion in glowing terms.

This wonderful perception of Hawaii was somewhat ironic given the point to which Maya consistently returned in our dinner interview about Obama: her brother's burning desire to leave Hawaii to seek something—knowledge, adventure, racial identity—away from the confines of Honolulu. "Hawaii has ideas and cultures and what have you," Maya said. "But basically you felt cut off from the world, much more so than you do now. I think moving to the mainland he found himself for the first time in the midst of arguments and, you know, defense. Here, I think he loved it here, and he was happy, and he was well-adjusted—he was all of those things. But there were just so many big questions, you know, that couldn't be answered here that lingered. He sort of had no one to help him understand or to urge him to pursue those questions, for instance, when he felt someone was being racist or when he was struggling with issues of identity."

In his memoir, Obama makes composite characters of the few blacks who dispensed advice on racial issues, such as the strident Ray, who seemed almost as lost as Obama in that respect. Obama wrote that his grandparents were ill-equipped to speak about race, and Maya said their mother's lack of experience with the African-American community was particularly evident. There were few family disagreements, Maya said, but Obama and his mother debated emotionally over one topic: Obama's connection to Malcolm X. His mother tried but failed to understand the anger brewing within the 1960s black nationalist leader, who perhaps was best known for urging his followers to seek black freedom, justice and equality "by any means necessary." "She just didn't understand—and she wanted to," Maya said. "She was a little naive and she was very sweet and she was very liberal and she just didn't understand how anybody who had been subject to discrimination could then be exclusionary in their tactics. She didn't understand the anger. . . . And I think, although she tried, she was not equipped to help Barack deal with what it is like to be a black man in this country. I don't think she knew at all how to help him. I think she just wanted to make him feel better. She was such a loving mother. She just wanted to make him feel like everything was okay."

But to Obama in these years, everything was not okay and black militancy was an alluring notion. Consider this passage from *Dreams* about his feelings of black anger: "At best, these things were a refuge; at worst, a trap. Following this maddening logic, the only thing you could choose as your own was withdrawal into a smaller and smaller coil of rage, until being black meant only the knowledge of your own powerlessness, of your own defeat. And the final irony: Should you refuse this defeat and lash out at your captors, they would have a name for that, too, a name that could cage you just as good. Paranoid. Militant. Violent. Nigger."

Maya said her brother realized that he must experience African-American culture up close if he were to cleanse himself of this internal racial confusion and bubbling anger. Even so, after being accepted by several colleges, Obama chose to attend a small liberal arts school in suburban Los Angeles that had relatively few blacks on campus. Still not clearly focused on adult goals, he decided on Occidental College for the same reason that many young men make decisions—not necessarily based on a rational, long-term plan, but because he was fond of a girl. The Occidental coed who caught his eye was from tony Brentwood, California, and she had vacationed on Oahu with her parents. While still in high school, Obama was inclined to use his charm and good looks to try to connect with college women visiting Hawaii on vacation. His intelligence and broad smile made him quite appealing to women older than himself, and his introspective manner made slightly more mature college women more appealing to him. Also, by his senior year, he was shedding the thicker body frame that helped him on the football field as a high school freshman. He wasn't quite as thin as he would become in his twenties, but he was now on his way toward the ultraslender physique of his adult years.

OBAMA WON A FULL SCHOLARSHIP TO THE WELL-MANICURED liberal arts college of fewer than two thousand students. At Occidental, Obama gravitated toward the school's small contingent of black students and instantly discovered that blacks in the United States were not at all the monolithic tough urban chic guys he saw on his television screen in Hawaii. They were as diverse among themselves as whites—with a range of viewpoints and divergent behavior patterns. At Occidental, many blacks were from middle- to upper-middle-class backgrounds, and Obama decided they were a lot like him—suburban kids "whose parents had already paid the price of escape." "They weren't defined by the color of their skin, they would tell you. They were individuals," he wrote. "In their mannerisms, their speech, their mixed-up hearts, I kept recognizing pieces of myself." Those characteristics of Obama were also evident to blacks in other parts of the country. Chicago native Don Terry, a writer for the *Tribune* whose mother was white and father was black, told me that Obama reminded him of "California blacks," projecting a more laid-back, less urban personality.

Obama wrote that he consciously chose politically active black students as his friends because he feared being labeled a "sellout." In trying to convey an image of being a true black, he would sometimes overreach to gain acceptance among his black peers. When Obama described an African-American student who lacked overt black mannerisms as a "Tom," another student chastised him for the comment, saying, "Seems to me we should be worrying about whether our own stuff's together instead of passing judgment on how other folks are supposed to act." When I first read this passage in his memoir, it struck me that in my experience with him, Obama still had a tendency to overreach in order to fit in with some urban blacks, even if in general he seems at ease in black settings. Occasionally during his Senate campaign, he would slap a black man on the back and exclaim something like "What's up, brother?" Obama's feigned enthusiasm and overly eager manner could sometimes be a tad off-putting. I recall one surprised black man responding, "Huh?" and another not responding at all. "Obama is a great candidate, just a great candidate. But that 'hey, brother' thing can be just a little too much sometimes," Scott Fornek, a political reporter for the *Chicago Sun-Times,* remarked during a campaign stop at an African-American church.

Besides blacks at Occidental, Obama found a diverse set of friends in "foreign students, Chicanos, Marxist professors, structural feminists, punk-rock performance poets." He would encounter intellectual discourse focused on political rebellion and engage in deep-into-the-night intense discussions in his dorms about such heavy topics as neocolonialism, Eurocentrism, decolonization and the Caribbean black activist Frantz Fanon. "When we ground out our cigarettes in the hallway carpet or set our stereos so loud that the walls began to shake, we were resisting bourgeois society's stifling constraints," Obama wrote.

Obama became involved in a popular campus movement of the day—urging divestment of university money from South Africa because of its policy of apartheid. It was through this activism that Obama first learned the power of words—and his own power with the spoken word. "I noticed that people had begun to listen to my opinions," he wrote. "It was a discovery that made me hungry for words. Not words to hide behind but words that could carry a message, support an idea." His first public-speaking moment occurred when he opened a staged rally in which he was to begin talking to an afternoon crowd only to be yanked from the stage in a physical metaphor for the voiceless black South Africans. But when his cohorts began pulling him from the stage, Obama sensed a connection with the audience and physically resisted. "I really wanted to

stay up there, to hear my voice bouncing off the crowd and returning back to me in applause," he wrote in *Dreams*.

Occidental was a powerful intellectual growth experience for
Obama, even if he still did not perform heroically in his studies. But
it was here that he began to slowly drift away from concentrating
on basketball and partying. He practiced with Occidental's Division II team for several weeks but said the team consumed too much
time and he didn't see basketball leading anywhere. So he quit the
team and instead played pickup games with students and faculty.
In the classroom, Obama found deep-thinking peers and professors
who challenged him to reconsider his own self-absorption and to
view the world in global terms. The somewhat idle yet capacious
mind of a teenage basketball player and bodysurfer was being fed
intellectual nutrients that it had never ingested before. Rather than
hanging out at the beach, Obama was now spending time in intellectual discourse in coffeehouses. Some of that was again driven by
male hormones, Obama conceded. "The schools I went to weren't
driven by athletics," he said. "To get girls, you had to be the smartest guy in the coffee shop, not the best shooter on the court." Nevertheless, being forced to rely on his intellect and charm served to
change him and focus him. "He seemed to have gotten some purpose in life during those two years at Occidental," his grandmother
told me. Maya noticed her brother growing as well: "Moving to the
mainland, he felt for the first time in the midst of vigorous discussion. He was always smart, but he found himself a meaningful par-
ticipant in all of that. It was the first time he was really confronted
with black culture. He really had to find his place in the world. And
that requires a great deal of reflection and a great deal of study."

Obama assumed a new persona at Occidental. During one coffee-
house conversation, a young woman, whom Obama was obviously
attracted to, inquired about his given first name: Barack. He explained that it meant "blessed" in Arabic, and that his paternal grandfather was a Muslim. "Do you mind if I call you 'Barack'?" she asked.

Obama smiled, and some readers of his memoir can see the lightbulb pop on over the young man's head with the likely caption "Hmmm. Women might just be into the name Barack over Barry." And so Barry Obama was forever discarded.

After his sophomore year, in 1981, Obama transferred from Occidental to Columbia University in New York City under a program assembled by the two schools. He said he wanted to experience New York and find a "community" of blacks where he could "put down stakes and test my commitments." "Occidental was so small," he said, "that I felt that I had gotten what I needed out of it and the idea of being in New York was very appealing."

Ironically, when he arrived in New York, it was not the hustle and bustle and the millions of cosmopolitan people that primarily consumed him, but the solitude and isolation of being a newcomer in a big city. He chose to live a monastic existence, spending time alone in his spare Manhattan apartment and digesting the works of Friedrich Nietzsche, Herman Melville and Toni Morrison, as well as the Christian Bible. He began an exercise routine, running several miles each day, and this routine would attach itself to his psyche and remains there to the present day. His sour moods can often be attributed to a missed workout or two. "I had two plates, two towels," Obama recalled. "My mother and sister, when they came to visit me, just made fun of me because I was so monklike. I had tons of books. I read everything. I think that was the period when I grew as much as I have ever grown intellectually. But it was a very internal growth." Obama's intellectually elitist nature also began to sprout during this period, and he perhaps began taking himself a bit too seriously. Maya, who was in her mid-teens at the time, recalled that her brother chastised her for reading a copy of *People* magazine and watching television instead of delving into the novels that he had given her. Maya said she questioned whether her brother was going off the deep end, but her mother convinced her it was simply another phase in Obama's life. To Julian Green, Obama's former

Illinois press secretary, these years were the most instrumental in developing Obama's present level of intellectualism. "Barack put himself through something that most of us never do—you know, that whole period of just reading and growing your mind. That's the difference between him and a lot of people who haven't succeeded, I think."

Obama focused on himself and did not give much credit to his college professors in his book, but a friend of Obama's said that one Occidental instructor in particular played an enormous role in Obama's intellectual evolution. Roger Boesche taught two classes taken by Obama: American Political Thought and Modern Political Thought. The latter had a lasting effect.

Boesche's personal hero is Thomas Jefferson, as well as others who advocated for participatory democracy. Yet Boesche said he did not push his own political predilections on his students. "My philosophy of teaching is for them to think for themselves. I've always thought that in a good political theory class, students come to their own conclusions," Boesche said. "I don't really care how they end up. I don't believe in converting students to my way of thinking. Whether they come out and become radicals or liberals or conservatives, I want them to rethink their assumptions and come out on the other end. I can be Plato one week or Aristotle for a week."

Boesche actually had only a vague recollection of Obama, which was interesting, considering that Obama told his friend how much Boesche had influenced his thinking and life choices. When Obama was his student, Boesche had been a college teacher for only a few years. In his early thirties at the time, Boesche was young enough that he would occasionally mingle with Obama and his peers in a campus burger-and-fries joint called The Cooler. And though his memories of Obama were scant, he vividly recalled one lunchtime chat in which he urged Obama to apply himself more vigorously to his studies. He said Obama showed great promise, but that promise went unfulfilled at Occidental. "You can often see the potential in a student's papers and in questions in class," Boesche said. "I would put Barry into that category. It was my feeling that his performance hadn't matched his immense talent. I was pushing him, I guess, to develop that talent."

When I told Boesche that just a couple of years later, Obama had devoured the writings of Nietzsche and other political thinkers during his solitary spell in New York City, Boesche said that Obama certainly must have gone through a philosophical growth experience. "Nietzsche calls everything into question. You have to call everything into question. He says God is dead," Boesche said. "So if he kept reading Nietzsche, he went through a whole reasoning process in which he reevaluated all the core beliefs that he had—and then came out on the other side."

Obama affirmed Boesche's observation. He considered Boesche's class on modern political theory his favorite course. He went so far as to say that Boesche's influence and this course, more than anything, spurred his following period of deep self-reflection and study. Boesche had a way of drawing out Obama's intellect, which had not yet been fully engaged. He did this by, of all things, giving Obama a B on an exam that Obama was certain deserved an A. Obama said that he was still "partying pretty hard" but found Boesche's classes special. "I knew that, even though I hadn't studied, that I knew this stuff much better than my classmates," Obama said. "I went to him and said, 'Why did I get a B on this?' And he said, 'You didn't apply yourself.' He was very much grading me on a different curve. And I was pissed."

Obama said that Boesche was among a number of people who helped him mature, who guided him out of his self-absorbed, angst-ridden teenage years into a serious adulthood. "There were people who recognized my potential and who were willing to challenge me on some of my less productive behavior, and I think that helped increase a sense of seriousness," Obama said. "Most importantly, it got me to recognize that the world wasn't just about me. There was a bigger world out there and I was luckier than most and I had an obligation to take not only my own talents more seriously but also see what I could contribute to others."

Indeed, Boesche's toughness tactic worked, and Obama worked even harder at stimulating and challenging himself intellectually. That evolved into his growth phase in New York City.

"THOSE TWO YEARS WERE EXTREMELY IMPORTANT FOR ME," HE said. "I just stripped everything down and sort of built things back up. For about two years there, I was just painfully alone and really not focused on anything, except maybe thinking a lot."

Obama graduated from Columbia in 1983 with a bachelor of arts degree in political science. He stayed in Manhattan and took a position with the Business International Corporation, a firm that published newsletters on global business and offered consulting to American companies operating overseas. Obama was an editor and research assistant in the firm's international financial information division. For a year, he "researched, wrote and edited articles, reports, and how-to manuals on international business and finance for multinational corporations," according to his résumé. Obama told me along the Senate campaign trail that this job did little to excite him, but it gave him a quick education in modern business, international finance and the "coldness of capitalism." In *Dreams,* Obama wrote about feelings of ambivalence toward this job and the progression of his life. His Manhattan apartment was on the trendy Upper West Side. He recalled that he barely recognized himself in suit and tie and briefcase, although he also had some visions of grandeur, of being a powerful businessman cutting major international deals. These observations of his own personal privilege ate at Obama. "I would imagine myself as a captain of industry, barking out orders, closing the deal, before I remembered who it was that I had told myself that I wanted to be and felt pangs of guilt for my lack of resolve."

Indeed, Obama had told his grandmother Toot that he wanted "to leave the world a better place." Becoming a wealthy and powerful captain of industry would not fulfill that mission. He had begun toying with the idea of becoming a neighborhood activist who would help empower the poor, although he did not know exactly how to go about doing that. He grew more unfulfilled in his day job and ultimately quit. Obama said he wanted to work close to the "streets," and took a couple of part-time organizing jobs in Harlem and Brooklyn that barely paid the rent. He sent out résumés and searched for a regular community organizing post, but none appeared. He had begun considering a return to the business world when he received a call from a man named Jerry Kellman, a community organizer in Chicago. Kellman had placed an ad in a trade publication seeking new hires for an organizing drive he was launching in a poor black neighborhood on Chicago's Far South Side. Being white and Jewish, Kellman and his partner had difficulty gaining the confidence of black residents. When Obama answered the ad, Kellman thought about Obama's name and wondered about its racial ancestry. "Could this be Asian?" he asked his wife, who was Japanese American. She said it might be. The two met in a coffee shop in Manhattan; Kellman was immediately impressed. First, Obama was black. It was especially hard to find a young, college-educated black man willing to work in community organizing for just thirteen thousand dollars a year, which was all Kellman could pay. Second, Obama was intelligent and hungered to serve society in a positive way. For Obama, whose soul was ever restless, Kellman offered an opportunity to experience a new city and a new culture. He also offered Obama something more: an opportunity to chase that unselfish life mission that he had confided to his grandmother.

The Organizer

He had this really refreshing dream and I was like, "Barack, no, no, no. Not going to happen."

—THE REVEREND JEREMIAH A. WRIGHT

Barack Obama arrived in Chicago in June 1985 at a tender twenty-three years of age, still wildly idealistic about the intrinsic goodness of mankind. But Obama's first Chicago experience would open his youthful eyes to how cynical, complicated and unjust the world can be—especially when politics, race and power are as inextricably intertwined as they are in the largest middle-American city. But these Chicago years also educated Obama about more than the shortcomings of human beings. For a mixed-race young man, it was his first deep immersion into the African-American community that he had longed to both understand and belong to. The means by which he was immersed in this society—through community organizing—also made a lasting imprint on Obama, an impression that would greatly affect how he later conducted himself as a politician.

After he rolled into the City of Big Shoulders in a beat-up old Honda, Obama still kept a daily journal of his observations and emotions, still read philosophy and literature extensively and still led a rather isolated personal life. His daily existence so resembled the solitary life of a writer that some of his fellow community organizers wondered whether he had undertaken organizing simply to find grist for his version of the great American novel. They would not be too far off, for a few years later Obama would vividly recount his time in Chicago when writing *Dreams,* although the prose would be a loose form of nonfiction. According to Jerry Kellman, "One of the things Barack thought he might want to be is a writer. When he came to Chicago, he was still contemplating that. He wrote short stories. That's a very different side from where he eventually went. He talked about how his father had made some wrong moves and wound up destitute, and he didn't want that for himself."

Obama was drawn to community organizing because it forced him directly into neighborhoods of poverty and despair. In New York, a former activist whose nonprofit had grown into a powerful entity reviewed Obama's résumé and offered him a job that seemed to fit his well-educated skill set—organizing conferences and lobbying politicians on behalf of poor black communities. But Obama wanted to be closer to the real lives of the dispossessed and dispirited, and he turned down the offer. Community organizing in Chicago placed Obama in the midst of a unique American culture. Chicago's South Side is filled with neighborhood upon neighborhood of African Americans, the largest single grouping of blacks anywhere in the nation. In the mid-1980s, the South Side ranged from communities of the stable middle class to neighborhoods of dire poverty. In the poor neighborhoods, violence, drugs and crime infected nearly every aspect of daily life. But even in some of the predominantly middle-class sections, a significant class dichotomy between middle and low income was present, creating tension between neighbors, between the haves and have-nots.

Kellman assigned Obama to the Roseland and West Pullman neighborhoods on the city's Far South Side. Kellman was Obama's tutor and sounding board through his many frustrations of organizing people who were financially poor, confused about authority and, in some cases, illiterate. Kellman cautioned Obama that his task would not be easy and that he should be prepared for utter failure, but that if he concentrated on a few specific problems, he could make a difference.

"All I had to do was to teach him not to be idealistic and he did the rest," Kellman said. "You can't go out there and do any kind of significant political work or organizing and be idealistic. And he was idealistic, almost ridiculously so. You know, it is in his nature. He was a dreamer. But at the same time, you can't perceive people through rose-colored glasses—right away, you have to get a sense of them."

Roseland, once a Dutch farming community named for the bushes that thrived in its soil, reflected a schism in prosperity. The neighborhood is nearly all black and is adorned with street after street of well-manicured lawns and well-maintained houses. But it is also home to the occasional block of blight, poverty, unemployment and crime. That makes the lives of all its residents vastly more precarious than those of, say, the city's whites on the Northwest Side. "For the middle class, it's hard to maintain the high lofty goals you have for yourself when this abject poverty is all around you," said Pat DeBonnett, executive director of a nonprofit group devoted to bringing business to several Far South Side neighborhoods. During the migration of southern blacks to Chicago from 1914 through the 1950s, Roseland suffered extreme disinvestment as nearly all its white inhabitants departed. White flight, in turn, translated into depressed property values, which translated into mortgage defaults, business failures, housing foreclosures, crime and unemployment. Then, in the 1980s, the decline of the South Side industrial base stung Roseland. By the mid-1980s, about one in six Roseland residents lived below the poverty line. The ubiquitous presence of the poor created an emotional paradox for middle-class blacks. Residents of moderate wealth were mindful that their neighbors lived in deprivation, a situation that instills a sense of compassion for those less fortunate. But middle-class blacks also harbored some resentment toward their poor neighbors because they dragged down the community's overall standard of living. "It creates a real division in a community—and I'd even say a sense of division within the race," DeBonnett said.

Into this uncomfortable, complicated stew dropped Obama. Kellman had hired him to launch the Developing Communities Project, an ecumenically funded group whose mission still today is to empower the poor and disenfranchised through grassroots organization. The group is based in the community organizing tradition of Saul Alinsky. Alinsky's activism in the first half of the twentieth century ran parallel with the labor movement, although he expanded the theory of organizing people to include grievances other than employment. Born in 1909, Alinsky grew up in Chicago's gritty Back of the Yards neighborhood during the Great Depression. He made his first political efforts in his home community of blue-collar European immigrants, organizing meatpackers to agitate for better working conditions. He believed in public confrontation in the form of sit-ins and boycotts, but his success stemmed mainly from instilling people with the belief that they had the power to redefine their own lives, through both activism and personal behavior. In his manifesto *Rules for Radicals,* Alinsky wrote, "When we talk about a person's lifting himself by his own bootstraps, we are talking about power. Power must be understood for what it is, for the part it plays in every area of our life, if we are to understand it and thereby grasp the essentials of relationships and functions between groups and organizations, particularly in a pluralistic society. To know power and not fear it is essential to its constructive use and control." Alinsky taught organizers to work behind the scenes, listening to residents for hours upon hours to decipher what their community needed and what it could realistically achieve.

Alinsky's life mission and his methodologies are both central to Obama's modern political message. As noted before, a recurring passage in many of Obama's speeches is his mission of "giving voice to the voiceless and power to the powerless." But Obama also speaks frequently about self-reliance as the most effective means of ultimately pulling oneself out of financial and social distress. Alinsky himself was politically active, traveling the country and attempting to politicize the masses. In the process, he established an institute that trained, among others, Cesar Chavez, founder of the United Farm Workers union. Alinsky's work also influenced the civil rights movement and anti–Vietnam War protestors. Obama wrote that he was drawn to Alinsky's form of direct action. "Once I found an issue people cared about, I could take them into action. With enough actions, I could start to build power. Issues, actions, power, self-interest. I liked these concepts. They bespoke a certain hardheadedness, a worldly lack of sentiment; politics, not religion."

One of Obama's initial discoveries about Chicago's broad swath of South Side African-American communities was the power and influence of the church in these neighborhoods. Whether it was the Catholic Church, African Methodist, Baptist or something else, the larger churches and their pastors exercised great influence over the local politics. And this was the first hard lesson that Obama learned about these institutions: Most of these ministers operated independently of the others, to the degree that they competed for congregants—and competed for power. Obama had idealistic notions of uniting these various conflicting personas and personal agendas into a coherent whole, to work on behalf of the entire area. But it did not take long for him to learn just how discouraging that mission would be in reality. "It wasn't until I came to Chicago and started organizing that all this stuff that was in my head sort of was tested against the reality," Obama said. "And in some cases, it didn't always work out. That's why I say that the best education I ever received was as an organizer because it reminded me that you can look at a map, but that's not the actual territory."

ONE OF THE MOST INFLUENTIAL OF THESE SOUTH SIDE PASTORS was Jeremiah A. Wright of the Trinity United Church of Christ. Wright tried to explain to Obama that the terrain on the map was treacherous. In fact, Wright counseled the young Obama that his dream of unifying the Chicago pastors was impossible:

> The first day we met, I said, "Do you remember the story about Joseph in the Old Testament, where he had these dreams and he shared them with his brothers? Then one day, Joseph was coming across the field toward his brothers and they said, 'Behold, the Dreamer.'" I told him, "That's what I feel like when I listen to you, Barack. You are a dreamer." He came up with this dream of organizing the churches of Chicago like you would organize the churches in New York or on the East Coast. His vision and view was to implement significant change and meaningful change that benefited people, real people. Not politics, not self-aggrandizement, not worrying about which politicians would look good. But how can we bring about change in our communities that the people want and need? I said to him, "Oh, that sounds good, Barack, real good. But you don't know Chicago, do you?" [Wright then broke into a hearty laugh.] Barack said to me, "You are a minister. Why are you sounding so skeptical?" . . . And I said, "Man, these preachers in Chicago. You are not going to organize us. That's not going to happen." He had this really refreshing dream and I was like, "Barack, no, no, no. Not going to happen."

Wright's pessimism was born in a cold reality, a lesson Obama would soon take to heart. In *Dreams,* he recounted a community police meeting that he organized in which one minister derided him for being a naive pawn in the hands of Chicago's whites—lakefront Jewish liberals and the Roman Catholic Church. Indeed, individuals from both of those groups were benefactors of Kellman's Developing Communities Project. The meeting "was a disaster," Obama said. At this juncture in his life, Obama's religious beliefs were similar to his mother's—a secular humanist who was largely agnostic. In *Dreams*, Obama described how religion was being used in Chicago's churches, and he questioned the true power of religious faith and religious institutions to make positive change: "For there were many churches, many faiths. There were times, perhaps when those faiths seem to converge—the crowd in front of the Lincoln Memorial, the Freedom Riders at the lunch counter. But such moments were partial, fragmentary. With our eyes closed, we uttered the same words, but in our hearts we prayed to our own masters; we each remained locked in our own memories; we all clung to our own foolish magic."

A main tenet of the Alinsky organizing philosophy was attention to listening—to pull together the masses for a common cause, the organizer must hear and understand the limitations, the fears and the experiences of the people being assembled. Working out of a small office in a church, Obama was assigned to conduct twenty to thirty interviews each week. Residents initially were wary of this overly serious young man from out of town, but Obama slowly gained their trust. "I'm here to do serious work," he told them. Over time, residents began calling him "Baby Face" for his youthful looks. Obama was known for his detailed and calculated planning, a trait he would carry into politics. He did not like to be surprised in a meeting and he especially loathed being unprepared. Kellman's first major project for Obama was assisting the people of a housing project called Altgeld Gardens. Altgeld's sprawling apartment complex was inhabited by two thousand residents, nearly all of them black. Situated just outside Roseland, it was built in the 1940s to house manufacturing laborers for surrounding factories. The complex sits in relative physical isolation amid a huge garbage dump, a noxious-smelling sewage plant, a paint factory and the heavily polluted Calumet River. Altgeld is in a constant state of disrepair, and circumstances have improved only slightly a generation later. Obama's group was one of two working in Altgeld to organize parents and others, with the mission of pushing the Chicago

Housing Authority to repair toilets, windows and the heating system. To some extent, Obama's efforts were successful. He organized residents to successfully lobby city hall to open a job bank, learning along the way that blacks still viewed the labor movement with suspicion. Chicago's trade unions historically had excluded blacks from their ranks, especially as manufacturing jobs dried up.

Of Obama's pursuits in Altgeld, a campaign to remove asbestos drew the most public attention, even if Obama himself garnered practically no publicity. He helped organize two meetings in downtown Chicago. One resulted in the housing authority testing Altgeld for asbestos and persuading an alderman, Bobby Rush, to hold city council hearings on the matter. A second meeting saw a group of several hundred travel downtown for a raucous visit with city officials. That confrontation prompted the housing authority to hire workers to seal off the asbestos. "He was our motivator," resident Callie Smith said of Obama. Kellman suggests that Obama downplayed his own significance because, as an organizer, he wanted to remain behind the scenes and allow the residents to be the public faces of discontent. For instance, Kellman said it was actually Obama who noticed a newspaper ad for asbestos removal services in the main administration building at Altgeld, even though in his book Obama credits an activist he called "Sadie." The ad prompted residents to question whether asbestos was in other parts of the complex too.

When writing his first memoir, Obama made Kellman one of his main Chicago characters, bestowing on him the pseudonym "Marty Kaufman." "Jerry Kellman is whip smart," Obama told me. "One of the smartest men I've ever met." In *Dreams,* Obama described Kellman (or Kaufman) as a bespectacled, ordinary-looking white man in his middle thirties who typically needed a shave and seemed a bit too sure of himself. When I met Kellman in March 2006, he seemed much the same man, only now in his mid-fifties, a bit more put together but still quite certain of his own words. A

New York Jew who converted to Catholicism, Kellman described himself as a "recovering organizer," giving me the impression that he did not view his organizing years with joy. He still worked in a modest-paying job that assisted the poor and disaffected, serving as director of a Roman Catholic social services project in an inner-ring suburb just north of Chicago.

Physically, it was difficult to recall specifics of his face only hours after our interview. But the substance of the interview was etched in my mind for days. Kellman's voice would occasionally drop into such a low pitch that I was drawn to it, not only because I had to strain to hear it, but because it seemed as if he was earnestly imparting a piece of wisdom that he deeply believed—and deeply wanted his listener to believe. His insight into Obama was extraordinarily intimate and accurate, not surprisingly since the two had been daily confidants for almost three years. They talked about everything from community organizing to Obama's vision for his future. Kellman's main mission seemed to be to keep Obama from burning out. He urged him to pursue a more active social life to take a break from the depressing moments of organizing. Obama, who initially was spending most of his weekends reading books, eventually heeded this advice. He started dating and had a live-in girlfriend for a while. But when that relationship ended, it was painful for Obama, who conceded that his own lack of maturity partly contributed to its demise.

Like Obama's grandmother, Kellman described Obama as a man on a mission to serve society. Obama told Kellman that he had been extremely fortunate in his own life, and so he felt a passion to contribute to the betterment of others' lives, especially by pushing for economic and social progress in the black community. "Barack wanted to serve; he wanted to lead," Kellman said. "And he was ambitious, but never just for ambition's sake. It was always mixed in with a sense of service. He is motivated by [a desire for] significant structural change for people he cares about. And he found himself in the African-American community, historically, intellectually and, finally, geographically."

The two men fell out of touch after Obama left Chicago for Harvard Law School, and thereafter Obama's life grew busier and Kellman's grew somewhat erratic. Kellman had not had an extended conversation with Obama since his rise to celebrity status, but in following his public career, he said, it seemed clear to him that the core of Obama's character had remained essentially intact. He felt so devoted to Obama's political cause, in fact, that he contributed fifteen hundred dollars to his Senate campaign. ("That's not much, but it is a great amount for me, someone paid from the Catholic Church," he said sheepishly. Not long after our interview, Kellman called Obama and the two men had an extended conversation.)

Before sending Obama off to Chicago's projects, Kellman urged him to read *Parting the Waters,* the first of author Taylor Branch's rich and thorough trilogy on the civil rights movement of the 1950s and '60s. After finishing the book, Obama confessed to Kellman, "This is my story." It would be yet another indication to Kellman of the seriousness with which Obama took both himself and his public work. But mostly it would be a sign that this man who had suffered such profound racial confusion was finally finding comfort as an American black man. Obama's physical features, after all, give him this appearance. He has joked that cab drivers who passed him by on the streets of New York failed to see his maternal white relatives. Nor have most blacks looked at him and questioned his identity. "I think there are a lot of black and white intellectuals who like to use me as a Rorschach test for their own confusion about race," Obama said. "And the truth of the matter is, you know, when I'm walking down the South Side of Chicago and visiting my barber shop and playing basketball in some of these neighborhoods, those aren't questions I get asked."

Kellman met with Obama on nearly a daily basis and observed as Obama grew more comfortable weaving in and out of the black community in those organizing years. After seeing how Obama's political career unfolded, Kellman made a bold proclamation: Despite chatter in some quarters of the black community that Obama hadn't lived the typical African-American experience, Kellman predicted that he would be the most likely heir to Martin Luther King's legacy as both the chief advocate and the moral voice of black Americans. He said Obama saw this role for himself years ago, even if he is reluctant to admit it publicly today for fear of sounding immodest and perhaps distancing himself from his non-black constituents. Kellman also predicted that Obama would assume this mantle with thoughtfulness and a full understanding of its gravity.

"If you look at the King analogy and you look at Barack," Kellman said, "Barack has become the expectation of his people, and in that sense he is similar to King. As I know Barack, he will carry that as a weight, but he will carry that burden with great seriousness. And obviously, that can cause someone to kind of lose perspective. You can get overly inflated. But it is also cause for some sobering loss of your own importance at the same time. I think he knows that if he wants to go where he wants to go in politics, he has to speak for more than the black community. But I think the rest of his life, he will take on that burden of being that person who changes the situation for African Americans."

A SECOND MAN PLAYED A VITAL ROLE IN OBAMA'S CHICAGO experience—Reverend Wright. Obama noted the influential role that the church played in the lives of the African Americans he was organizing. "So I figured I better attend some services myself and see what it was all about," he said. It was Wright's church and his charismatic preaching that most closely spoke to Obama's budding spiritual nature. In seeking a permanent American identity, Obama discovered that, in a religious land, his agnosticism relegated him to a place of isolation. "I came to realize that without a particular commitment to a particular community of faith," he wrote in his book *The Audacity of Hope,* "I would be consigned at some level to always remain apart, free in the way that my mother was free, but also alone in the same ways that she was ultimately alone."

Wright was among the most liberal of African-American preachers—he could be all fire and no brimstone. When Obama knocked on Trinity's door, Wright was in his mid-forties and in the midst of growing his Trinity congregation to its present membership of nearly eight thousand. Burly and light-skinned, Wright is the son of a Baptist minister in Philadelphia. His intellectual sermons sometimes more resemble left-wing political rants than religious preaching. Startling for a preacher, he can be both profane and provocative. Despite advancing a multicultural agenda, like Obama, Wright's church is rooted in Afrocentrism. Wright himself often dons colorful African dashikis and is not shy about laying historical and modern-day blame on whites for much of the social and economic woes in the African-American community. His sermons frequently denounce Republican politics, and he has called people who voted for George W. Bush "stupid." Trinity United is considered among some Chicago blacks to be the church of elites, attracting celebrities like the rapper Common and TV talk mogul Oprah Winfrey to its congregation.

In some ways, Obama and Wright seem a mismatch because of their distinctively different styles. But in other ways, they seem like a perfect fit—an attraction of opposites. In contrast to Obama's cautious style, Wright is bombastic, rebellious and, in his own estimation, unafraid to speak truth to power. Wright earned bachelor's and master's degrees in sacred music from Howard University and initially pursued a Ph.D. at the University of Chicago Divinity School before interrupting his studies to minister full-time. His intellectualism and black militancy put him at odds with some Baptist ministers around Chicago, with whom he often sparred publicly, and he finally accepted a position at Trinity. The church's motto was "Unashamedly black and unapologetically Christian." Wright unquestionably took that motto to heart.

Wright remains a maverick among Chicago's vast assortment of black preachers. He will question Scripture when he feels it forsakes common sense; he is an ardent foe of mandatory school prayer; and he is a staunch advocate for homosexual rights, which is almost un-heard-of among African-American ministers. Gay and lesbian couples, with hands clasped, can be spotted in Trinity's pews each Sunday. Even if some blacks consider Wright's church serving only the bourgeois set, his ministry attracts a broad cross section of Chicago's black community. Obama first noticed the church because Wright had placed a "Free Africa" sign out front to protest continuing apartheid. The liberal, Columbia-educated Obama was attracted to Wright's cerebral and inclusive nature, as opposed to the more socially conservative and less educated ministers around Chicago. Wright developed into a counselor and mentor to Obama as Obama sought to understand the power of Christianity in the lives of black Americans, and as he grappled with the complex vagaries of Chicago's black political scene. "Trying to hold a conversation with a guy like Barack, and him trying to hold a conversation with some ministers, it's like you are dating someone and she wants to talk to you about Rosie and what she saw on *Oprah,* and that's it," Wright explained. "But here I was, able to stay with him lockstep as we moved from topic to topic. . . . He felt comfortable asking me questions that were postmodern, post-Enlightenment and that college-educated and graduate school–trained people wrestle with when it comes to the faith. We talked about race and politics. I was not threatened by those questions."

Wright also played an assisting role in another part of Obama's evolution—from a questioner of religion to a practicing Christian. Along his Senate campaign trail, Obama would never fail to carry his Christian Bible. He would place it right beside him, in the small compartment in the passenger side door of the SUV, so he could refer to it often. When I first questioned Obama about his religious faith and ever-present Bible in October 2004, he seemed just a bit hesitant to answer. He was also uncharacteristically short in his responses. Obama, without fail, would mention his church and his Christian faith when he was campaigning in black churches and more socially conservative downstate Illinois communities. But in speaking to a reporter, it seemed that he had something to say about religion and politics, although he had yet to turn that inclination into a coherent message. He told me that he referred to his Bible a "couple times a week." "It's a great book and contains a lot of wisdom," he said simply. When I pried further, he said he was drawn to Christianity because its main tenet of altruism and selflessness coincided with his own philosophies. "Working with churches and with people of faith, I think, made me recognize that many of the impulses that I had carried with me and were propelling me forward were the same impulses that express themselves through the church," he said.

But more than that, Trinity's less doctrinal approach to the Bible intrigued and attracted Obama. "Faith to him is how he sees the human condition," Wright said. "Faith to him is not . . . litmus test, mouth-spouting, quoting Scripture. It's what you do with your life, how you live your life. That's far more important than beating someone over the head with Scripture that says women shouldn't wear pants or if you drink, you're going to hell. That's just not who Barack is."

Overall, Obama's first Chicago experience proved powerful in his later development as a politician. His Christianity would be well received among blacks and some rural whites. Community organizing taught him that idealism must be coupled with pragmatism and hard realism. This would help him most prominently in his legislative work, when he realized that compromise was often necessary to move a bill forward. Organizing also turned this once self-absorbed teenager and college student into an attentive listener to other people's concerns, a trait that enabled him to perceive the world through other people's eyes and then communicate those concerns to voters.

Obama never got very far in his effort to organize Chicago's pastors, but a magazine essay he wrote upon returning to Chicago a couple of years later showed that he had not lost hope for that dream. In the 1990 article, Obama called the African-American church "a slumbering giant in the political and economic landscape of cities like Chicago." He continued: "Over the past few years, however, more and more young and forward-thinking pastors have begun to look at community organizations . . . as a powerful tool for living the social gospel, one which can educate and empower entire congregations and not just serve as a platform for a few prophetic leaders. Should a mere 50 prominent black churches, out of the thousands that exist in cities like Chicago, decide to collaborate with a trained organizing staff, enormous positive changes could be wrought in the education, housing, employment and spirit of inner-city black communities, changes that would send powerful ripples throughout the city."

After some losses, some victories and new spirituality, Obama summed up his experience working for the Developing Communities Project by strongly endorsing community organizing as an effective means of advancing American society toward the ideal of equity and justice for all:

"In helping a group of housewives sit across the negotiating table with the mayor of America's third largest city and hold their own, or a retired steelworker stand before a TV camera and give voice to the dreams he has for his grandchild's future, one discovers the most significant and satisfying contribution organizing can make.

In return, organizing teaches as nothing else does the beauty and strength of everyday people. Through the songs of the church and the talk on the stoops, through the hundreds of individual stories of coming up from the South and finding any job that would pay, of raising families on threadbare budgets, of losing some children to drugs and watching others earn degrees and land jobs their parents could never aspire to—it is through these stories and songs of dashed hopes and powers of endurance, of ugliness and strife, subtlety and laughter, that organizers can shape a sense of community not only for others, but for themselves.

6
Harvard

We had the sense . . . that he genuinely cared what the conservatives had to say and what they thought and that he would listen to their ideas with an open mind. And so there was just a much greater comfort level with the notion of Barack . . . than some of the others.
—BRAD BERENSON, A HARVARD LAW SCHOOL CLASSMATE

Community organizing is high-burnout work. Battling a harsh world that often ignores the poor and dispossessed gave rise to perhaps the first extended period of disillusionment in the life of the young idealist from Hawaii. Before that point, Barack Obama had had little contact with failure. At a minimum, he had found modest success at whatever he tried, from pickup basketball to college studies to attracting young women.

Initially, one of his biggest organizing projects—the asbestos campaign in Altgeld Gardens and its nice run of media coverage in Chicago—seemed to be a small victory for Obama and other Altgeld activists. Ultimately, though, the victory proved ephemeral. Budget priorities in Washington soon sapped much of the activists' sense of lasting accomplishment. Federal officials told the public-housing residents that they had a choice between repairing Altgeld's antiquated plumbing and leaky roofing or cleaning up the poisonous asbestos—there was not enough funding for both. This left many residents wondering what all their hard work and activism had wrought. "Ain't nothing gonna change, Mr. Obama," one dispirited resident complained. Moreover, squabbling between members of

Obama's Developing Communities Project and another group did not help instill optimism about the long-term success of organizing efforts in Altgeld. Thus, a couple of residents told Obama they were so busy trying to make ends meet in their households that they no longer had time for public activism, especially since it appeared to be leading to scant advancement.

As these frustrations mounted for Obama, an abrupt external event radically altered the political and racial landscape in Chicago. In November 1987, the city's first black mayor, Harold Washington, suffered a massive heart attack at his desk in city hall and died. Washington's election in 1983, the year before Obama arrived, was a seminal moment for Chicago's African-American community, even if it cracked wide open the city's deep and lingering racial divide. It was also an object lesson for Obama in the real power that American democracy bestows on its duly elected political leaders.

Organizers had registered more than one hundred thousand new black voters, and Washington squeaked through a three-way Democratic primary in which the vote was largely divided along racial lines. The two white candidates, incumbent Jane Byrne and Richard M. Daley, son of the longtime mayor Richard J. Daley, had split the white vote. Washington, meanwhile, won nearly all the black vote and captured just enough support from liberal, reform-minded whites. Typically in Chicago, where the Cook County Democratic Party reigns supreme, the Democratic primary winner is a lock in the fall general election. But Washington was a different story. He barely eked out a November victory against Republican Bernie Epton, who suddenly found support among white Democrats and powerful ward organizations in white neighborhoods. Even after Washington took office, he encountered stubborn white resistance on the city council and could not win enough votes to enact his programs and appoint his chosen nominees. The city council descended into a bitter, racially polarized battle dubbed the "Council Wars," after the then-popular science fiction film *Star Wars*. In 1986, two years after Washington took office, special elections were called and enough Washington-endorsed candidates were victorious to allow his agenda to go forward.

Obama wrote poignantly in *Dreams* of the significance of Washington's ascendance in the lives of Chicago's more than one million blacks. When Washington died, it was a devastating blow to that community. Thousands attended his two-day wake in the lobby of city hall. "Everywhere black people appeared dazed, stricken, uncertain of direction, frightened of the future," Obama wrote.

With Washington's abrupt death, and considering his own modest organizing successes, Obama grew dispirited. This Chicago adventure—to follow his sense of mission in assisting poor African Americans—was worthwhile in many ways. It opened his eyes to both the complexities and the shortcomings of a unique racial culture, as well as to the cruel reality of American priorities. But after three years of working close to the "streets" he had so longed for, Obama felt shackled by the limited power of a small nonprofit group to create expansive change.

Bobby Titcomb, his friend from the Punahou Academy, visited Obama around this time and found his teenage chum far less optimistic and readily enthusiastic than during their Hawaiian childhoods. One night, an extremely frustrated Obama arrived at his apartment near the University of Chicago from a meeting of unhappy residents at a church. "I just can't get things done here without a law degree," Obama told Titcomb. "I've got to get a law degree to do anything against these guys because they've got their little loopholes and this and that. A law degree—that's the only way to work against these guys." It had not gone unnoticed by Obama that Mayor Washington had a been a graduate of the Northwestern University School of Law, and he had parlayed that lofty degree and his own personal charisma into a highly successful political career. Indeed, Obama watched closely as Washington unabashedly wielded his power to pour resources into the city's ailing minority communities. The mayor led the fight to redistrict council wards to give more representation to Latinos and blacks; he issued an executive order to increase city contracts with minority-owned businesses; he upgraded city services in poor black and Latino neighborhoods. Washington could do more for Chicago's poor blacks with the wave of his veto pen than Obama could in countless days and nights of community meetings in Roseland and Altgeld.

Titcomb was not prone to Obama's activism and was grappling with a set of family problems at the time, so he had only a partial understanding of Obama's personal disheartenment. But he could see the resolve in his friend's eyes. Obama would soon be accepted at the most prestigious law school in the country, lifting him even higher into a world of intellectual elites and setting him on a course to the kind of political power that even Harold Washington could only dream of possessing.

OBAMA ARRIVED AT HARVARD LAW SCHOOL WITH A UNIQUE pedigree. At twenty-seven, he was several years older than his typical classmate, who most likely had come straight from a highly regarded undergraduate school. Like many of the students, however, Obama had been privileged to receive an elite primary and secondary education. Punahou Academy, Occidental College and Columbia University—all were excellent private schools. Yet Obama's life was already much different from that of the typical Harvard Law student. He was a black man from Hawaii who had been a community organizer in Chicago for three years; by now, he had also taken his first trip to Kenya to explore his father's roots. After his several years of a minimalist life in New York, and several years of diligent work in Chicago's neighborhoods, Obama had attained a maturity level and an extraordinary degree of self-discipline that would greatly abet his success at Harvard. He was now committed to his studies as never before, and his grades reflected this—he would graduate magna cum laude. A classmate, Michael Froman, who would later work in Bill Clinton's Treasury Department, said it was readily evident that Obama was operating on a different plane from everyone else. "He was mature beyond his years in being able to approach issues in the way that he did. I saw people who were much older and more seasoned who have similar attributes as Barack. And he was doing this in his twenties. This is a temperament and a style that he clearly developed ahead of his peers." (In 2004, Froman would reconnect with Obama and serve as a key adviser in organizing his Senate office.)

As in New York and, to some degree, Chicago, Obama spent a vast amount of time by himself while in Cambridge. In his initial year, his daily routine included carving out a spot for himself in one of the sections for first-year students at the library and burrowing in for several hours of intense study.

Obama again made friendships with the small number of black students on campus. But after largely coming to terms with his mixed racial ancestry, he also reached out and secured several close white friendships. He researched and wrote articles for the *Harvard Civil Rights–Civil Liberties Law Review.* He was active in the anti-apartheid movement on campus, gave a speech at the annual dinner hosted by the Black Law Students Association and served on the association's board of directors.

Harvard was the forum in which Obama's long-term public message of unity and altruism would take its first oratorical shape. At the Black Law Students' dinner, Obama asked his audience to remember that their privileged education meant that they now had the means, opportunity and, yes, the responsibility to use their prestigious law degrees to return something to the less privileged. Friends and professors recalled that Obama invoked similar rhetoric in his often fiery and inspirational speeches concerning the importance of cultures and ideas mixing on campus—his hope being that if students with differing philosophies interacted more often, they would be less wary of opinions the opposite of their own. This bridge-building approach to racial and partisan politics was the first indication of the tightrope he would walk throughout his career. He would give voice to minorities who felt slighted but in a conciliatory tone that did not threaten whites; in addition, he would give conservatives the impression that he was willing to listen to their arguments. In his political career, Obama would deliver numerous speeches along these lines—about the value of helping those less privileged and the ideal of multiculturalism. And, of course, Obama would center his presidential campaign on the message of uniting a bitterly divided country.

His message of racial and intellectual unity may not necessarily have been controversial, but it certainly flew in the face of events on the Harvard campus in the early 1990s. Black students, in particular, were agitating for more minority representation among the faculty ranks. Toward that end, a group of blacks sued the school for discrimination; and a black professor, Derrick Bell, resigned over the matter. Obama largely steered clear of the fray, but he did give a speech calling for greater faculty diversity and heaping praise on Bell for his defiant stand.

Besides these racial tensions, Harvard Law was in the midst of a bitter ideological war. Liberals and conservatives waged fierce intellectual battles in classrooms, lunchrooms, at parties and, of course, in the offices of the prestigious *Law Review*. Despite these differences, Harvard Law is considered by most to be a bastion of liberal thought. This left-leaning history is rooted in the Vietnam War era when progressive activists at Harvard Law railed against school policies and, most vociferously, the war. In recent years, liberal professors like Laurence Tribe and Alan Dershowitz have gained national reputations, pushing the school's image even farther in that direction. *The Economist* magazine went so far as to dub the school "the command centre of American liberalism."

Obama immediately made a mark at Harvard among his peers and professors. He went to work for Tribe as a research aide and so thoroughly impressed the renowned legal scholar that Tribe would later call Obama his "most amazing research assistant." In a 2006 Harvard admissions blog written by an administrator, Tribe added this incredible statement: "He's a guy I hope will be President someday."

Obama made enduring friendships at Harvard, one of the most significant being with a black woman named Cassandra Butts, who would later become a senior adviser to Representative Richard Gephardt of Missouri. The two met in the financial aid office the first week of classes. Obama's personal charm, interesting background and unique perspective intrigued her. "But you know, once you get past the charm, what keeps you, what engages you—or at least, what engaged me—was his decency and the intellectual curiosity. And you know, that combination, and the experience that he had from his work as an organizer, the international experience that he had— he just saw the world in a different way than anyone I had met, to that point, and definitely anyone who was in law school with us. And so that, you know, that just made him interesting. [He] was a wonderful filter through which to see what we were learning and how you apply what we were learning to the outside world. He came to discussions with much more life experience than most of the students. I mean, we all had big ideas, but Barack had the experience." This experience often gave his opinions a greater weight, although she said Obama did not offer them as such.

THE MOST GLARING EXAMPLE OF OBAMA'S DIFFERENT PERSPECTIVE came during a large organized discussion among black students of Harvard's law, medical and business schools. The meeting was convened to discuss a hot topic of the day among educated blacks: Should they refer to themselves as "blacks" or "African Americans"? This debate would not end in the 1980s. In an essay in the *Chicago Tribune Magazine* nearly a decade later, columnist Clarence Page explored the source of the issue. He asserted that the label debate was a way for black Americans to define their own identities rather than having the white majority define it for them. But he conceded that this led to great confusion among people of all races. "Diversity is enriching," Page wrote, "but race intrudes rudely on the individual's attempts to define his or her own identity. I used to be 'colored.' Then I was 'Negro.' Then I became 'black.' Then I became 'African American.' Today I am a 'person of color.' In three decades I have been transformed from a 'colored person' to a 'person of color.' . . . Changes in what we black people call ourselves are quite annoying to some white people, which is its own reward to some black people. But if white people are confused, so are quite a few black people."

At Harvard, that identity confusion came forth passionately in the organized conversation among the black students, with participants arguing heatedly for each side. But when Obama stood up to speak, he didn't take sides. Instead, he looked at the discussion from the pragmatic viewpoint of a former community organizer. He said the whole issue was immaterial to the real world. As Butts recalled it, Obama told the crowd: "You know, whether we're called black or African Americans doesn't make a whole heck of a lot of difference to the lives of people who are working hard, you know, living day to day, in Chicago, in New York. That's not what's going to make a difference in their lives. It's how we use our education in these next three years to make their lives better. You know, that's what's going to have an impact on making the U.S. a more just place to live, and that's what's going to have an impact on their lives."

OBAMA'S MOST IMPORTANT EXPERIENCE AND DEFINING ROLE AT Harvard would be his tenure as a writer, editor and, finally, president of the *Harvard Law Review,* the most influential legal publication in the country. It was hard for him to see the significance of this role at the time, but the *Review* presidency would provide him with his first lessons in managing both bitter electoral politics and the personal agendas of individual people.

The *Review* is edited by students, allowing them a high-profile public forum in which to display their research and writing skills. But it also permits scholars and others to voice their arguments in the most highly read publication in the legal community. A *Law Review* post, especially the presidency, is an immediate attention-grabber on any résumé. Even so, the top job held little appeal for Obama. As his first year of law school wound down, he garnered a position on the *Review* through high grades, a writing competition and endorsements from other students and professors. Almost immediately, friends urged him to run for the presidency, but Obama expressed no interest. He told them that overseeing the legal journal would do nothing to enhance his future as a lawyer. After all, he did not envision clerking for a federal judge or seeking a position at a prestigious law firm. Obama wanted to return to Chicago and use his degree to help the city's disadvantaged, or perhaps follow Harold Washington's model and take a stab at politics. Therefore, the credential of *Law Review* president seemed of little value.

In 1990 the *Review*'s staff of about seventy-five students was riven by intense partisan feuding. Large factions of liberals and small bands of conservatives were engaged in titanic ideological struggles, with each side trying to push the *Review*'s leading viewpoint in its own direction. Since the liberals outnumbered the conservatives, they mostly dominated the discussion. But a growing movement of conservatives on campus had gained a foothold and was pushing for a larger voice. The Federalist Society, a national group of attorneys and law students that describes itself as "conservatives and libertarians dedicated to reforming the current legal order," was gaining more adherents among students of Harvard Law. (Over the next two decades, the Federalist Society would grow exponentially and become a force of conservatism at the school. And the group would gain national prominence for exerting great influence in the Republican administration of George W. Bush, with society members appointed to cabinet posts and throughout the Department of Justice.) Brad Berenson, a classmate of Obama who later served in Bush's Justice Department, said he had never seen more vicious political infighting and backbiting than during his Harvard Law days. He said the law school campus "was populated by a bunch of would-be Daniel Websters harnessed to extreme political ideologies." He added that "political rivalries and personal divisions inside the *Review* were just ridiculously bitter, given how little was at stake."

This heated battle of ideology and personalities played out fiercely as the *Review* conducted elections for its 1990–91 president. The electoral process itself is like no other. Inside the *Review*'s cluttered, cramped offices in a three-story Greek Revival building that was formerly a single-family home, candidates spend election day preparing food for their roughly seventy-five colleagues as ballots for the presidency are cast. Over hours of spicy chili and less spicy spaghetti, candidates are eliminated and the field is winnowed down. At the last moment and at the urging of his friends, Obama cast his hat in the ring. He was one of nineteen editors who ran for the presidency—about one in four of the staff. As such, the balloting session was egregiously long that year, lasting from a Sunday morning into the wee hours of the following Monday morning. Candidates eliminated from the next round of balloting typically sit down for a meal and start casting ballots themselves.

Finally, after the last conservative was voted out of the competition, that faction threw its support behind Obama, tilting the election in his favor and bestowing on him the honor of being the first African American to hold the presidency in the more than a century of the *Review*'s existence. As the final outcome was announced, another black student with tears in his eyes, Kenneth Mack, threw his arms around Obama. The decision would provide Obama with his first bit of national media exposure, profiles quickly appearing in the *New York Times* and several other publications. That publicity, in turn, opened up the opportunity for Obama to publish *Dreams*.

Berenson, a member of the conservative group, said that Obama won over his peers for various reasons, but race was not among them: "He was picked completely on the merits." Obama was a devoted liberal, but conservatives believed he would give their opinions a fair hearing. Obama seemed less ideologically rigid and more evenhanded than the other progressive candidates, Berenson recalled. "Barack always floated a little bit above those controversies and divisions. Barack made no bones about the fact that he was a liberal, but you didn't get the sense that he was a partisan—that he allied himself with some ideological faction on the *Review* and had it in for the other ideological factions on the *Review*. He was a more mature and more reasonable and more open-minded person. We had the sense, and I think it was borne out by the experience of his presidency, that he genuinely cared what the conservatives had to say and what they thought and that he would listen to their ideas with an open mind. And so there was just a much greater comfort level with the notion of Barack as president than some of the others."

The conservatives were indeed correct in their assessment. Obama was an avowed social and economic liberal, but his reasonable tone and attentive listening skills gave him a nonthreatening appeal to partisans on the right. Obama, in fact, used some of his appointment power to place conservatives in key editorial positions on the *Review*. He asserted that each viewpoint deserved a fair hearing—a magnanimous sentiment that would produce some criticism from people in his own progressive crowd, as well as from minorities who wanted him to put their advancement at the top of his *Law Review* agenda. But Obama was more interested in making his publication run smoothly and convey diverse opinions than in pleasing everyone in the liberal and black contingents. His tenure as *Review* president, in fact, would foreshadow his future political style: a belief in giving attention to people with views other than his own; a desire to reach across the aisle to form consensus; a tendency to disappoint people in his own crowd—blacks and progressives— by not being more strident in his demeanor or behavior.

Despite this grumbling, however, the *Review* ran rather peacefully under Obama, especially considering how bitter the partisan feuding had been. "He did show great political deftness as president of the *Review* in maintaining good relations with most, if not all, of the editors of the *Review*," said Berenson, the devout conservative. "He made people feel generally included and valued and he got everybody in harness, working toward a common goal, notwithstanding a lot of the other problems and fissures that existed. I remember marveling at the amazing set of interpersonal and political skills that he had. It was a fractious, headstrong bunch. And he led the group with considerable skill and finesse."

One group disenchanted with Obama's *Law Review* performance was some campus blacks who criticized him for not filling more management posts with African Americans. This is a criticism that Obama concedes has dogged him, and would continue to beset him throughout his public career. Obama has consistently advocated racial diversity and affirmative action, but he has also advocated promotion based on merit. So when he appointed some conservatives to the *Review*'s upper editing ranks and bypassed some minorities and women, the criticism rang forth. Butts defended Obama's choices, even though she wasn't as close to the *Review*'s internal workings as he was. She said he was fixed on making the best personnel choices based on talent, dedication and temperament.

Obama, for his part, would not talk in specifics about his personnel choices. But in the weeks leading up to his inauguration as a U.S. senator, he told me that the *Law Review* experience was a precursor to what he expected to confront in Washington—that he would anger some minorities and liberals by concentrating on serving a constituency of all races, ethnicities and political affiliations. "On the *Law Review*, that was the first time I had to deal with something that I suspect I'll have to deal with in the future, which is balancing a broader constituency with the specific expectations of being an African American in a position of influence," he said. "I had to manage a *Law Review* with seventy students who all want their own positions and who all want their advancement. And I had to make decisions about promoting diversity but also ensuring that people feel that I am being fair. So as for the criticism, I'm not sure there was anything all that surprising about that."

As his Harvard schooling wound down, Obama began setting his sights on the mission that he had been training for—politics back in Chicago. All through his time in Cambridge, he had never mentioned running for any office but mayor of Chicago. Harold Washington's tenure in that position had so impressed Obama that, in his mind, overseeing Chicago City Hall was the top political job in the country. "He wanted to be mayor of Chicago, and that was all he talked about as far as holding office," Butts said. "He never talked about the U.S. Senate; he never talked about being governor. He only talked about being mayor, because he felt that is really where you have an impact. That's where you could really make a difference in the lives of those people he had spent those years organizing. He could have gone on to great acclaim, but those people still were his mission."

Sweet Home Chicago

*Barack was like, "Well, I wanna be a politician. You know, maybe
I can be president of the United States." And I said, "Yeah, yeah,
okay, come over and meet my Aunt Gracie—and don't tell
anybody that!"*

——CRAIG ROBINSON, BARACK OBAMA'S BROTHER-IN-LAW

Barack Obama seemed to know almost immediately upon meeting
a round-eyed, statuesque African-American lawyer named Michelle
Robinson that she was his choice for a spouse; the young Miss Robinson was far less sure about her future husband. And that in itself
says much about the two people: Barack is the romantic dreamer;
Michelle is the balanced realist. Upon meeting her, he was swept off
his feet; she took some convincing.

Obama and Robinson met after Obama's first year at Harvard
Law, in 1988, when he was a summer intern in the Chicago office
of the high-brow law firm now called Sidley Austin. Robinson, a
young lawyer at the firm, was assigned to be his mentor. Initially,
Robinson was skeptical about Obama because, even before he arrived for the summer, he had been talked up by so many others at
the firm—too many others, she thought. Secretaries gossiped about
how handsome he was. Associates marveled at his magnificent
first-year performance at Harvard. Senior partners hailed an intro-
ductory memo by Obama as nothing short of brilliant. "He sounded
too good to be true," Michelle recalled. "I had dated a lot of brothers who had this kind of reputation coming in, so I figured he was

one of these smooth brothers who could talk straight and impress people. So we had lunch, and he had this bad sport jacket and a cigarette dangling from his mouth and I thought, 'Oh, here you go. Here's this good-looking, smooth-talking guy. I've been down this road before.' Later I was just shocked to find out that he really could communicate with people and he had some depth to him. He turned out to be an elite individual with strong moral values."

Obama suffered no false preconceptions about Michelle and was immediately taken with her. Nevertheless, at first she resisted his amorous advances. She thought it would be improper to date an employee she was assigned to train. In addition, they were the only two African Americans at the law firm. "I thought, 'Now how would that look?'" Michelle said. "Here we are, the only two black people here, and we are dating. I'm thinking that looks pretty tacky." Michelle tried to set up Obama with a friend, but he showed no interest in anyone but her. Eventually, she relented and agreed to a date, and, over chocolate ice cream at a Baskin-Robbins shop near the University of Chicago, he won her affection. When Obama returned to Harvard in the fall, the two carried on a long-distance relationship that Obama conceded would have been impossible just a few years before. "Before I met Michelle, I was too immature to hold something like that together," Obama told me, acknowledging that as he approached thirty he gained a different perspective on a secure romantic relationship. As a community organizer, Obama had had a serious girlfriend (and a pet cat), but all three parted amicably when he went to Harvard.

During his late twenties, Obama shifted his thinking toward the value of marriage and family. Even though his restless mind and ambitious energy made him fear a static life, he was beginning to desire a stable relationship and family. He considered the upheaval of his father's family life and longed for a different outcome for himself. On trips back to Chicago from Harvard, Obama often visited Jerry Kellman and his wife at their house in the Beverly neighborhood, one of Chicago's few racially mixed communities; he looked out at Kellman's backyard and confided that he wanted "this kind of stability."

Michelle Obama grew up in a tightly knit, working-class family in the South Shore neighborhood of Chicago's sprawling South Side African-American community. Obama, who sometimes described his own childhood as that of "an orphan," is fond of chiding his wife for "being raised by Ozzie and Harriet," a reference to the idyllic American family from the 1950s sitcom.

The Robinsons lived in a small apartment on the top floor of a classic low-slung Chicago bungalow. Michelle's father, Frasier Robinson, worked odd-hour shifts overseeing the inner workings of boilers at the city's water filtration plant. Her mother, Marian, did not work outside the home until Michelle reached high school, when she took a position as an administrative assistant in the trust division of a bank, a job she still held in 2007. Michelle has one sibling, a brother, Craig, sixteen months older than she. A talented basketball player, he ultimately left a lucrative job in high finance to coach in the college ranks. (In 2006 he became head coach at Brown University.) Their father suffered from a debilitating illness that family members believe was multiple sclerosis, although he never received an official diagnosis. Both children were indelibly shaped by their father's unstable physical condition—and the strong will he showed in coping with it. He was devoted to setting a sturdy paternal example and sufficiently providing for his family. He rarely missed work or time with his children, even as his physical state deteriorated. "We always felt like we couldn't let Dad down because he worked so hard for us," Craig Robinson said. "My sister and I, if one of us ever got in trouble with my father, we'd both be crying. We'd both be like, 'Oh, my god, Dad's upset. How could we do this to him?'"

Much of Michelle's childhood was spent scurrying on the heels of her older brother and, to some degree, living in the tall shadow he cast. Craig was a good student and popular athlete, and Michelle had a deeply competitive spirit. Long limbed and extremely tall at five feet eleven inches, Michelle showed great athletic prowess in the neighborhood, often holding her own on the basketball court with her brother and his friends. But to fight comparisons with her high-achieving brother, she decided against playing organized sports. Instead, she immersed herself in pursuits of her own—learning the piano, writing short stories in her spiral notebook, serving as student council treasurer and excelling in school. She skipped the second grade and consistently made the honor roll at Whitney Young High School, one of the premier public institutions in the Chicago system. Her academic excellence secured her acceptance at Princeton University. She graduated cum laude from Princeton and then, like her husband, went to Harvard Law. Her brother also attended Princeton, at the urging of their father. Craig was gifted enough on the basketball court to attend a school with a strong Division I program, and he was offered full-ride scholarships at several colleges with top basketball programs. But his father said education was more valuable and sent him to Princeton, even though he had won only a partial scholarship. "Dad said it didn't matter about the cost—it was the education that was important," said Craig, who went on to be one of the top players in the history of the Ivy League.

At Princeton, Michelle underwent a racial identity crisis similar to what Obama experienced in his formative college years. For the first time in her life, she had stepped into a nearly all-white cultural and academic setting. And even though she was popular and quickly acquired a handful of good friends, she admitted in a thesis, titled "Princeton-Educated Blacks and the Black Community," that she felt racially isolated as one of the few black women on the campus. "I have found that at Princeton, no matter how liberal and open-minded some of my White professors and classmates try to be toward me, I sometimes feel like a visitor on campus, as if I really don't belong. Regardless of the circumstances under which

I interact with Whites at Princeton, it often seems as if, to them, I will always be Black first and a student second." Yet as she progressed through Princeton, Michelle said she realized that as an alumna of an elite college, she probably would move in predominantly white circles later in her professional life. She had wanted to use her education to serve the black community in some way, but she wrote that "as I enter my final year at Princeton, I find myself striving for many of the same goals as my White classmates—acceptance to a prestigious graduate or professional school or a high-paying position in a successful corporation. Thus, my goals are not as clear as before."

The position at the firm now known as Sidley Austin certainly fit the description of a high-paying position in the upper echelons of the legal community. Yet a few years later, after marrying Obama, Michelle found herself moving out of that predominantly white world of high-powered law and into the public service sector that she had always envisioned for herself. First she left Sidley Austin to work for a deputy chief of staff to Mayor Richard M. Daley. Then, in 1993, she was hired to launch the Chicago office of Public Allies, a program established under Bill Clinton's AmeriCorps to help young people find employment in public service. Michelle, exceedingly efficient and excessively organized, built the program from the ground up. Over her three-year tenure as executive director, she assembled a solid board of directors and raised enough cash to establish the program for the long haul.

I first met Michelle Obama during her husband's U.S. Senate campaign, in January 2004, as I researched my first profile of her husband for the *Chicago Tribune*. At the time, she was directing community affairs for the University of Chicago Hospitals, just blocks from their town house and children's school in Chicago's Hyde Park neighborhood. Her credentials made her seem considerably overqualified for the position of community liaison for a hospital, and as I walked to her office on a blisteringly cold morning,

I noted mentally that she had made a career sacrifice for the sake of the family. Her small office was situated in a difficult-to-find back corner section of a sprawling medical building. The structure was designed with a mind-bending maze of hallways that, each time I would visit, gave me the feeling of the proverbial rat in a science experiment. Her office was much as I would find her to be—highly functional with few frills. It was simply decorated with the same mass-produced wooden furniture found in her secretary's waiting room. Attractive family photos of her children and her husband were atop seemingly every square inch of desk or cabinet space, and no visitor needed to guess about her top priority in life.

Having previously interviewed wives (and husbands) of political candidates, I was uncertain what to expect. Some spouses, fearful of a verbal gaffe, are heavily scripted by campaign aides, equally fearful of a blunder. Others are more at ease with their own words but are still generally cautious in their approach. Michelle Obama exuded neither quality. She was not the least bit scripted. She was open and relaxed and greeted me as if we had met many times before. As with her husband, one of her strengths is the ability to put others at ease in her presence. She answered my questions in such an unhurried and relaxed fashion that she seemed only mildly calculating, if at all. I found her to be comfortable with herself, personable, intellectually engaging and deeply committed to her husband. She knew that he badly wanted to win this election. Toward that end, she was adept at pointing out his positive traits; and yet she seemed to have little interest in putting an artificial gloss on her husband's foibles and faults. "I call him 'The Fact Guy,'" she told me. "He seems to have a fact about everything. He can argue and debate about anything. It doesn't matter if he agrees with you, he can still argue with you. Sometimes, he's even right." She broke into a playful grin at this last remark.

She told me how Obama's competitive streak sometimes led him to be overly boastful at winning family games, such as Scrabble or Monopoly. She stressed that her husband had made many sacrifices in his career, particularly financial sacrifices, in order to serve the public. The emphasis on this part of her husband's character certainly would have been considered helpful to his campaign, but she spoke so openly about his faults that neither observation came across as overtly scripted.

Our interview ran almost two hours, at which time I ended the session. I stepped back into a maze of hallways with the impression of a woman who was confident in her own skills, confident in her marriage and her own career and also highly respectful of her husband's abilities.

Obama has publicly portrayed Michelle as a reluctant political wife, and that has unmistakably been true. His immense personal ambition and good political fortune have pushed his career into overdrive, sometimes leaving his family breathing exhaust fumes. And this has caused friction in the marriage. But initially, her attraction to Obama was wedded, at least in part, to his mission to serve the public. When Obama first mentioned his desire to enter politics, she was encouraging. "I told him, 'If that's what you really want to do, I think you'd be great at it,'" she said. "'You are everything people say they want in their public officials.'"

Their four-year courtship seemed rather effortless to outsiders. Both clearly were devoted to each other, although Michelle's family expressed some initial hesitancy. Michelle's mother, Marion Robinson, was fond of Obama but was concerned that his biracial ancestry might evoke a clash of cultures, or that their union might not be readily accepted by others. Michelle's brother, meanwhile, wondered if Obama could live up to the rigorous standards that his exacting younger sister had placed on previous boyfriends. "My mom and I and my dad, before he died, we were all worried about, 'Oh, my god, my sister's never getting married because each guy she'd meet, she's gonna chew him up, spit him out.' So I was thinking, Barack says one wrong thing and she is going to jettison him. She'll fire a guy in a minute, just fire him. . . . Unfortunately for these guys, and I don't want this to sound conceited, but my dad was my dad. And so she had a definite frame of reference for a guy. She had an imprint in her mind of the kind of guy she wanted. And my mom used to tell her—and I used to tell her after she got older—I was like, 'Look, you're not gonna find guys that are gonna be perfect, because they didn't have Dad as a father. So you've gotta sort of come up with your framework.' But she was hardheaded and refused to let that go."

Craig's first impression of Obama was positive: "He was tall," Craig said with a chuckle. Michelle had dated men shorter than she, and one surmises that her confident demeanor and extreme tallness for a woman could be intimidating to any suitor, especially one deprived of height. Craig's most indelible impression of his sister's new boyfriend came on the basketball court. Obama was not nearly as talented as Craig, who had played professionally in Europe after being drafted and cut by the NBA's Philadelphia 76ers. But Obama had never been reticent about his own basketball skills and he was eager to step on the court with the former college star. That moxie impressed Craig. "Barack's game is just like his personality—he's confident, not afraid to shoot the ball when he's open. See, that says a lot about a guy," Craig said. "A lot of guys wanna just be out there to say they were out there. But he wants to be out there and be a part of the game. He wants to try and win and he wants to try and contribute."

This extreme level of confidence is something that was ingrained in Obama's psyche early in life. His maternal grandfather would tell Obama that this was the greatest lesson he could learn from his absent father: "confidence—the secret to a man's success." That is how Obama's father led his life; and even in times of self-doubt, Obama has hearkened back to that wisdom.

Though Obama and Craig bonded on the basketball court, Craig was taken aback at one of their first holiday gatherings when Obama confided what profession he might pursue after Harvard Law—politics. And not only that, but Obama seemed to hint that he was destined for great things in this often poisonous field of endeavor. Obama speculated to his future brother-in-law that he just might be president one day. "Barack was like, 'Well, I wanna be a politician. You know, maybe I can be president of the United States.' And I said, 'Yeah, yeah, okay, come over and meet my Aunt Gracie—and don't tell anybody that!'"

Michelle, in contrast to her husband, was more circumspect about Obama's lofty ambitions. She thought he had great talent, but looked at him as something of a dreamer as far as national politics was concerned. He might be a star in that realm one day, but that was of little interest to her. Beyond Obama's keen intellect and personal charm, what sealed Michelle's love for him was his civility and human compassion. She said she was intimately affected by his treatment of one of her uncles who had a drinking problem. Obama was studying at Harvard Law at the time and clearly had a bright future ahead of him. So he could easily have dismissed her uncle. Instead, "Barack treated him with respect and dignity, like an equal," Michelle said. (Perhaps Obama had compassion for her uncle because Obama's own family was not immune to alcohol issues. His maternal grandfather, who helped raise Obama, drank to excess, as did Obama's Kenyan father.)

LOVE AND ATTRACTION ARE SUCH INTANGIBLE NOTIONS THAT IT IS impossible to definitively analyze what drew Obama to Michelle Robinson so intensely. But people close to him believe that race probably played a factor. After coming to the mainland from Hawaii, he sought to find not only his own cultural identity, but a comfortable human community in which to live. One can surmise that in choosing an African-American woman as his wife, he consciously (or subconsciously) decided to root himself in the black community. Kellman, for one, believes it was no fluke that Obama married a black woman. He said Obama was attracted to the black experience in America. "If you are biracial, I think as a kid, you begin to identify with the underdog, the people who have injustice thrust upon them," Kellman said. "I think that has great appeal to you, and that is what Barack began to care about, intellectually. You can see that in college and in other places. I mean, blacks in America, this is the great injustice of our history. So why not opt for that and choose that path? I think it is very natural, in that sense, for him to do that and to be inspired by that. . . . In writing more about his dad than his mother, the person he knew and [who] reared him, it makes the case that this is what he chose for his future—the fact he chose to marry Michelle, the ideal person who could help him develop those kinds of roots, and the person to share this career with. And personally, it just seems to have worked out wonderfully for him."

The Obamas settled in Chicago's Hyde Park neighborhood along the lakefront on the city's South Side. With the University of Chicago as its epicenter, Hyde Park is one of just two or three communities in racially segregated Chicago that is populated by significant numbers of both blacks and whites, many with college educations. A particularly trendy place for upwardly mobile blacks, Hyde Park is also in vogue for mixed-race couples. So when Obama married Michelle in 1990, he also married into her budding network among Chicago's community of successful, white-collar African Americans. Indeed, as two attractive Harvard Law graduates, the Obamas made for a striking black professional couple. But despite this status, neither pursued financially rewarding careers. Michelle left Sidley Austin in the wake of losing two people close to her—her father (shortly before she and Obama married) and a college roommate who died suddenly at twenty-five—which prompted her to take a close look at how she conducted her own life. She interviewed for a Chicago City Hall post under Valerie Jarrett, then chief of staff to Mayor Richard J. Daley. When Jarrett offered her the job, Michelle had one rather odd request: She asked that Jarrett first meet with Obama, then her fiancé. It turns out that Obama had concerns about his impending bride going to work in Daley's city hall. He worried that Michelle might be too straightforward and outspoken to survive in such an overtly political environment; and suspicions about Daley's administration, which had been criticized as an updated version of Chicago's machine politics and a vehicle designed primarily to serve Daley's political interests over community interests. Jarrett promised Obama that he would protect Michelle from political backstabbing, and Obama eventually agreed to the job.

When Obama returned to Chicago from Harvard, he put off his legal career for six months to take a position directing a voter registration and education campaign targeting Chicago's low-income blacks. Illinois Project Vote registered nearly one hundred and fifty thousand new voters for the 1992 presidential election. "It's a power thing," declared the project's radio commercials and brochures. The effort was critical in electing two Democrats. It helped Bill Clinton win Illinois, and it greatly assisted an African-American state lawmaker, Carol Moseley Braun, in becoming the first black woman to serve in the U.S. Senate. But Obama was also working on another project at the time, and this dual workload caused strains in his marriage. Running Project Vote by day, Obama was writing his first memoir at night, leaving Michelle feeling rather lonely. Michelle is religious in her routine of going to bed early and rising at 4:30 A.M. to hit the treadmill, but her husband's disappearance into his writing hole until the wee hours took some adjustment on her part. This was a pattern that the ambitious Obama would fall into throughout his political career: heavily burdening himself with work duties, much to the chagrin of his wife. Obama admits that this trait is a personal shortcoming. "There are times when I want to do everything and be everything," he told me. "I want to have time to read and swim with the kids and not disappoint my voters and do a really careful job on each and every thing that I do. And that can sometimes get me into trouble. That's historically been one of my bigger faults. I mean, I was trying to organize Project Vote at the same time as I was writing a book, and there are only so many hours in a day."

Michelle has not been shy about grumbling in public regarding her husband's busy career. Much like Obama's no-nonsense grandmother, Michelle has consistently played the role of stabilizing influence in the Obama home, particularly after their two daughters were born. "I cannot be crazy, because then I'm a crazy mother and I'm an angry wife," she said. "What I notice about men, all men, is that their order is me, my family, God is in there somewhere, but 'me' is first. And for women, 'me' is fourth, and that's not healthy." Michelle qualified the statement to say "all men," which left her husband off the hook. But I think it would be fair to say that "all men" do not put themselves ahead of their family. More often than not in American households, it is indeed the mother who makes the career sacrifice for the sake of children, but not always. The man Michelle married, however, is prone to placing his professional ambitions at the top of his priority list. Indeed, someone who writes a nearly four-hundred-page memoir—at the age of just thirty-three—might be accused of self-indulgence. And Michelle was not the first woman in his life to accuse Obama of a certain level of self-absorption. A female friend at Occidental College told Obama spitefully, "You always think it's about you."

AFTER THE NOVEMBER 1992 ELECTION AND PROJECT VOTE'S conclusion, Obama set out to practice law in Chicago. As a magna cum laude graduate of Harvard Law and the first black president of the *Harvard Law Review,* he had his choice of top law firms. He picked Miner, Barnhill & Galland, which specialized in civil rights and discrimination cases. The firm was in many ways the antithesis of the corporate Sidley Austin. Miner, Barnhill was an activist firm that strove to rectify social and economic injustice through the courts. In this sense, it fit Obama's mission agenda perfectly. Miner, Barnhill had picked up Ivy League graduates before, but when senior partner Judson Miner had called the *Law Review* the year before to inquire about Obama, the response gave him little cause for optimism. "Leave your name and take a number," Miner was told. "You are caller number six hundred and forty-seven." Obama's *Law Review* presidency had garnered him national media attention and put him at the top of the list of Harvard graduates.

Over the nine years that Obama's law license was active in Illinois, he never handled a trial and mostly worked in teams of lawyers who drew up briefs and contracts in a variety of cases. He was one of the lawyers representing an activist group in a successful lawsuit that accused the state of Illinois of failing to apply a federal law designed to help the poor register to vote. In another case, Obama wrote a large portion of an appeals brief for a whistle-blower who exposed misconduct by Cook County and a private research institute in the handling of a five-million-dollar federal research grant. The grant money was used to study the treatment of pregnant substance abusers, and the whistle-blower was a doctor fired from the program after raising questions about expenditures. Obama was also among a group of attorneys who sued on behalf of black voters and Chicago aldermen who alleged that new ward boundaries drawn up after the 1990 census were discriminatory. An appeals court ruled that the new ward map violated the Voting Rights Act, and another set of boundaries was drawn.

But beyond the firm's legal work, it was Judson Miner himself who appealed to Obama, for another reason. Miner had been corporation counsel in Harold Washington's administration. Miner, in fact, was one of the lawyers who helped Washington lead the fight against the white political machine on the city council. From those days, Miner had a bevy of contacts in Chicago's political circles. And as Obama had mentioned to friends and family, politics greatly interested him. He had seen how effective Washington had been in quickly altering the racial and social dynamics of the city. The speed of this change impressed Obama. "The courts are generally very slow and they are generally pretty conservative, not ideologically conservative necessarily, but conservative as institutions," he said. "Law school and practicing law put the framework around how this country works, but it also drove home that social change through the court system is a very difficult thing. There are very few moments in our history, *Brown v. Board of Education* being a singular exception, where substantial change was initiated through the court system. . . . So it was at this point that I started thinking more seriously about political office."

CHAPTER

Politics

> *I am surprised at how many elected officials—even the good ones—spend so much time talking about the mechanics of politics and not matters of substance. They have this poker chip mentality, this overriding interest in retaining their seats or in moving their careers forward. . . .*
>
> —BARACK OBAMA

Barack Obama's first foray into electoral politics revealed both the burning intensity of his personal ambition and his deeply held desire to press for social change, particularly in poor African-American communities. To Obama, it displayed once again the thorny thicket of intramural politics that besets Chicago's black community. The city's African-American political universe is lighted by a select group of insiders who have amassed power and prestige, and they are loath to relinquish it. It is also a society of unyielding internal rancor, which often impedes overall black progress. For these reasons, as Obama entered the public political sphere, the eternal optimist began to express pessimism about the state of black Chicago.

"Upon my return to Chicago," he wrote in *Dreams from My Father,* which was published around this period, in 1995, "I would find the signs of decay accelerated throughout the South Side—the neighborhoods shabbier, the children edgier and less restrained, more middle-class families heading out to the suburbs, the jails bursting with glowering youth, my brothers without prospects. All too rarely do I hear people asking just what it is that we've done to make so many children's hearts so hard, or what collectively we might do to right their moral compass—what values we must live by. Instead I see us doing what we've always done—pretending that these children are somehow not our own."

Running Project Vote and working for Judson Miner provided Obama with entree to the diverse constellation of black politicos on Chicago's South Side. So when an opportunity to run for public office presented itself in 1995, Obama seized it. A respected state senator in her first full term, Alice Palmer, had decided to run for Congress, which led to Obama cutting his first deal to advance his career in politics. Palmer was a progressive African American in the vein of Obama, and she threw her support behind Obama as her replacement.

But that is only where the story begins.

In Obama's version of events, Palmer agreed that even if she was not successful in the three-person race for Congress, she would retire from politics and he would run for her seat. But when Palmer began to falter badly in the congressional contest, she changed course dramatically. Her supporters met with Obama and asked him if he would step aside if Palmer was to lose. This would allow her to run again for the Illinois Senate seat she was holding, but it would leave Obama out in the cold. After pulling together a campaign, Obama had no interest in ditching the effort. And he did not equivocate in expressing that sentiment to Palmer's representatives. At thirty-four, Obama was eager for this next step in his evolution. This was, after all, a man who had mused to his brother-in-law a couple of years earlier that, *hey, you never know,* he might be the president of the country one day. Palmer, in Obama's view, was reneging on an agreement that they had negotiated in good faith. So he told Palmer's people that she had promised to relinquish her seat and support him, and that he would not withdraw from the race.

Predictably, this did not sit well with Palmer. She, indeed, lost the congressional primary contest in November 2005 to Jesse Jackson Jr., and then quickly filed to run for her old seat in the March 2006 Democratic primary against Obama—even though she had publicly supported him for the seat. "Since she endorsed me, I can always use, 'Even my opponent wants me' as a campaign slogan," Obama quipped to the *Chicago Tribune*.

But humor aside, this established a problematic election for the upstart Obama. Suddenly, instead of running at the head of the pack with the incumbent's blessing, he was running *against* the incumbent herself. He had worked the hustings and accumulated a great deal of support for his candidacy, but he would have been hard pressed to match Palmer's power of incumbency: a ready-made army of supporters to distribute literature and get people to the polls on election day, as well as the endorsement of an array of established black politicians. Indeed, Palmer called a press conference and accepted a petition from more than one hundred supporters urging her to seek reelection. Also in her corner was the new congressman, Jesse Jackson Jr. In fact, fresh from the congressional victory, Jackson's field organizer attended Palmer's press conference and pledged full assistance. Palmer also possessed an incumbent's most potent weapon—name recognition from her previous election—while Obama's name was not a plus to his campaign. When it came to his name, the debate generally revolved around which was odder—his first name or his last.

But Obama had one card up his sleeve. He could not envision how Palmer's supporters, even as solidified as they seemed to be, had gathered the necessary number of voter signatures on her nominating petitions in such a short time. Palmer herself confessed at her press conference that the nearly sixteen hundred petitions she had filed with the state elections board had been accumulated in just ten days. So a volunteer for Obama challenged the legality of her petitions, as well as the legality of petitions from several other candidates in the race. As an elections board hearing on the petitions neared, Palmer realized that Obama had called her hand, and she acknowledged that she had not properly acquired the necessary number of signatures. Many of the voters had printed their names, rather than signing them as the law required. Palmer said she was desperately trying to get affidavits from those who had printed their names, but time was running out. She had no choice but to withdraw from the race. The other opponents were also knocked off the ballot, leaving Obama running unopposed in the primary.

Publicly, Palmer claimed she held no grudges; privately, she was extremely bitter. This turn of events was embarrassing to her, especially after her poor showing in the congressional contest. But more important than embarrassment, it effectively ended her once promising political career. So she refused to support Obama in the primary or the fall election, telling a Chicago journalist, "I've since discovered that he's not as progressive as I first thought." Nevertheless, since the Republican Party is almost nonexistent in African-American districts on the South and West Sides of the city, Obama cruised to victory in the fall election.

For Obama, the saga pointed up several things. Rather than winning a position in the Illinois General Assembly by ousting an incumbent or taking an open seat, he appeared to have slipped in the back door on a technicality. And by challenging Palmer, who was highly regarded in black political circles as a fighting progressive, he had left a bad taste in the mouths of many black political leaders, influential people whom he would have to work with in the state capitol and in his district. Palmer, for her part, seemed more than happy to see this bitterness and resentment toward Obama spread to as many people as possible.

But most significantly, the whole episode showed that Obama was an extraordinarily ambitious young man willing to do whatever it took to advance not only his agenda of community empowerment but his own political career.

★ ★ ★

OBAMA'S STATE SENATE DISTRICT ON CHICAGO'S SOUTH SIDE WAS one of the most economically diverse in the state. It included some of the poorest African-American neighborhoods and public housing projects in the country, but it was also home to middle- and upper-middle-class residents, most living in Hyde Park. In addition, his district ran up into the South Loop area, just south of the main downtown commercial district of the city. As development spread outward from the city's high-rises from the 1980s through today, the South Loop and neighborhoods to its near south experienced intense development pressures. The South Loop was undergoing gentrification, with new condominium buildings and town-house communities drawing downtown professionals and others back to city living. Attendant stores, businesses and restaurants sprouted alongside these developments.

Obama's fratricidal state senate race provided some inside-page grist for the Chicago media, but it was still a relatively obscure contest in a city filled with higher-stakes political intrigue. Yet the emergence of Obama onto the city's political scene did not go completely unnoticed. Obama's Harvard pedigree, the release of his first memoir and his own personal eloquence combined to draw some interest from the press and from influential people outside the South Side environment.

He was named one of the city's top movers and shakers by *N'DIGO* magazine, a publication tailored to black professionals in Chicago. But the most substantial interest came in the form of a long front-page profile in the city's leading weekly alternative newspaper, the *Chicago Reader*. The four-thousand-word article was highly laudatory of Obama, and the last third of the piece was largely an account in Obama's own words of his views on the complicated intersection of race and politics. The article served as a manifesto of sorts for Obama. This was the first time he was able to voice his political philosophy to the broader Chicago community, especially the all-important liberal set. His vision for a "new kind of politics" was still in its nascent stage, and he was speaking primarily to the more narrow concerns of his future black South Side constituents. But the overall principle was essentially identical to what he preaches politically today: If America is to progress as a free society, people must learn to work together to build healthy communities rather than fight with each other over parochial, self-serving interests. As Obama put it:

> The political debate is now so skewed, so limited, so distorted. People are hungry for community; they miss it. They are hungry for change. What if a politician were to see his job as that of an organizer, as part teacher and part advocate, one who does not sell voters short but who educates them about the real choices before them? As an elected public official, for instance, I could bring church and community leaders together easier than I could as a community organizer or lawyer. We would come together to form concrete economic development strategies, take advantage of existing laws and structures, and create bridges and bonds within all sectors of the community. We must form grass-root structures that would hold me and other elected officials more accountable for their actions. The right wing, the Christian right, has done a good job of building these organizations of accountability, much better than the left or progressive forces have. But it's always easier to organize around intolerance, narrow-mindedness, and false nostalgia. And they also have hijacked the higher moral ground with this language of family values and moral responsibility. Now we have to take this same language—these same values that are encouraged within our families—of looking out for one another, of sharing, of sacrificing for each other—and apply them to a larger society. Let's talk about creating a society, not just individual families, based on

these values. Right now we have a society that talks about the irresponsibility of teens getting pregnant, not the irresponsibility of a society that fails to educate them to aspire for more. . . . I am surprised at how many elected officials—even the good ones—spend so much time talking about the mechanics of politics and not matters of substance. They have this poker chip mentality, this overriding interest in retaining their seats or in moving their careers forward, and the business and game of politics, the political horse race, is all they talk about. Even those who are on the same page as me on the issues never seem to want to talk about them. Politics is regarded as little more than a career.

One of Obama's central themes was the powerful potential of multiculturalism in American society. Rather than continually fighting white interests and castigating whites for an oppressive history of mistreating blacks, Obama suggested, blacks would do better if they infiltrated the mainstream power structure and worked from there to effect social change.

"Any solution to our unemployment catastrophe must arise from us working creatively within a multicultural, interdependent, and international economy," Obama said. "Any African Americans who are only talking about racism as a barrier to our success are seriously misled if they don't also come to grips with the larger economic forces that are creating economic insecurity for all workers—whites, Latinos, and Asians."

At Harvard, Obama's practice of patiently listening to all sides of a debate made him popular and defused conflict with members of the school's conservative ranks. But his steadfast beliefs about multiculturalism and race made him less than a unifying force in Chicago's black community. The idea of building bridges to people of all races was anathema to many old-school black leaders and so-called black nationalists who still sounded a voice in Chicago's African-American community. These individuals practiced a form of identity politics. They were not only pessimistic about the capacity of whites to share economic wealth with minority groups, but they were especially wary of blacks like Obama who were educated in elite white institutions, had assimilated into mainstream white society and now espoused these kinds of multicultural notions. It did not help Obama's cause in these circles that, in addition to his law practice, he was lecturing in constitutional law at the University of Chicago School of Law. The university was situated in Hyde Park, and to some South Side blacks the campus represented the encroachment of the white establishment into what had historically been claimed as African-American turf. The school was considered a physically forbidding and intellectually impenetrable institution of white elitism sitting in the middle of struggling black communities—and there was Harvard Law–trained Obama, teaching classes at this establishment.

Most of this grumbling occurred behind the scenes, although some was aired in public. An African-American political science professor at Northwestern University criticized Obama for his "vacuous-to-repressive neo-liberal politics." The head of a task force on black empowerment was more direct and more cynical, charging that Obama was little more than a tool of forces beyond the black community. These kinds of concerns from blacks were nothing new to Obama, of course. He had rankled some African Americans at Harvard for trying to promote harmony among partisans by appointing conservatives to editorial posts at the expense of pushing specific blacks up the ladder.

SOME OF THESE ISSUES IN THE BLACK COMMUNITY FOLLOWED Obama into the Illinois General Assembly, where he received a cool reception from some members of the Illinois Legislative Black Caucus. This group of black lawmakers was not immune from internal squabbling. Chicago has two significant communities of African

Americans—West Siders and South Siders—and legislators representing districts from each part of town were often at odds, in part
because they were competing in Springfield for the same state resources. And Obama did not necessarily ingratiate himself with the
legislators by making some privately disparaging comments about
the disorganized, free-for-all atmosphere in Springfield. While some
lawmakers acknowledged that Obama's criticisms were on the mark,
others wondered, Just who is this hotshot new guy from Harvard
Law who shows up in Springfield and immediately takes potshots at
us? "At the swearing-in, there was already some heat from certain
people," a Springfield politico recalled.

In fact, the man who soon became Obama's chief political adviser acknowledged that he did not care for Obama upon their first
meeting. Dan Shomon was a hardworking, gregarious, bespectacled
wire service reporter who had left journalism to become an aide to
Democrats in the state capital. By 1997, Shomon had worked in
Democratic Party politics in Springfield for eight years and he knew
the terrain. Obama was displeased about the amount of time that
his assigned aide was devoting to his staff work and noticed that
Shomon was a dogged worker for the senator in the office next
door. So Obama pressed the Black Caucus leader, Emil Jones Jr., to
have Shomon assigned to him, as well.

Shomon was initially reluctant because he had heard some of
the scuttlebutt about Obama. He had also met Obama a couple of
times the year before when Obama helped campaign door-to-door
for other Democratic legislative candidates. On one of his volunteer
walking efforts, Obama ran into another volunteer apparently assigned to walk the same precinct and grew miffed because he felt
his time had been wasted. This gave Shomon the impression that
Obama could be rather testy and elitist. So when Shomon was
approached about working for Obama, his first reaction was: "I am
thinking that I am really busy. He wants to change the world and
that is great, but I don't really like the guy that much." Still,

Shomon agreed to meet with Obama, and when Obama took Shomon out for dinner, the two wound up hitting it off. So Shomon went to work for Obama, mostly in the capacity of helping the freshman lawmaker with legislative strategy. Shomon offered to help with media affairs, since that was his specialty as a former reporter. But Obama told Shomon that he did not need a press aide because he dealt with reporters on his own. Obama wanted full control over his own public message, and he didn't want a spokesperson interpreting his language and ideas.

(Thus began Obama's cordial, but not chummy, relationship with the news media. He was known for an occasional thoughtful quote, but capitol reporters were not so charmed by him that they consistently sought him out. Obama joked in later years that only one reporter, Dave McKinney of the *Chicago Sun-Times,* paid him any attention at all. "And McKinney only talked to me to be nice," Obama said.)

One of Shomon's first tactical moves would later prove pivotal in pushing forward Obama's budding political career—and it would convince Obama that he could win a statewide political contest. Shomon told Obama that, because he would be voting on issues that affected Illinoisians outside the Chicago region, he should consider traveling to the southern reaches of the state to gain an understanding of Illinois's vastly divergent cultures. Obama thought the trip would be instructive for that reason and for another: Springfield was not his final political destination. He harbored notions of running for statewide office one day. This was 1997, before Obama's first child was born, and there were fewer encumbrances on his time. So, according to Shomon, Obama responded without hesitation: "Let's do it." (In Obama's second book, *The Audacity of Hope: Thoughts on Reclaiming the American Dream,* he takes credit for the idea of this first far downstate trip.)

Behind Shomon's thinking was this: Illinois suffers from a constant geographic tussle between the Chicago region and the rest of the state. About two-thirds of the state's population lives in the Chicago area, which, at roughly eight million people, is the third-largest metropolitan region in the United States, behind New York City and Los Angeles. The remaining third live in rural and small urban communities across Illinois, and these communities consistently complain that they are overwhelmed by Chicago and its collar counties, especially when resources are divvied up by lawmakers in Springfield. Because of this dynamic, most politicians from Chicago are viewed with some suspicion by voters in Chicago's immediately surrounding suburbs and with even greater suspicion outside the metro region altogether.

But in central and southern Illinois, another factor might play a role in how Obama would be considered: race. Obama was not only a Chicago politician but an African-American one. This was potentially a second strike against him in this part of the world. Southern Illinois, in particular, more resembled the Deep South than the Midwest. It had endured a long history of racial intolerance. At Illinois's southern tip, where the Ohio and the Mississippi rivers meet, the small town of Cairo has a particularly unsettling past that includes the public lynching of black men in the early 1900s and a race riot in 1967. The riot was so ugly that it helped to spur the departure of thousands of residents. On several trips into this potentially hostile environment, Obama would not only measure his overall reception but also seek to make political and fundraising contacts. The key question: Had the racial equation changed enough that he would be viewed as a friend or, at the very least, a nonthreatening outsider? "I said that we will go for a week and play golf and we will see the state," Shomon recalled.

On this first trip, Obama's eyes were opened to this largely white part of the world—the so-called red state world. The two men disagreed over some logistical and political aspects of the trip, but overall it was a resounding success, signaling to both of them that Obama probably could compete in that rural, small-town environment.

Obama may have written four hundred serious pages in *Dreams* about the painful odyssey of finding comfort with his mixed racial ancestry, but he seemed comfortable with his identity by this time.

"We are driving through Perry Township and we pass the Pinckneyville Coon Club," Shomon recalled, spelling out the word "C-O-O-N" for special emphasis. (*Coon* in this case was a truncated colloquialism for *raccoon,* but historically in the United States it has been used as an epithet for a black person.) "And, you know, this guy has never really been in the South. So Barack looks at the sign, and looks at me, and he says, 'I don't think they're going to let me join the Pinckneyville Coon Club.' He then starts laughing and he is laughing so hard he almost fell off his chair."

While traveling in another small town on a later trip, Obama and Shomon were pulled over by a police officer in DuQuoin County when they inadvertently turned the wrong direction down a one-way street. When the officer questioned the state license plate on Obama's green Jeep Cherokee and Shomon told him that Obama was a state senator, the officer "looked flabbergasted," Shomon said. "He's not a state senator from these parts!" the officer told Shomon.

Still, in setting after setting on these downstate trips, Obama discovered that everyday people reacted warmly to him. In some cases he was treated like a dignitary because people were not accustomed to a state senator visiting them. "It was just a great trip because it really did open Barack's eyes," Shomon said. "He thinks these people are really cool, and they could relate to him—although, you know, they couldn't pronounce his name."

The trips were also helpful in yet another regard, allowing Shomon to school Obama in another lesson of politics: Try to adapt as best you can to the culture you are stepping into. Obama did not always take well to Shomon's unsolicited advice. While Obama's family never had a great deal of money, his private schooling and Harvard degree meant that he had been immersed in an elite culture, especially during his adult years. One morning, Shomon met Obama coming out of their hotel and noticed that Obama was wearing khaki pants and a silky black shirt with a flat collar and buttons all down the front, a shirt more typically worn to a picnic than to a round of golf. Most significantly, it was not the universal golf uniform of a two- or three-button polo shirt. (Obama shows his minimalistic tendencies by consistently wearing khaki pants and black shirts in casual settings. "Buy him a black shirt for Christmas and he is a happy man. He's not flashy. That's all he wears," his wife once told me.) "So I asked him what was going on with the shirt? And he asked what I meant," Shomon said. "I told him that we have the golf outing today. He didn't think anything was wrong with the shirt. I told him that he needed to wear a golf shirt like everyone else. And he asked me why, because he only had one golf shirt and he had worn it playing golf the day before. I said, 'Well then, wear it again.' So he put it back on. I reminded him that this was southern Illinois and you don't want to look too 'uptown.' He got mad and kind of frustrated about that. He didn't have a problem with changing after I explained it to him, but he really had not been exposed to that stuff. He is fairly sophisticated."

Obama recounted a slightly different version of this anecdote in his second book. He wrote that Shomon told him to wear polo shirts and khakis throughout the trip, in order to fit in. Obama also recalled the story of Shomon advising him in a downstate restaurant to eat regular yellow mustard rather than the more pretentious Dijon mustard. Obama portrays Shomon's advice as rather petty. These disagreements might appear insignificant on the surface, but they reveal something about each man. Both could be stubborn and opinionated, but each also had legitimate points. And the discussion itself crystallizes a dilemma that political aides have consistently had with Obama: his occasional air of elitism coupled with his skin color. Rightly or wrongly, Shomon was concerned that Obama's urbane sophistication, coming as it did from a black man, would alienate working-class white voters—and Shomon sought to temper that image. Obama, however, was wise enough to know that if he tried to morph into something that he was not, he would be perceived as inauthentic.

In any case, Shomon and Obama would take more of these trips downstate, and the excursions satisfied Obama that he could win votes among this electorate. These mostly white, middle-class, down-to-earth midwestern people reminded him of his grandparents and he felt completely comfortable in their midst. And that comfort obviously was reciprocated. "I understand these folks," Obama told me on a downstate campaign trip during his U.S. Senate race. "My grandmother was Republican. I grew up with these people."

The Legislator

*For Barack, it's not a constant flow of glorious defeats. He has
good attention to ideals and core principles, but a recognition that
it is good to get things done from year to year.*
  —JOHN BOUMAN, ANTIPOVERTY ADVOCATE

For the first six of Barack Obama's almost eight years in the Illinois
General Assembly, he suffered from an aggressive legislator's most
debilitating affliction—membership in the minority party. In national politics today, Illinois is a pure blue state, considered an immediate loser for a Republican presidential candidate. But Illinois
voters have a long history of electing moderate Republican gover-
nors, as well as turning over state legislative chambers to the Grand
Old Party. Rural areas of Illinois are known for sending Republicans to the state legislature, as are the suburban counties that encapsulate Chicago. The Illinois House of Representatives has been
dominated by Democrats for some time, but Republicans captured
control of the Illinois Senate in 1992 and held on to it for a decade.

No red carpet unfurled for Obama when he arrived in the
state capital of Springfield in January 1997. If anything, his fellow
senators cast a cold eye. Many of his colleagues viewed him in the
same visceral way that his aide, Dan Shomon, first reacted to him:
Here comes an aloof Ivy League good-government type who too
often mentions his years of sacrifice as a community organizer and
his Harvard Law pedigree. "The fact that he had a Harvard Law
degree and was a constitutional law professor—that made some

eyes roll," said Kirk Dillard, a Republican senator from suburban Chicago.

Despite his time organizing the poor on the city's Far South Side, Obama's ability to connect on a broad scale with urban blacks outside his collegial Hyde Park neighborhood was highly questionable. And even though he had spent several years as an organizer in the city's economically depressed communities, he exhibited far more comfort in university-type settings where he could woo the white lakefront liberal set—environments that highlighted and celebrated his cerebral, mission-oriented, policy-wonk nature. His public speeches tended to be policy-heavy and overly intellectual— fine for Harvard Law students, but fairly dry for everyday people. Indeed, some of his senate colleagues, especially in the Legislative Black Caucus, saw in him an Ivy League elitist unwilling to sully himself with the unseemly universe of Chicago ward politics or the muck of the legislative process. "It wasn't like Barack took Springfield by storm," Shomon said. "The first few years he was thought of as intelligent, thoughtful, bright. But he certainly wasn't considered to be a major player." Rich Miller, the publisher of *Capitol Fax,* a statehouse newsletter, was less diplomatic. "Barack is a very intelligent man," Miller said in 2000. "He hasn't had a lot of success here, and it could be because he places himself above everybody. He likes people to know he went to Harvard."

Miller's characterization might be unduly harsh, but in charitable terms, Obama initially toiled in relative obscurity in the statehouse. His intellectualism initially did not translate well to Springfield, where most grunt legislative success is accomplished over spirits in a local tavern or between iron shots on the fifth fairway. In addition, Obama's palpable disdain for some of the petty antics of his fellow legislators did not endear him to those colleagues. At the same time, Springfield had its share of self-absorbed, monosyllabic hacks with little use for an urbane sophisticate like Obama.

Nevertheless, with Shomon as his main tutor, Obama eventually adapted to this environment. He picked up golf. He established close working and personal friendships with a number of fellow legislators, and these associations would greatly assist him as his political career unfolded. By the time his tenure in Springfield ended, it could be argued that he had evolved into a very successful state lawmaker. Viewed warily by some of his fellow blacks, Obama primarily formed alliances with downstate Democrats and a handful of Republicans. These friendships would show that while some people, especially some blacks, regarded Obama as elitist, he could mix with other everyday legislators quite comfortably.

Perhaps most significantly, Obama joined a weekly poker game that included fellow senators Terry Link and Denny Jacobs, white Democrats who represented districts outside Obama's home base of Chicago. The son of a man who studied economics at Harvard, Obama was not surprisingly a skilled poker player. As when playing board games at home, the highly competitive Obama took the game seriously. He played carefully and was adept at not revealing his hand. Senator Link joked that Obama took his money at the card table and Link won it back on the golf course, where Obama was still learning the game.

The loose poker atmosphere, however, did not always fit with Obama's exceedingly strict sense of propriety and morality, traits instilled primarily by his mother. During one poker game, Obama was irked when a married lobbyist arrived with an inebriated woman companion who did not acquit herself in a particularly wholesome fashion. Without offending his buddies, Obama registered his displeasure with the situation. "He didn't think much of that," an Obama associate said. "He didn't see the purpose of bringing her."

Legislatively, Obama managed to pass a decent number of laws for a first-term lawmaker in the minority party. His first major legislative accomplishment was shepherding a piece of campaign finance reform in May 1998. The measure prohibited lawmakers

from soliciting campaign funds while on state property and from
accepting gifts from state contractors, lobbyists or other interests.
The senate's Democratic leader, Emil Jones Jr., a veteran African-
American legislator from the South Side, offered Obama the opportunity to push through the bill because it seemed like a good fit for
the do-good persona projected by Obama. Obama was also recommended to Jones by two esteemed Chicago liberals who had taken
a liking to him: former U.S. senator Paul Simon and former congressman and federal judge Abner Mikva. Working the bill was an
eye-opening experience for the freshman senator. It was a tough assignment for a new lawmaker, since he was essentially sponsoring
legislation that would strip away long-held privileges and perks
from his colleagues. In one private session, a close colleague angrily
denounced the bill, saying that it impinged on lawmakers' inherent
rights. But Obama worked the issue deliberately and delicately,
and the measure passed the senate by an overwhelming 52–4 vote.
"This sets the standard for us, and communicates to a public that is
increasingly cynical about Springfield and the General Assembly
that we in fact are willing to do the right thing," Obama told reporters immediately after the bill's passage. The bill was not a wa-
tershed event anywhere but Illinois. It essentially lifted Illinois, a
state with a deep history of illicit, pay-to-play politics, into the
modern world when it came to ethics restrictions. The bill gave
Obama a legislative success, but his public criticism of Springfield's
old-school politics did not sit well with some of his colleagues, who
already considered the Ivy League lawyer overly pious.

Furthermore, this highly publicized success stood rather alone.
Obama was learning that, like the court system, the legislative branch
of government could move at a snail's pace. Indeed, as his time wore
on in the minority party, he became increasingly frustrated by the
workings of the Illinois legislature. He was especially irritated that
too many grandstanding, feel-good measures seemed to go forward
easily, while bills that offered structural change were bottled up in

committee or never got off the ground. For instance, lawmakers easily passed a bill that strictly outlawed graffiti painting, but were reluctant to take on deeper juvenile justice issues. The graffiti bill garnered its sponsors a political gold mine of media attention.

IN SPRINGFIELD, OBAMA RAN INTO SOME OF THE SAME ISSUES WITH blacks that he experienced at Harvard. He was not shy about criticizing black leaders and their legislative strategy; also, he did not necessarily follow his caucus's talking points. Moreover, he worked closely with white Democrats and even conservatives to pass his own bills. Obama, to be sure, had allies in the black caucus, but he had his share of critics as well. His chief antagonists were Rickey Hendon, who represented a district on the city's West Side, and Donne Trotter, who would run against Obama for Congress.

Hendon and other African-American lawmakers from the West Side often found themselves at odds with their South Side brethren, but the rift between Hendon and Obama was particularly acute. Hendon and Trotter would "just give Barack hell," said Senator Kimberly Lightford, an Obama ally in the black caucus. Hendon, nicknamed "Hollywood" because he once aspired to produce films, was a flamboyant personality in Springfield, known for his smart-aleck humor and occasionally inappropriate public manner. In one legislative session, the two nemeses nearly came to physical blows when Obama, apparently inadvertently, voted against a bill that included funding for a project that assisted Hendon's district. Years later, details of the incident remain in the eye of the beholder. Obama supporters say that Obama had stepped away from his seat and asked someone else to vote for him, not an uncommon practice considering the thousands of votes cast each session. His proxy, however, accidentally voted against his wishes. When Obama asked that the record reflect that he voted the wrong way, Hendon publicly accused Obama of duplicity. Hendon has never been shy about holding back his feelings, and he had a special way of penetrating Obama's usually smooth exterior. Soon, the two men were shouting at each other on the senate floor. They took their disagreement into a nearby room, and a witness said that Obama had to be physically restrained. Neither man cares to discuss the incident today, but Hendon remains unconvinced of Obama's explanation that his vote was accidental. Individuals close to the situation say Hendon still believes that Obama voted against his project in order to pacify North Side fiscal conservatives who were leery of some West Side projects. For his part, the rarely reticent Hendon won't discuss the altercation, except to confirm that it occurred. "I have been advised to leave Barack alone and that is what I am going to do," Hendon said. "I am going to let things stay in the past. It happened. That's all I can say. It happened."

Though Obama alienated some colleagues like Hendon and gained scant public attention, he nevertheless accrued a rather impressive record as a first-term legislator for the out-of-power Democrats. Much of it was like his community-organizing success—low-key and behind the scenes. In his first two years, Obama introduced or was chief cosponsor of fifty-six bills, with fourteen of them becoming law—not bad rookie and sophomore seasons. Besides the campaign finance and ethics bill, other Obama-led legislation that became law included measures that compensated crime victims for certain property losses, prevented early probation for gun-running felons, streamlined administrative processes when municipalities adjudicated ordinances and increased penalties for offenders who used date-rape drugs on victims.

In his third year, 1999, Obama was even more successful. He cosponsored almost sixty bills and eleven became law. They included measures that established a state-funded screening program for prostate cancer (a disease that disproportionately afflicts blacks), provided a training program in the use of heart defibrillators, strengthened hospital testing and reporting of sexual assaults, increased funds for after-school programming, increased investigation of nursing home abuses and hiked funding for lead abatement programs (another large issue in the black community).

Most of Obama's legislative efforts were rooted in his discussions with special interest groups, and he was particularly attentive to activists who sought his help for a cause he supported, such as fighting poverty or protecting civil liberties. Yet he declined to take up issues simply for the sake of garnering attention. He was highly results-oriented. "Obama is interested in sponsoring bills that will pass and is [uninterested] in symbolic legislation," Don Wiener, an opposition researcher hired by two later campaign opponents, wrote in a report assessing Obama's early senate career. "He does not seem to sponsor legislation just because he is encouraged to do so." Obama concentrated his efforts on pursuing legislation that reflected his basic sense of social justice. "He is idealistic but practical," said John Bouman, a director of the Chicago-based National Center for Poverty Law. "For Barack, it's not a constant flow of glorious defeats. He has good attention to ideals and core principles, but a recognition that it is good to get things done from year to year. He is willing to hammer out a good compromise, but he doesn't compromise for the sake of it."

A major reason for his high rate of bill passage was his deftness in reaching across the aisle to Republicans. After working in peace with conservatives on the *Harvard Law Review,* Obama was not the least bit uncomfortable taking the concerns of Illinois conservatives into account and placating their fears. His relentlessly polite, inoffensive manner was key to his success in this regard. "I always found him to be a true gentleman in all the negotiations and dealings that I had with him," said Joe Birkett, a hard-nosed Republican prosecutor in conservative DuPage County who was Obama's ideological opposite.

In the course of his legislative career, Obama persuaded Republicans to go along with an array of initiatives, winning bipartisan support on potentially polarizing legislation like reforming welfare and battling racial profiling by police. "The most important thing that you do in Springfield is you bring all sides of an issue to the table and you make them feel they are being listened to," said Obama, hearkening back to lessons learned as a community organizer. This reasonable tone and genuine attentiveness to Republican concerns made Obama a key swing legislator for both Republicans and Democrats, even if his voting record was decidedly liberal. "Members of both parties listened closely to him," said Dillard, the Republican senator from suburban Chicago who frequently cosponsored legislation with Obama.

As OBAMA'S FIRST TERM WOUND DOWN, HE MADE HIS FIRST MAJOR political miscalculation. The cause of this misjudgment: unbridled ambition.

Obama had returned to Chicago from Harvard Law with an eye on the mayor's office after witnessing Harold Washington's historic tenure at city hall. But while Obama was away, Richard M. Daley, the son of iron-fisted longtime mayor Richard J. Daley, had brokered a peace with the black community and won the city hall job. By 2000, Daley had built a massive political army and held such a firm grip on the city council that unseating him looked akin to dethroning a king.

So Obama looked at Congress instead, deciding to challenge Representative Bobby Rush in the 2000 Democratic Primary. To Obama, Rush looked vulnerable in his South Side district. Rush, in fact, had tried to oust Daley in 1998—but he was stomped by the mayor. For this reason, Obama saw Rush as an aging politician ready to be replaced by a younger man with a fresh vision and new enthusiasm for tackling the ills of the black community. Obama figured that if he could spread his message of community empowerment and social justice, he could upend Rush. This was a circle-of-life philosophy, since Rush had won his seat in Congress in 1992 in exactly that fashion. Rush had alienated some Democrats by unseating a well-regarded black congressman, Charles Hayes, who needed just one more term to earn a comfortable pension. Despite this political history, Rush was not the least bit understanding of Obama's challenge.

Through the 1990s, Rush had been neither a stellar success nor a failure in Congress—he had just sort of existed comfortably inside the Beltway. This was a bit surprising considering his somewhat radical past, which had gained him star status with black voters. Rush was a 1960s civil rights veteran, having been a member of the Student Non-Violent Coordinating Committee and an acolyte of the outspoken Black Power advocate Stokely Carmichael. After coming under the influence of Carmichael, Rush left SNCC and served as head of the Illinois Black Panther Party, whose mantra was "Power to the People." The Panthers' philosophy was in direct contradiction to the nonviolent tenets of SNCC, and they derived power from fiery eye-for-an-eye rhetoric and from arming themselves to defend against white-led urban police forces, most notably in Chicago and Los Angeles. After the Panthers faded from the public scene, Rush became a Chicago alderman in the Harold Washington era, and he served as one of Washington's lieutenants on the city council.

With this history, Bobby Rush had the consummate résumé for a black politician representing a black congressional district. Indeed, the most enduring image of Rush remains a photograph of him as a Black Panther, clad in a black leather jacket and clutching a rifle.

So when thirty-eight-year-old Obama made a run at fifty-three-year-old Rush, the black political world did not embrace him. For the most part, blacks wondered what Obama was doing trying to unseat this black elder statesman; and some African Americans privately began to question not only Obama's motives but his black credentials.

This questioning of Obama's blackness gained some resonance

for several reasons. Primarily, this came about because Obama had not been brought up in a traditional African-American setting. He had been active in Chicago's poor black neighborhoods for a few years, sure, and that had gained him some street credibility. But that was nothing compared to being a former Black Panther and one of Harold Washington's trusted foot soldiers. Obama was a beneficiary of the work by Rush and others to advance the black cause. And his running against Rush raised questions that perhaps this young upstart was an ungrateful beneficiary of that labor.

Much of this sentiment percolated beneath the surface or among Chicago's black nationalist crowd. That group largely consisted of an aging generation that vividly recalled the civil rights struggle, but it was still an influential voice in Chicago's black community. The charge that Obama lacked black authenticity appeared in its sharpest form in a March 2000 story in Chicago's alternative weekly tabloid, the *Chicago Reader.* Another candidate in the Rush contest, State Senator Trotter, viciously accused Obama of being secretly controlled by powerful white interests. Before long, that doubt was reverberating through segments of Chicago's South Side black community. Feeding into this perception were Obama's personal résumé, which included ties to two institutions of white power (Harvard and the University of Chicago), his lack of connection to the broader black community while growing up in Hawaii and his campaign's substantial financial support from wealthy white liberals—people like Abner Mikva, a former federal judge, and Bettylu Saltzman, a former aide to Illinois senator Paul Simon.

In addition, Obama's state senate district on Chicago's South Side was undergoing a demographic and socioeconomic shift during this period, a trend that has continued into the 2000s. Once blighted areas just south of downtown were experiencing the first waves of gentrification. Obama had formed a political alliance with the local alderman, Toni Preckwinkle, and both politicians were the beneficiaries of substantial campaign donations from developers eyeing these suddenly attractive South Side areas. These financial donations were both a boon and a detriment to Obama. The campaign cash from developers helped him remain independent of the Daley political machine, but they also caused him problems with the older African-American nationalists. (One developer who actively courted Obama, Antoin "Tony" Rezko, would cause Obama significant problems in later years when Rezko was indicted amid a federal investigation of pay-to-play politics.) There was considerable tension among these new developers, some of whom were white, and long-time black activists in the area. And there was virtually no way to reconcile their interests. Trotter not too subtly raised these concerns when he told the *Chicago Reader,* "Barack is viewed in part to be the white man in blackface in our community. You just have to look at his supporters. Who pushed him to get where he is so fast? It's these individuals in Hyde Park, who don't always have the best interests of the community in mind."

THE ACCUSATIONS OF BEING LESS THAN BLACK TOOK HOLD IN CIRCLES that were open to such talk. With this political history working against Obama in the black community, some of his advocates began wondering in earnest about his political strength among Chicago's blacks. Said the *Reader* during Obama's 2000 challenge of Rush: "There are whispers that Obama is being funded by a 'Hyde Park mafia,' a cabal of University of Chicago types, and that there's an 'Obama Project' masterminded by whites who want to push him up the political ladder."

Contributing to this perception was his loving treatment by the *Chicago Tribune,* the region's most influential newspaper. The *Tribune* has historic ties to the Republican Party, and its readership is still

•

disproportionately white compared with the diverse population it serves in the Chicago region. The *Tribune,* gushing that Obama was "a rising star in the Democratic Party," endorsed Obama over Rush and Trotter in the 2000 primary race. Any black leader with the imprimatur of the editorially conservative *Tribune* had to be viewed with wary eyes on the South Side, where home-delivered *Tribune*s can go untouched for days while copies of the rival *Sun-Times* must be picked up from doorsteps early in the morning, lest they be swiped. "What was interesting," observed Jerry Kellman, the Chicago community activist who brought Obama to Chicago, "Barack is in the black community and he is being attacked for not being black enough, not because he is half-white, but because he went to Harvard, and because he talks like he could be white and he moves like he does. There are thousands of black people who move and talk like he does, but they are not running for Congress from the South Side of Chicago."

For his part, Obama maintains that the charge of not being black enough was little more than a canard drummed up by his political foes. "What's fascinating is it's never been an issue among regular folks on the street," he said. "You know, it's never an issue with the bus drivers or the teachers or the guys on the street corner who I'm talking to. This is always an issue that's been brought up in the context of a political situation by professional politicians. And it's a handy shorthand to try to create some separation between me and what's a strong part of my base. But it really is not something that has ended up being a problem for me with respect to the voters." Privately, however, Obama told me that he has encountered blacks who question his commitment to the black cause and his self-identified status as an African American. "What I've found," he said, "is they are usually going through identity issues themselves and they project those issues onto me."

The issue of Obama's black authenticity would follow him from this time forward—rearing its head again in his race for the U.S.

Senate and then again in his presidential bid. But it was in this congressional contest that it was first broached.

OBAMA'S CAMPAIGN FOR CONGRESS WAS STAR-CROSSED ALMOST
from the very beginning.

On October 18, 1999, five months before the primary, Rush's
twenty-nine-year-old son, Huey Rich, was gunned down on a
South Side street. For days, the Chicago media overflowed with the
story of Rush's son fighting for his life, spurred in no small part by
public appearances of a solemn-looking Rush in which he talked
about the incident. By the time Rich died four days later, the congressman had been transformed in the public eye from the 1998
mayoral candidate who had been thrashed by Mayor Daley into a
grieving parent who had suffered the tragic loss of his son.

The death all but silenced Obama's campaign. As he was trailing
Rush badly in the polls (one survey found that Obama had only 9
percent name recognition in the district compared with the incumbent's 90 percent), Obama's only hope was to go negative on Rush.
This was a strategy Obama had been considering before the shooting. Opposition researchers for Obama's campaign had found that
Rush's relations with his grown son had been strained. Rush had
fathered Huey in the late 1960s with Saundra Rich, another member of the Black Panthers. Rush, in fact, had named his son after
Panther founder Huey Newton. But Obama's research showed that
Rush had touch-and-go involvement in his son's upbringing, with
Huey being raised primarily by an aunt. One of the men convicted
of the shooting told police investigators that he sought out Huey
because he believed the congressman's son was carrying cash for a
drug dealer.

If Obama were to unseat Rush, he would have to use something
along these lines to portray Rush in a negative light. Now that
Huey Rich had died tragically, that morsel of potentially negative

information was off the table. Attacking Rush for being an absent parent now was completely out of the question. In fact, attacking Rush with any ferocity would now look in bad taste.

"I'll never forget getting a phone call from Reverend [Jesse] Jackson," Obama said. "He said to me, 'You realize, Barack, the dynamics of this race have changed.'" Added Shomon, who was managing Obama's campaign: "There was just no way to beat Bobby after his son was shot."

But that was not the only misfortune to beset Obama's congressional campaign.

Throughout Obama's tenure as a state lawmaker, Shomon would stress nothing more vigorously to Obama than the importance of not missing votes in the General Assembly. This continual preaching from Shomon could irritate—and sometimes infuriate—Obama. On many of his three-hour commutes from Chicago to Springfield, an impatient Obama would phone Shomon and complain that he was making another pointless trip to the state capitol when it was clear that little would be accomplished in that particular session. Shomon would explain that it didn't matter whether Obama's presence made any difference to the capitol debate or to pending legislation, it was crucial to Obama's future political career to have a high-percentage voting record so that a political opponent down the road could not accuse him of shirking his public duties. "You gotta show up for work," Shomon told him. Obama, who loathed having his time wasted, balked and rebelled verbally—but he heeded Shomon's advice and missed few votes during his time in Springfield.

There was, however, one instance in which Obama did not follow this plan—and it cost him dearly. As he did every holiday season, in December 1999 Obama gathered up Michelle and Malia, now eighteen months old, to visit his grandmother in Honolulu. These annual trips to his childhood home have served as a respite for the hardworking Obama, who looks forward all year to spending a couple of weeks decompressing in the serene tropical environment of his youth. But the 1999 trip was not stress-free, by any means. That holiday season, a showdown was brewing in the state capitol in Illinois. Governor George Ryan was pushing for reenactment of a gun control law known as the Safe Neighborhoods Act, and the General Assembly was locked in a heated debate over the bill. The legislation, among other things, would have raised the penalty for illegally transporting a firearm from a misdemeanor to a felony. Obama, a staunch gun control advocate who represented a district containing neighborhoods ravaged by gun violence, wholeheartedly supported the legislation. Opposition mostly came from rural conservatives, who argued that hunters and other law-abiding gun owners could be hit with felony counts for accidentally carrying a rifle or other firearm from place to place.

As Obama vacationed in the Pacific, the bill moved closer to a vote in a special session of the senate called the week between Christmas and New Year's. Shomon jumped on the phone with Obama during that week, advising his boss that he ought to seriously consider returning to Illinois for a possible vote. Obama, however, had been spending a good deal of time away from Michelle and their young daughter in recent months, as he had two jobs—state senator and college law lecturer—in addition to campaigning. In fact, he had already truncated his usual two-week Hawaii trip to just five days because of his hectic campaign schedule. So at this point, leaving the family vacation early for the sake of work, especially during the holidays, would put him in a bad spot with Michelle. She was already displeased with her husband's prolonged absences for the sake of his political career and, according to Obama, was barely on speaking terms with him. Moreover, Obama said, his daughter had come down with a bad cold, which worried Michelle. Obama was torn between his duty as a public official and his duty as a father and husband. In the end, he chose to remain in Hawaii.

Back in Illinois, Shomon was left to defuse a potential public relations bomb. The governor's office and reporters were calling asking the whereabouts of Obama and if he could be counted on to vote should the bill come before the senate. Ryan, desperate for every vote, even offered to fly Obama from Chicago to Springfield. Shomon was in a no-win situation—and he stonewalled. For a day or two, he quit returning calls, hoping the issue would go away. How could he tell the governor and the media that his boss was on vacation in Hawaii as this important matter was debated? Shomon crossed his fingers that the bill would not come up for a vote. Unfortunately, that strategy failed, as no compromise was brokered in the waning hours. On December 29, the bill came to the floor and fell just three votes short of the thirty-six needed for passage. Obama was among three lawmakers Governor Ryan had relied on but who had missed the vote. "I'm angered, frankly, that the senate didn't do a better job," Ryan told reporters tersely.

As Obama's plane touched down in Chicago the following week, he took another phone call from Shomon, who was not shy about letting him know of the political disaster awaiting him upon his return to Illinois. He warned Obama that this political ad was in his future: an image of Obama sipping a tropical drink on the Waikiki Beach while legislators fell just a few votes short of passing a tough anticrime bill, all of this transpiring as Chicago was suffering the highest murder rate in its history. Besides Governor Ryan, criticism came from various segments of the media. The *Tribune*, which, despite its conservative leanings, was a staunch advocate of strict gun laws, took Obama to task on its editorial page: "Sen. Barack Obama (D-Chicago), who has—had?—aspirations to be a member of Congress, chose a trip to Hawaii over public safety in Illinois." Bobby Rush also piled on, telling the *Tribune*, "This vote was probably the most pivotal vote, one of the most important votes in memory before the General Assembly, and I just can't see any excuse that Mr. Obama could use for missing this vote."

It was during this episode that I first encountered Obama. About two weeks after the holidays, I was assigned a weekend reporting shift at the *Tribune*. And that Sunday afternoon, Obama assembled a group of senior citizens in his Hyde Park neighborhood to tout a proposal instituting price controls on prescription drugs. But with Obama's missed vote still fresh in the public consciousness, reporters covering the event had little interest in the health care proposal and instead focused on seeking an explanation directly from Obama for the missed vote. Instead of fielding questions about health care, Obama found himself explaining how his daughter had become ill and how he felt duty-bound to stay with the family in Hawaii. He handled the questioning with a confident demeanor, but it was obvious that he felt chastened by the experience. Before a spread of news photographers and television cameras, he shifted from one leg to the other and maintained that he had put his family's interests above his own political career. He cited his high-percentage voting record and said this occurrence was not the norm, but there were extenuating circumstances. "I cannot sacrifice the health or well-being of my daughter for politics," Obama told reporters. "I had to make a decision based on what I felt was appropriate for my daughter and for my wife. . . . If the press takes my absence as the reason for the failure of the Safe Neighborhoods bill, then that's how the press is going to report it. . . . I have the track record of someone in Springfield who takes his legislative duties seriously."

Most of the roughly fifty seniors who listened to Obama seemed to buy his explanation of a sick child. When Obama maintained that the real culprit for the botched legislation was a feud between the governor and the senate president, many seniors nodded in support. I, however, was more cynical. Using the excuse of a sick toddler while vacationing in Hawaii seemed a bit hard to swallow. Thus, I began my story the next day in the *Tribune* this way: "Proving the political principle that one memorable public decision can define a politician for some time, state Sen. Barack Obama called reporters to a senior citizens' home Sunday to unveil a health care proposal but soon found himself explaining his controversial Christmastime vacation."

Rather than a rising star, Obama looked like a withering candidate. His performance reminded me of a comedian dying onstage.

The incident was indicative of Obama's almost uncanny run of bad luck in the Rush contest. The missed vote, the questioning of his racial authenticity, the death of the incumbent's son—Obama's first major campaign in the public eye had become an utter disaster. Obama would later say that this period was one of the low points of his life. "Less than halfway into the campaign, I knew in my bones that I was going to lose," Obama wrote in his book *The Audacity of Hope*. "Each morning I awoke with a vague sense of dread, realizing that I would have to spend the day smiling and shaking hands and pretending that everything was going according to plan."

Over the next couple of months, Obama fared only slightly better. He managed to pick up the *Tribune*'s endorsement, after all. And he won a few high marks for his performance in public debates, although the major news media largely ignored them. The candidates' only televised appearance, perhaps fittingly, went poorly for Obama. He resembled a fighter struggling on the ropes, desperately trying to prove he belonged in the ring in the first place. The rather informal panel discussion was hosted by Chicago's public television station and featured a single moderator, Phil Ponce, asking questions of the candidates. By now, the three men had been rivals for some time, and the discourse tended to be combative. Obama and Rush's other main challenger, Donne Trotter, accused Rush of failing to lead in Washington, and both argued that the district needed fresh, aggressive leadership. Rush, meanwhile, portrayed Obama as an overly ambitious young state lawmaker with modest accomplishments who had the gall to ask voters to send him to Washington. When Obama impatiently tried to interrupt Rush and plead his case, moderator Ponce cut him off and asked him to give Rush his time to speak—an unpleasant exchange that only furthered Obama's growing image as a self-centered young upstart. To make matters worse, Rush batted away the interruption by averting his eyes from Obama and attacking Obama's newcomer status: "Just what's he done? I mean, what's he done?" Obama responded by citing his history of organizing voters and poor residents, but Rush's line cut through Obama's message like a steak knife through butter. Obama's candidacy was in tatters, and he lost the election by 30 percentage points.

The New Rochelle Train

He always talked about the New Rochelle train, the trains that took commuters to and from New York City, and he didn't want to be on one of those trains every day. The image of a life, not a dynamic life, of going through the motions . . . that was scary to him.

—JERRY KELLMAN, CHICAGO COMMUNITY ORGANIZER

Barack Obama's ill-fated race against Bobby Rush taught the young state lawmaker a host of crucial lessons, not the least being that, in politics, no matter how appealing you are as a candidate and no matter how impressive your credentials, you are never completely in control of your own destiny. Something as random as a drug-related shooting can alter an election in such a profound way that not even the most sagacious political forecaster could predict it.

Obama had learned a similarly hard lesson about the vagaries of life once before—when despite his hard work and intense desire, he was benched from the basketball team at Punahou Academy after arguing with the coach about playing time. Yet for an incredibly competitive politician like Obama, the electoral loss to Rush stung sharply. Obama had entered the contest knowing it would be difficult to unseat an incumbent. But with his optimistic streak, he believed that if everyday voters could just hear his message, experience his intellect and feel the passion in his heart, he could win them over. But politics is about more than just delivering an appealing message or being on the right side of the issues. It's about shrewd calculation, raising money, catering to the necessary special interests.

It's about assembling a coherent strategy to win and then executing that plan.

When Obama stood outside a polling place and shook hands with voters on election day, he realized that he had gravely miscalculated. One after another, voters told him that they liked him, that they thought he offered a lot to public life and had a bright future, but they couldn't vote for him. The reason was summed up by one elderly woman who explained to Obama succinctly: "Bobby just ain't done nothin' wrong." Obama said it became clear to him that he had put himself ahead of the electorate, that his own time frame for advancement was not necessarily the same time frame that voters saw for him. "It made me realize that, you know, they were right, that there was no great external imperative for me to be a congressman at that stage," he said. "It really had more to do with me feeling anxious to be in the mix."

Obama learned that success in politics, as in life, requires balancing fierce ambition with due patience. On the campaign trail, as in the television debate, Obama looked too much like a man in a hurry. By then, he had published a book, led the most prestigious law journal in the country, been profiled in national publications and been embraced by various elites in Chicago, from the *Tribune* editorial board to guardians of the city's liberal establishment like Abner Mikva. In Springfield, he had amassed some success as a first-term lawmaker, but he was perceived by some as ineffectual because he had not curried enough favor with the right interests—the influential political reporters, the legislative powers-that-be, the union leaders, the political insiders. "Barack, you didn't have enough of the people in the party with you—you were kind of out there on your own," his friend and counselor the Reverend Jeremiah Wright told him.

But even if the right people were behind him, that could be of little worth if the voters didn't feel connected to him as an individual. It might be hard to imagine as the years have unfolded and Obama's skills as a politician have been seen worldwide, but he was an extremely poor political candidate in that race. By most accounts, in the Rush contest Obama was too fond of reciting his impressive résumé, too often mentioned that he had forsaken a high-priced law firm for public office and too often spoke in the high-minded prose of a constitutional law lecturer, all of which could make him appear condescending to his audience. That tone might have played well with a majority of voters in his state senate district, anchored by the college neighborhood of Hyde Park, but in the broader South Side black community, it could be alienating. Working-class voters gathering in neighborhood church basements want to know specifically how a candidate is going to work to change their lives. These black voters, in particular, want to feel that a candidate is committed to their cause and not to furthering his own career. "In his race with Bobby Rush, it really taught Barack how to connect to a black audience," Mikva recalled.

In this respect, Obama's story is similar to that of one of the country's most talented and charismatic politicians, a man to whom he is often compared—John F. Kennedy. As with Obama, it might be hard to believe that JFK was initially a poor stump campaigner. But at the outset of his political career, instead of working a room with handshakes and grins, the introverted Kennedy would disappear into a small group and rarely come up for air. Given his family's wealth and his elite education, Kennedy, too, was initially viewed by some as a condescending elitist. But slowly, as his campaign for Congress unfolded, Kennedy learned the value of pleasing oratory and press-the-flesh connections with everyday voters, until he grew into one of the most skilled practitioners of this aspect of politics. If Obama were to experience success down the road, he needed to absorb the same kinds of lessons about politics from his Rush experience.

When Obama returned to Springfield, his colleagues and friends noticed a changed man, a more chastened figure. He was no longer the bright young guy from Harvard Law School beloved by the liberals and primed for a big political office; he was the brash guy who aimed a little too high, too fast, and had come up way short. He had not just lost to Rush, but in Obama's own words, he had been "spanked." His poker crew was comforting but not surprised by the pounding Obama had taken at the polls. State senator Terry Link and others had warned him that dislodging a sitting congressman like Rush would be close to impossible. When Obama sat down with these buddies, he started off by saying he knew that they had told him so—and it need not be said again.

Yet instead of sulking, Obama impressed his colleagues and friends by putting his head down and diving back into the trenches of the General Assembly. And rather than holding grudges against those who had been less than supportive of him through the congressional contest—such as Donne Trotter, the state senator who accused him of lacking black authenticity; and Rich Miller, the Springfield scribe who had criticized his performance in the legislature—Obama sought them out and worked to improve relations. (The one burned bridge that could not be repaired, at least not for years, was with Rush, who now harbored a deep grudge against Obama.)

Obama conducted some soul searching upon returning to Springfield and to his college lecturing. He wondered if politics truly was the avenue he wanted to keep traveling. In 2001 his second child, Sasha, was born, making his commitment to family that much stronger. He had been appointed to the boards of several nonprofit organizations—the prestigious Joyce Foundation perhaps the most important of these—and maybe it was time to think about seeking a full-time post as the director of one of these groups. After all, he was still in the minority party in the senate and there were no guarantees that Democrats would regain power anytime soon. How effective could he be? Moreover, with a second child to tend to, Michelle longed to have a husband with a more stable working life. And financially, the congressional campaign had not been kind. His campaign had spent nearly five hundred and fifty thousand dollars on the race, with thousands coming out of pocket from Obama.

He and Michelle were living a middle- to upper-middle-class, white-collar existence, going home to a spacious town house in Hyde Park and employing a caregiver to help with child care. But despite their combined incomes, which topped $250,000 a year, Obama had personal debt. He had maxed out his credit card, partly on campaign expenses, and the couple were both repaying student loans from Harvard. He had no immediate future in terms of higher office and no law clients because he had suspended his legal practice to challenge Rush full-time. When a friend encouraged him to attend the Democratic National Convention in Los Angeles in the summer of 2000, he said he lacked the resources to go. Ultimately, he found an inexpensive Southwest Airlines flight and made the trek. But upon arrival at the Los Angeles airport, the rental car agency rejected his credit card. "I was broke," Obama recalled. "And not only that, but my wife was mad at me because we had a baby and I had made this run for Congress. I tried to rent a car in Los Angeles and my credit card wasn't accepted. It wasn't a high point in my life." He could not get a floor pass to the convention, made few networking connections and, in a state of dejection, he wound up returning home by midweek.

This was perhaps the first time that Obama began to consider the importance of money to his personal life. He had always been generous to his employees and friends when it came to money, making it a point to provide holiday bonuses to his staff. But personally, he had never been money-motivated, and he had never sought expensive possessions. In fact, quite the opposite. "He is motivated by people and getting things done. He does not think about money," Shomon said. "He used to forget to put in his expense reimbursements for his state trips. I would say, 'What the fuck is wrong with him?' His wife would thank me for being their accountant! But he was never cheap. A staff member was getting married and Barack wrote her a generous check. He is not a cheap guy at all. If I went to Barack and asked to borrow five thousand dollars, he would do it. He is really generous."

Obama's lack of emphasis on money extends to his personal tastes and his other minimalist tendencies. As he was growing up in Hawaii, it was only his grandmother who placed any real emphasis on money, and she was very pragmatic about it. "I'm sure Michelle would have been happier if I would have emphasized that a little more," his grandmother told me with a small smile. Obama's minimalistic nature can be seen in his physical appearance during that period. After entering politics, he always looked sharp in a crisp blue or black suit, and he certainly had a smart sense of how to look presentable. But he had only four or five suits in his closet. When spring rolled around, his one khaki suit would be added to his weekly wardrobe—perhaps twice weekly. He loathed buying new clothing, telling Michelle to pick out a couple of new shirts and ties for him at Christmas. His socks were worn in the heels, and his charcoal wool winter coat was a decade old with frayed lining. With his ultra-thin physique, most clothing looked good on Obama, and he was particularly fond of plain black polo shirts or mock turtlenecks paired with khaki pants. In casual settings, he gave the appearance of a walking Gap advertisement, except that his lagging fashion sense generally would have made the ad a few years old. Michelle, later to be joined by his children, would push Obama to wear more colorful garb, or at least patterns and stripes. But he was satisfied with plain whites, blacks and khakis. True to his nature, Obama refrains from cologne and wears little jewelry—just a gold wedding band and his one modest wristwatch with dark leather strap.

Being somewhat financially bereft after the Rush contest, Obama began thinking about seeking a tenured position at the University of Chicago or with a nonprofit group. But he could not see himself making the transition to a more traditional career. Instead, he put off such a move and tried to satisfy himself by knuckling down and working hard in the General Assembly, as well as teaching. "There were a range of options that I examined," Obama said. "But, you know, I continued to enjoy just the day-to-day work of drafting bills and, you know, framing debates. And so it was a time of reflection, but it wasn't a time of depression."

The most important relationship Obama improved upon when he returned to the senate was with the senate's Democratic leader since 1993—Emil Jones Jr., a street-tough African American who had risen from Chicago sewer inspector to enter the corridors of power in Springfield. Jones is one of the least eloquent speakers among politicians of his success. He has a thick voice that sometimes comes across as a deep mumble, making him difficult to understand. But Jones has been one of the most influential black politicians in the state over the past two decades, building a grassroots political operation on the city's South Side that could not be challenged. His power among Democrats in the legislature is undisputed and also rarely challenged. If Obama wanted to get his name on key pieces of legislation, Jones was the man to convince. Obama had actually met Jones while he was still a community organizer. He had organized a neighborhood meeting near Jones's home, and when the meeting expanded into a small march, Jones stepped outside to see what was going on. The two men could not have come from more different backgrounds: Obama was raised in a white family in Hawaii and educated in elite institutions; Jones, twenty-five years Obama's elder, was one of eight children of a truck driver and a homemaker on Chicago's South Side who found employment with the city's sanitation department, presumably with the help of his father, who was also a formidable precinct captain for the Democratic Party. Jones grew up in the belly of Chicago's machine politics. But as Obama's career evolved in Springfield, Obama and Jones would grow so close that Jones would talk about the fatherless Obama as a blood son. Obama consistently paid Jones the utmost respect. While most people pronounce Jones's first name with a midwestern flatness (E-mul), Obama is always careful to pronounce it correctly (E-meel). "Emil is driven by a sense that the African-American community has not been given its fair share and he is trying to make up for that—and I respect that mission," Obama said. Jones puts his fondness for Obama in more personal terms. "I am blessed to be his godfather and he feels like a son to me," Jones said.

As 2002 approached, the political dynamics of Illinois began to alter dramatically. It appeared as if the Democrats could retake the Illinois senate—and by the November election, that scenario became a reality. Jones became senate president, and Obama's career as a legislator took a decisively sharp turn, veering out of the wilderness of the minority party and into the bright lights of the majority.

THROUGHOUT 2002, OBAMA'S AMBITION BEGAN NAGGING AT HIM again. Suddenly he was in a position to pass laws, but he still had his sights set on higher office. The question: Which office to seek? It was about this time that Obama began to think about the U.S. Senate race in 2004. It looked as if the Democratic nominee would face an incumbent, Republican Peter Fitzgerald. Unseating a sitting senator would typically be considered a difficult task. But Fitzgerald had been such an outspoken, go-it-alone maverick in his first term that he had alienated established members of his own party, both in Washington and Illinois. For that reason and others, he looked particularly vulnerable, giving Obama and many other Democrats hopes of reclaiming the seat that, before Fitzgerald, had been held by the first African-American female senator, Democrat Carol Moseley Braun.

Obama said he "put out feelers" to colleagues in the Illinois senate, asking if they would support him in a run for the U.S. Senate, and he "got a pretty favorable response." Yet nearly everyone close to Obama was unified in their counsel: Don't run.

In addition to his wife, Michelle, chief among people dispensing this advice was his top aide and good friend, Shomon. He offered this viewpoint to Obama as a friend, not as a paid consultant. Sure, Obama would be a long shot in the Senate race and probably wouldn't win, but Shomon believed that Obama would regret getting into the race not because of another political failure, but because it would strain relations with his family. In 2002 the couple's second child, Sasha, was just a year old and the eldest, Malia, was only four. Obama had been regularly absent from the household for months in 1999 and 2000 in his campaign against Bobby Rush, something that still did not sit well with his wife. Now, with two children, Obama just two years later was considering an ambitious statewide contest that would consume even more time than a district congressional race. Shomon said he felt that another hectic campaign would overwhelm Obama's family life.

Little did Shomon know at the time, but Obama had been seriously exploring the Senate race through the year and had all but made up his mind. To Obama, this was his best, and likely his last, shot at advancing his career in politics. It was clear by now that Bobby Rush would hold his congressional seat for as long as he wanted. So where else was Obama to go? Perhaps he could run for a state office, but other Democrats were in line ahead of him for those positions. Obama had interviewed for private sector jobs, as head of nonprofit foundations. But his restless soul and driving ambition had given him an intense fear of winding up in such a prosaic societal position—a nine-to-five office job that lacked excitement and adventure.

"He always talked about the New Rochelle train, the trains that took commuters to and from New York City, and he didn't want to be on one of those trains every day," said Jerry Kellman, the community organizer who enticed Obama to Chicago from his Manhattan office job. "The image of a life, not a dynamic life, of going through the motions . . . that was scary to him."

Shomon, on the other hand, was ecstatic at the prospect of Obama landing a position as the executive of a nonprofit agency. Obama, in fact, interviewed for such a job—as head of the Chicago-based Joyce Foundation. "I am thinking that my relationship with a politician is going to pay off! I am going to get this hundred-and-fifty-thousand-dollar-a-year policy job!" Shomon said with a laugh. "So he calls me . . . and he tells me that he was literally shaking when he went in to the interview for fear that he would get the job—because he did not want it. I said, 'What the fuck is wrong with you?! This is a dream. You can build up money, build up relationships and run again.'

"We were on the side of the road on Illinois 4 and I told Barack, 'I don't think you should run,'" Shomon recalled. "I said I thought it was a bad idea because of Michelle and the kids. Barack feels tremendous guilt. He has a conscience. I thought he would wish he hadn't done it afterward. But he just looked at me and said, 'I'm running anyway.'"

Winning a U.S. Senate seat was perhaps Obama's last chance at leading that dynamic life that he had told Jerry Kellman he craved. It was his last chance not to take the New Rochelle train home from work every day.

11
The Candidate

The older I get, the more I realize it is not always what you say,
but the way you say it. And that is particularly true in politics.
—HOTEL MAGNATE PENNY PRITZKER

Barack Obama's first thoughts about running for the U.S. Senate came as early as mid-2001, less than a year after his stinging defeat by Bobby Rush. After that race, Obama had been approached about running for Illinois attorney general, but discarded the notion for the same reasons Dan Shomon warned him away from the Senate run. Said Obama, "I put Michelle and the family through such heck with the congressional race and it put such significant strains on our marriage that I could not just turn around and start running all over again, so I passed that by."

Coincidentally, the thought of Obama running for the Senate had also occurred to Eric Adelstein, a Chicago-based media consultant for Democrats. Adelstein had scheduled a meeting with Obama for September 2001 to discuss their mutual idea. But little could either man have known what would transpire on September 11 of that year. "So 9/11 happens and immediately you have all these reports about the guy who did this thing is Osama Bin Laden," Obama said. "Suddenly Adelstein's interest in the meeting had diminished! We talked about it and he said that the name thing was really going to be a problem for me now. In fairness to Eric, I think at that point the notion that somebody named Barack Obama could win anything—it just seemed pretty dim."

So Obama went back to concentrating on his two jobs as constitutional law instructor and state lawmaker. But by mid-2002, he again was growing restless in his political career. He began toying again with the idea of the Senate contest and he opened serious conversations with his Illinois senate colleagues about the notion. Several seemed receptive—Denny Jacobs, Terry Link, Larry Walsh and members of the black caucus, including senate president Emil Jones Jr. They promised to support him, even to lead an exploratory committee.

But there was one person whose affirmation was vital—Michelle. Convincing Michelle to support him through another campaign was the most significant hurdle Obama had to clear. Michelle knew that her husband's political career was of immense importance to him, but upon hearing his Senate idea, she began to wonder if this optimistic dreamer was going off the deep edge. The couple had two kids, a mortgage and credit card debt. After Obama's crushing loss to Bobby Rush, she worried that her husband was about to undertake another lost cause, although, realistically, she worried primarily about finances. Another political race could keep the family in debt, or perhaps plunge it deeper into debt. And even if he won the Senate seat, she thought, their financial condition was not likely to improve.

"The big issue around the Senate for me was, how on earth can we afford it?" Michelle told me. "I don't like to talk about it, because people forget that his credit card was maxed out. How are we going to get by? Okay, now we're going to have two households to fund, one here and one in Washington. We have law school debt, tuition to pay for the children. And we're trying to save for college for the girls. . . . My thing is, is this just another gamble? It's just killing us. My thing was, this is ridiculous, even if you do win, how are you going to afford this wonderful next step in your life? And he said, 'Well, then, I'm going to write a book, a good book.' And I'm thinking, 'Snake eyes there, buddy. Just write a book, yeah, that's right. Yep, yep, yep. And you'll climb the beanstalk and come back down with the golden egg, Jack.'"

But Obama was confident that he was destined for more than just a day job running a foundation or practicing law or languishing in the minority party in the Illinois senate. And for that golden destiny to come to fruition, he knew he had to do his most convincing sales job yet.

Explained Obama, "What I told Michelle is that politics has been a huge strain on you, but I really think there is a strong possibility that I can win this race. Obviously I have devoted a lot of my life to public service and I think that I can make a huge difference here if I won the U.S. Senate race. I said to her that if you are willing to go with me on this ride and if it doesn't work out, then I will step out of politics. . . . I think that Michelle felt as if I was sincere. I think she had come to realize that I would leave politics if she asked me to."

Added Michelle, "Ultimately I capitulated and said, 'Whatever. We'll figure it out. We're not hurting. Go ahead.'" Then she laughed and told him hopefully, "And maybe you'll lose."

Obama went back to Shomon and told him he was in. "So I told Dan that I had this conversation with Michelle and she had given me the green light and that what I want to do is roll the dice and put everything we have into this thing."

Still tapped out from his congressional race, Obama knew he needed start-up cash. So he held a small fund-raiser that netted thirty-three thousand dollars. That allowed him to begin paying Shomon as a full-time campaign manager, as well as hiring an office administrator and a full-time fund-raiser.

BEFORE THAT FUND-RAISER, HOWEVER, OBAMA HAD BEEN MAKING other vital moves. He had a crucial meeting with another important crowd—the key financial supporters of his unsuccessful congressional race in 2000. Obama knew that his fund-raising for a Senate campaign had to far exceed what he had raised in his race against Bobby Rush. So he invited a group of African-American professionals to the house of Marty Nesbitt, who had served as finance chairman of his congressional campaign. Nesbitt is the president of a Chicago parking management company and vice president of the Pritzker Realty Group, part of the Pritzker family empire. A tall, slender African American, he had been friends with Obama for years. Their wives were also close, and Nesbitt's wife, Anita Blanchard, an obstetrician, delivered both of Obama's children. The two men played basketball together and mixed with the same Hyde Park neighborhood crowd of young, successful black professionals. Nesbitt was soft-spoken, polite and amiable. He was also extremely active in civic affairs, serving on the boards of Chicago's Museum of Contemporary Art, Big Brothers/Big Sisters and the United Negro College Fund, among many others.

Not long after Obama's loss to Rush, Nesbitt had suggested that Obama might want to run statewide in his next contest. It seemed pointless to try to unseat Rush again after failing so miserably. So when Obama addressed the group of black professionals and mentioned his plans for a statewide race, Nesbitt quite naturally assumed Obama was considering a state office, such as attorney general or treasurer. Then Obama dropped the bomb.

"Barack says he wants to run for the U.S. Senate," Nesbitt recalled. "Blahhh!! I mean, I literally fell off the couch. And we all started laughing—and he said, 'No, really, I am gonna run for the U.S. Senate.'" Then Obama proceeded to make a rational argument that he could win such a race. He said that Senator Fitzgerald's approval ratings were so low that he was destined to lose to a Democrat in 2004. Obama said he believed he could bring together blacks and liberals into a coalition and come out on top in what was looking to be a crowded primary field. "He convinced us in the room that day that he could pull it off," Nesbitt said. "But he mostly convinced us because we were his friends and we wanted to support him."

Obama, who had learned the significance of money in politics during his Rush contest, also told them the hard fiscal truth that he was going to need not hundreds of thousands but millions of dollars to pull off a victory. He even broke it down into specifics, telling the group that with three million dollars, he had a 40 percent chance of winning; with five million dollars, he had a 50 percent chance; with seven million, 80 percent. And with ten million, Obama proclaimed, "I guarantee you that I will win."

Nesbitt said that Obama's supreme confidence, clear vision and attention to detail convinced them that it was doable. It certainly wasn't a sure win, but it was doable.

"So we all said, 'We've gotta make a grassroots, ground-level push and get this going,'" Nesbitt said. "But you know, at this point, I was very naive. Barack is not afraid to ask for money. But I didn't have any idea how far we had to take it, to the next level."

The next level meant that Nesbitt and Obama had to reach beyond his previous contributor base. In his congressional race, Obama successfully picked up cash from black business leaders and a smattering of lakefront liberals, but he needed to reach even deeper into those pockets and find more. So Nesbitt arranged a weekend getaway to help Obama reach inside the deepest pockets he knew— those of the Pritzker family.

THE PRITZKERS RULE A FAMILY EMPIRE THAT EMBODIES HIGH society and extreme wealth in Chicago. One of the richest families in the country, with a fortune estimated at twenty billion dollars, the Pritzkers began their American story in 1902 when Nicholas Pritzker, a Ukrainian immigrant, opened a law firm in Chicago's downtown Loop commercial district. Over four generations the family amassed its fortune through Hyatt hotels, financial services and numerous other enterprises. The clan is intensely private, even if they have also been extremely philanthropic. Nesbitt knew that if Obama could sell himself to Penny Pritzker, her support would not only reap huge immediate financial dividends but also be a crucial step in the foundation of a fund-raising network.

So in late summer 2002, Obama, Michelle and their two daughters drove to Penny Pritzker's weekend cottage along the lakefront in Michigan, about forty-five minutes from Chicago, to sell his candidacy.

Like many people at this point, Pritzker was impressed by Obama's intellectual heft but was unsure whether he could pull off what he had in mind. It would be up to Obama to sell his vision to a veteran businesswoman who was no easy sale. Pritzker, who has mostly supported Democrats but deems herself a "centrist," recalled the weekend as a "seminal moment." She and her husband, Bryan Traubert, were in training for the Chicago Marathon, and under a beautiful sunny sky the couple slipped away for a long run along Michigan's winding country roads near the lakeshore, all the while discussing whether Obama merited their backing. She described the discussion:

> We had known Barack and Michelle previously, but we hadn't made up our minds about supporting him for the Senate. And we had to make a decision. So we spent some time talking with him about his philosophy of life and his vision for the country. He is a very thoughtful human being in the way he articulates ideas and the way he thinks about the world. And the older I get, the more I realize it is not always what you say, but the way you say it. And that is particularly true in politics. . . . So Bryan and I had a long discussion about Barack and his values and the way he carries and expresses himself, his family and the kind of human beings he and Michelle are—what kind of people they are, as much as about lofty political ideas. Really, at that point

the question was, these other people are running and what
obligations do you have to the other people? It became clear to
us that if Barack could win, he had the intellect, the opportunity,
to be an extraordinary leader, not just because he would be an
African-American senator, but a male African-American senator.
Here is a guy with an amazing intellectual capacity to learn and
an interest in learning. . . . But how much did he know about
medicine, about business, about foreign affairs and the economy?
He was dealing with all these things in fragments and we asked
ourselves what was his capacity to deal with these things as a
whole? . . . He was someone who views himself as a healer, not
a divisive character, but someone who can bring disparate
constituencies together. . . . It became clear to us that . . . he is
not perfect, but he is bright and thoughtful and confident. He
possesses a lot of confidence. That was the seminal moment
when we simply decided after that weekend that we would
support them.

With Penny Pritzker on board, other influential Chicago-based
Democrats and philanthropists soon followed suit: Newton Min-
now, a Chicago lawyer who advised Senator Adlai Stevenson and
Presidents John F. Kennedy and Lyndon Johnson; James Crown and
members of the wealthy Crown family; John Bryan, then the chief
executive of the Sara Lee Corporation.

Obama had not yet announced his candidacy, but he realized he
needed more political talent behind him. By now, Adelstein had
signed on with another potential candidate, Gery Chico, the former
president of the Chicago Board of Education. So Obama turned to
the other major political consultant he knew, David Axelrod. A couple of years earlier, Axelrod and his wife, Susan, had thrown a quick
fund-raiser for Obama at their downtown Chicago high-rise condominium when Obama's senate district lines were redrawn to include

much of downtown, including their home. Obama was largely anonymous outside Hyde Park at that first gathering, which drew all of twenty people. "We were pulling in people from the pool, urging them, 'Hey, come meet your new state senator,'" Axelrod said.

After this tepid fund-raiser and Obama's collapse against Bobby Rush, the political consultant was less than enthused about a potential Senate candidacy for Obama. Axelrod told the hopeful Obama that he thought Obama was a "terrific talent" and "I consider you my friend," but running for statewide office was "probably unrealistic." Axelrod was blunt with Obama: "If I were you, I would wait until Daley retires and then look at a mayor's race because then the demographics would be working in your favor."

That disappointed Obama, but it did not dissuade him—he still thought the Senate was a good possibility for him. Even so, he soon encountered an even bigger stumbling block than Axelrod's caution. Carol Moseley Braun, who represented Illinois in the Senate from 1993 to 1999, announced that she might run to reclaim the seat she had lost to Republican Fitzgerald. Moseley Braun had made history as the first African-American woman to serve in the Senate but was defeated by Fitzgerald in 1998 after a tumultuous first term. Allegations arose that her then boyfriend, Kgosie Matthews, who ultimately took over her campaign, sexually harassed female campaign workers, although Moseley Braun said an investigation found no evidence. The couple was accused of spending campaign money on clothes and jewelry, and they were criticized for taking a month-long trip to Africa right after the election. But her major downfall was meeting with the former dictator of Nigeria, General Sani Abacha, without giving advance warning to the Department of State. Abacha had been accused of a host of human rights violations.

Fitzgerald looked extremely vulnerable, having feuded with the GOP power structures in Washington and Illinois. This was the major factor that led Obama to think that unseating him was possible.

But if Moseley Braun were in the race, Obama said he would have to defer to her and decline to run. The reality was, he would have no choice: she would gobble up both of his potential bases of support—African Americans and liberals. "There was no way to win," he said. So Obama asked to meet with her in his senate office in Springfield. "We . . . asked her how serious she was, and her basic attitude was that she had not made up her mind but obviously it is [her] prerogative to potentially run," Obama recalled. "I understood her position. She had been a trailblazer. It was frustrating for me to think that maybe this was one chance to go after something I really cared about and potentially [I] could not do it. But that is the nature of politics."

Moseley Braun presented not only a problem for Obama, but a major headache for the Democratic Party powers in both Illinois and Washington. Because she had been a national figure as the first black woman in the Senate, her alleged improprieties in office had been a major embarrassment to the party. And even though Fitzgerald appeared extremely vulnerable, Democrats feared that these past embarrassments would doom her in a general election. Eric Zorn, a liberal columnist for the *Chicago Tribune,* went so far as to predict that she would be "walloped" by almost any Republican. With a Senate seat seemingly up for grabs, Democrats could ill afford to run a candidate tainted with past scandals.

To keep Moseley Braun out of the race, various Democratic powers close to Obama set out to find her a job elsewhere. "But the problem was, they could not find her employment," said an Obama confidante. "Nobody could find her any work." While the job search persisted, Moseley Braun's indecisiveness about a Senate run began to wear thin on the potential candidates and on Democratic activists looking to support someone in the race. It especially wore thin on Obama, whose political career was hanging in the balance. But that apparently did not matter to Moseley Braun (who declined to be interviewed on the subject). "She felt that [Obama] was a young whippersnapper, a pretender, a cheat," David Axelrod said. "I think she took it personally. He was essentially messing in her territory. She made it pretty clear that she was not happy about Barack's entrance."

As Moseley Braun considered her options, Obama decided to travel to Washington in September 2002, to spend a weekend at the annual Congressional Black Caucus conference in hopes of garnering support for himself. He figured he would meet some influential black lawmakers and ask for their help and guidance. But the excursion was a major disappointment for the earnest Obama. He returned to Chicago significantly disillusioned about the ways of Washington.

A couple of weeks after returning home, he sought the counsel of his pastor and friend, the Reverend Jeremiah Wright. Visibly dejected, Obama slumped onto the sofa in the pastor's second-floor office at Trinity United Church of Christ. He told Wright that the Senate idea was thoroughly frustrating him because Moseley Braun would not make up her mind whether she was in or out. And not only that, but other names had begun to surface for the race—Illinois comptroller Dan Hynes, whose father was a powerful Chicago ward alderman; Blair Hull, a multimillionaire securities trader; and Gery Chico, formerly a top aide to Chicago mayor Richard Daley and school board president. None was a certain nominee or impinged on his bases, but each had strengths, and each was pulling ahead of him in organizing a campaign operation. Hynes had run statewide before and would have his father's political machine behind him. Hull would have tens of millions in personal wealth to lavish on a campaign. Chico was already raising significant money and was far along in assembling a campaign structure. Another name floating around was Representative Jan Schakowsky, a liberal from the North Shore suburbs who would have cut directly into Obama's lakefront support.

"My name should be out there," Obama told his pastor. "But Carol Moseley Braun won't say what she's going to do, and I'm not gonna run against a black woman. If she's gonna run, then I'm out. Until she says yes or no, I can't say anything."

But what truly struck Wright from that meeting was Obama's astonishment over the black caucus event in Washington. It opened Wright's eyes once again to just how innocent and idealistic Obama could be about the world of politics. The conference was nothing like what Obama had envisioned, but it was exactly the way Wright, a former adviser to Chicago's only black mayor, Harold Washington, recalled it.

"He had gone down there to get support and find out who would support him and found out it was just a meat market," the pastor said in an interview, breaking into a laugh. "He had people say, 'If you want to count on me, come on to my room. I don't care if you're married. I am not asking you to leave your wife—just come on.' All the women hitting on him. He was, like, in shock. He's there on a serious agenda, talking about running for the United States Senate. They're talking about giving [him] some pussy. And I was like, 'Barack, c'mon, man. Come on! Name me one significant thing that has come out of black congressional caucus weekend. It's homecoming. It's just a nonstop party, all the booze you want, all the booty you want. That's all it is.' And here he is with this altruistic agenda, trying to get some support. He comes back shattered. I thought to myself, 'Does he have a rude awakening coming his way.'"

A few months later, Moseley Braun was still dithering as Obama took his annual Christmas season sojourn to his native Hawaii. Deeply dispirited over the prospect of Moseley Braun's candidacy, Obama rested on Sandy Beach, a fifteen-minute drive along the rugged southeast shoreline from downtown Honolulu. This was the beach of Obama's youth, the childhood paradise where he spent mindless high school days body-surfing, drinking beer and seeking to convince college women that he was worthy of their attention. With the tree-covered hills of Oahu again cascading behind him, with the tall waves of the Pacific Ocean again crashing before him, the forty-one-year-old Illinois state senator contemplated his future.

At this moment, Obama's ambitious nature was eating at him. He was terrified that all his hard work was coming to naught. The frustrating years trying to organize programs that would assist the impoverished on Chicago's Far South Side, the decision to forgo a lucrative law career for a meagerly paid civil rights practice, the years foraging in the wilderness of a minority party in the Illinois General Assembly, the days and nights away from his devoted wife and two precious daughters to travel the campaign trail—what had this sacrifice done for him and for the world? That Christmas, Obama's once bright future seemed as if it was crashing in the surf. Suddenly his grand idea of winning an open seat on the world's most powerful lawmaking body looked less like a serious notion than a quixotic dream.

For the first time in his professional life, the supremely confident Obama was deeply frightened. He was terrified that the final story of Barack Obama's political career would be this: Another talented black man with grandiose dreams somehow flamed out and disappeared from public life. Most frightening of all: This story resembled all too much the tale of his father's unfulfilled dreams.

But then Obama peered out at his two young girls splashing in the ocean with Michelle. They looked playful and happy. Obama reconsidered his dreams and thought, well, perhaps he was not meant to be a senator or a mayor. "I didn't grow up thinking that I wanted to be a politician," Obama recalled. "This was something that happened as a sideline, as a consequence of or an outgrowth of my community organizing, and if this doesn't work out I am fine with it. Kids keep you grounded, and I had to remind myself that it was not all about me and my personal ambitions, that there is a set of broader issues."

But just as Obama was coming to terms with the notion that his grand political career might never happen, he received a phone call that would again put his personal ambitions front and center. He was still in Hawaii when the news came via cell phone: Moseley Braun had decided against running for the Senate and instead would seek the presidency. Obama knew exactly what he needed to do—and fast. He clicked the cell phone back open and dialed David Axelrod.

12
The Consultant

David was in his glory. He was the belle of the ball. Everyone wanted him and he was feeding on all the attention.
—AN ILLINOIS POLITICAL LOBBYIST ABOUT
CONSULTANT DAVID AXELROD

Perhaps the most significant event in Barack Obama's campaign for the U.S. Senate occurred more than two full years before Illinois voters went to the polls and many months before Obama even made a final decision to run for the seat. This occurrence was a business-like yet utterly strange final meeting between the highly regarded political consultant David Axelrod and a complete political neophyte with his eyes on the Senate seat, multimillionaire Blair Hull.

The discussions between Axelrod and Hull ultimately pushed Axelrod into the campaign of Obama, rearranging the dynamics of the race and instantly turning the relatively little-known state sena-tor into a serious contender. Without Axelrod as his marketing spe-cialist, it's unclear if Obama would have won the Democratic Party primary. And if Obama had still managed to pull off victory, there's absolutely no question that he would never have done it in such an astonishing fashion—in a way that propelled him to national star-dom even before he hit Washington. Obama possessed the innate talent for success, to be sure. But Axelrod was the coach who har-nessed that talent and massaged it into its present form, and then he was the publicity agent who sold Obama to Illinois voters. Obama would lean on Axelrod increasingly throughout the 2003–2004 campaign season—and thereafter. If Obama was on his cell phone, which seemed to be at every spare moment, there was a high probability that he was consulting with "Ax," the apt nickname for this cutthroat political operative. Obama would call Axelrod three or four times a day, at all hours of the day and night, seeking his counsel and his wisdom.

Axelrod is a former reporter for the *Chicago Tribune* who had slogged through several years of political campaigns as a private consultant before establishing himself as the preeminent message and media adviser for Democrats seeking office in Illinois. His talent for producing television commercials that highlight the best sale points of a candidate is unsurpassed. A native of New York City, Axelrod is a tall man who was then approaching fifty. He has a thick salt-and-pepper mustache, thinning hair, a perpetual slouch and dark droopy eyes that make him appear continually sad. He has a keen, somewhat scalding sense of humor. But even when tossing off an acidly funny line, he rarely smiles at his own wit.

Whenever I was assigned to cover a political campaign for the *Tribune,* I would never fail to schedule a breakfast or two with Axelrod near the downtown Chicago high-rise condominium he shares with his wife, Susan, whom he met playing coed basketball in college. It was a mutually beneficial source-reporter relationship. I would endeavor to pick his brain about the race at hand, while he would use the encounters to convey his candidate's message and try to spin a reporter from the influential *Tribune.* Without fail, a disheveled Axelrod would arrive for these meetings at least twenty minutes late. He was typically outfitted in a baseball cap and wrinkled sports attire, which he might or might not change in the course of his business day. Over one breakfast, he rubbed his eyes hard with the palms of his hands and confessed that he suffers from sleep apnea, a condition in which someone sleeps but fitfully and unrestfully. This made me think that Axelrod's scarily shrewd political mind is always at work, even during sleep. His sloppy eating habits are a well-worn inside joke among those who know him well—Obama foremost among these wags. During Obama's 2004 Senate primary contest, I recall a breakfast in which Axelrod fervently pitched his client's potential star power for minutes on end. But as he extolled the virtues of Obama, I was more in wonder at the seemingly endless length of time that a gooey string of yellow cheese from his omelet could extend from his mustache to his plate without snapping or being noticed by him.

Axelrod has handled media strategy for major Democrats across the nation, including Chicago mayor Richard M. Daley, Representative Rahm Emanuel of Illinois, Senator Christopher Dodd of Connecticut and a host of others. He initially oversaw media for Senator John Edwards's 2004 presidential campaign but mysteriously left that role amid squabbling among Edwards's advisers. Axelrod won't talk about that campaign in depth, except to say that "For whatever reason, John Edwards couldn't close the deal. The candidate has to close the deal with voters."

Rising from the streets of Chicago's bare-knuckles, fratricidal machine politics, Axelrod developed a specialty in urban races. He is especially masterful at helping to elect black candidates in broad geographic areas where some white support is essential, most notably Michael White for Cleveland mayor and Deval Patrick for Massachusetts governor. He has a special talent for tapping into the most compelling personal life stories of a candidate and composing a campaign script that enhances those qualities for public consumption. Despite this artistic side, Axelrod also has no fear of negative, go-for-the-jugular campaigning. Of all the operatives that Republican strategist Ed Rollins has gone up against, Rollins placed Axelrod at the top of his list of "Guys I Never Want to See Lobbing Grenades at Me Again."

Axelrod did his first political work at nine years of age, when the precocious youngster handed out leaflets for Robert Kennedy's Senate campaign in New York. "I just wanted to go out and do it and so I went over and volunteered—I was a weird kid," Axelrod said with a smirk. His mother was a newspaper reporter before running focus groups for a New York advertising agency. His father, a Russian immigrant, was a psychologist and, perhaps most significant to the sports-maniac Axelrod, an excellent baseball player who was an amateur teammate of the great hitter Hank Greenberg, known for making the National Baseball Hall of Fame after overcoming virulent anti-Semitism. Axelrod's father went to Long Island University on a baseball scholarship and studied art and philosophy before turning to psychology. Axelrod's parents, who would divorce during his childhood, were passionate about politics and "classic New York leftists," he said. This philosophy remains with Axelrod today. He is a potent blend of irascible left-wing idealist and smooth-talking, high-priced operative for the establishment. He talks with such deep conviction about using politics to promote the greater good that it is impossible to think he does not believe this at some level. On the other hand, one of Axelrod's most enduring clients is Mayor Daley, who lords over Chicago with a powerful political machine unmatched anywhere in the country. When Daley's political apparatus became the focus of a federal corruption investigation into widespread illegal patronage hiring at city hall, Axelrod rushed to defend the mayor with TV appearances and op-ed pieces. He explains this moral contradiction by saying that Daley's overall record is one of bringing harmony and development to a city that the mayor clearly loves.

Axelrod has been dealt two family tragedies that perhaps explain why I have always sensed a certain melancholy about him. His father committed suicide while Axelrod was in college, and his only daughter among three children became developmentally disabled after suffering epileptic seizures.

For years, Axelrod was loath to acknowledge his father's suicide. Whenever I tried to broach the subject, he would deftly skirt around it, and I would retreat. But in 2006, Axelrod finally came to terms with the loss. He wrote a poignant column in the *Tribune* about his father's chronic depression: "Dad never shared his anxieties or sadness with me or, so far as I could tell, anyone else. At his funeral, several of his patients told me that he had saved their lives. Yet he couldn't reach out for help to save his own. . . . We still have a long way to go before depression and other psychiatric conditions are fully accepted as illnesses, not defects of character or failures of self-discipline. I know, because it's taken me more than thirty years to say out loud that the man I most loved and admired took his own life."

Axelrod came to Chicago to study political science at the University of Chicago and, just days after graduation, found himself following his mother's profession, covering nighttime "mayhem"—cops, courts, killings—for the *Tribune*. He worked his way up to political reporter and was largely responsible for a great deal of good press coverage of the 1979 campaign of an outsider for Chicago mayor, Jane Byrne. In fact, Axelrod's reporting was instrumental in Byrne's upset victory to become the first woman to head city hall. After her election, Axelrod then dutifully chronicled every election promise that she broke, and essentially reversed her image from reformer to another go-along politician. Thus, Axelrod is intimately familiar with how the media often operates, having been a purveyor of its duplicitous nature. The press can carry a candidate into office on one day and knock that politician out of office the next. Set 'em up, then knock 'em down.

Axelrod's stay with the *Tribune* ended in 1984 after, in his version, he grew concerned about the newspaper's direction away from a "sense of mission" and toward "corporatization." He talks nostalgically of his reporting days at the *Tribune*. "It was just all about the journalism and the mission," he said. "You would have loved working then." Axelrod had a front-row seat for Harold Washington's historic mayoral run and the racial animosity it stoked in the city. But missions aside, Axelrod also departed the *Tribune* because his career hit a roadblock. He had sharp differences with more senior reporters and editors who stood in his path. Upon leaving the newspaper, Axelrod was immediately hired as communications director for the successful Senate campaign of then Representative Paul Simon, whose devotion and integrity "inspired" Axelrod. After a shakeup in Simon's staff, Axelrod quickly found himself comanaging the campaign of the bespectacled, bow-tied, populist Democrat. Still in his twenties, Axelrod went from newspaper reporting to overseeing a major Senate campaign almost overnight. It was in Simon's campaign that Axelrod formed a lasting friendship with another young Chicago operative—the brash, youthful, hard-driving Rahm Emanuel, who went on to become a top White House aide to Bill Clinton and then a leading Democrat in Congress. The merciless Emanuel was chief architect for the Democrats' takeover of the House of Representatives in 2005.

When Axelrod left the newspaper for political consulting, he carried with him a Rolodex of media contacts and intimate knowledge of how deeply politics and newspapers are intertwined in Chicago's culture and how the inside players of that world operated. It would take time, but this knowledge would help him build a lucrative media consulting business that is thriving more than two decades later—with his top client, of course, now being Obama.

THE MEETINGS BETWEEN AXELROD AND BLAIR HULL, WHICH occurred throughout 2002, provided Axelrod with a vivid glimpse into the private life of Hull, a former professional Las Vegas gambler who had amassed extraordinary personal wealth trading securities on Wall Street.

Hull was willing to spend tens of millions to win a Senate seat—and he wanted Axelrod on board his campaign. Axelrod was especially accomplished at spotting and assessing the modest or far-reaching talent possessed by a candidate, as well as at crafting a message and articulating that message through appealing television advertising. But just as important, as a former *Tribune* reporter, he offered those invaluable inside connections to Chicago's media elite. He could call nearly any political reporter or newspaper editor in the Chicago region, including those in the often impenetrable *Tribune* ivory tower, and immediately get an audience. Axelrod had cultivated reporters throughout Chicago and the rest of the state. (Full disclosure: I was among those reporters under cultivation.) From his earlier stint at the *Tribune,* Axelrod was also on a friendly, first-name basis with the editor of the *Tribune,* Ann Marie Lipinski, as well as with its editorial board editor, Bruce Dold. He throws an annual December holiday party in which the guests are split almost evenly between media members and political insiders. Thus, each serious candidate in the 2004 Senate contest ranked Axelrod the overwhelming first choice among media advisers to hire. "David was in his glory. He was the belle of the ball," one influential Illinois political lobbyist recalled. "Everyone wanted him and he was feeding on all the attention."

In being the sought-after adviser, Axelrod was dealing from a position of extreme power when selecting an employer. No one grasped the demographic and political dynamics of the entire state and urban Chicago as astutely as Axelrod. He knew the terrain and the people who populated that landscape. The candidate who won his services would have the sharpest, most ruthless political mind in the state at his disposal, immediately providing that person with a huge advantage amid a competitive field.

For those like Axelrod who ply their trade in the meat-grinder of campaign politics, Hull was the employer of choice. Hull not only brought immediate national attention to the race because of his excessive wealth—several hundred million dollars—but most important to itinerant political workers, he offered an unlimited bank account from which to fund a campaign. Hull had pledged to spend tens of millions of dollars—whatever it took to win the seat. And best of all, he was such a newcomer to politics that he was at the mercy of his paid staff about nearly everything, from what to say to reporters to how much to drop on television ads. His campaign manager, for example, struck a one-hundred-and-fifty-thousand-dollar-a-year deal—an over-the-top sum to run a Senate campaign. Hull was a cash cow all too willing to be milked prodigiously. "He was a meal ticket for everybody," in the words of Axelrod.

Axelrod also had a history of signing with a deep-pocketed first-time candidate like Hull. In 1992 he was chief adviser to free-spending multimillionaire lawyer Al Hofeld when Hofeld unsuccessfully sought the Democratic nomination against Senator Alan Dixon of Illinois. So when the 2004 Senate race rolled around, Axelrod looked first to Hull. By Axelrod's account, he met with Hull at least twice in the first half of 2002. Hull told Axelrod that he had already hired a highly recommended Washington-based media consultant, Anita Dunn. Yet, according to Axelrod, Hull promised him double Dunn's fee if he would come on board too. "I would have made far more money working for Hull in the primary than what I made from Barack in both the primary and the general [elections] combined," Axelrod said.

Still, none of the meetings between Axelrod and Hull went particularly well for the awkward Hull. And most important, the encounters provided Axelrod with extraordinary insight into Hull's character and his astounding lack of political talent. In their first chat, Hull misinterpreted Axelrod's concern about Hull's dismal history of voting in elections as a veiled threat to be used against him later. In subsequent discussions, Axelrod learned intimate and sordid details about Hull's private life that would later be of significant value to Axelrod's ultimate employer, Barack Obama.

Rumors had swirled in Democratic circles that Hull's divorce had been particularly messy, and that his ex-wife had made allegations against him, although the specifics of those allegations were unclear. Talk that Hull had a past alcohol abuse problem also floated through the political air. Learning of this through the grapevine, Axelrod could see a complicated future for a Hull candidacy. If these things were true, Hull could go up in flames, especially if the public relations surrounding these matters was not handled with expert perfection.

So Axelrod questioned Hull sternly and specifically about the rumors. He wanted to know exactly what kind of storm he might have to weather if he signed with Hull. And he wanted to know how Hull would handle himself should those rumors be proved factual and then leak into the public domain. Realizing the importance of enlisting Axelrod on his team, Hull decided to come clean with him—sort of. Hull acknowledged that he indeed had used cocaine in the past and had sought treatment for alcohol abuse and that, yes, his ex-wife had alleged physical and mental abuse.

But to Axelrod's astonishment, when he pried further about the ex-wife's allegations of abuse, Hull gave him an answer that ultimately would end their relationship. Axelrod explained to Hull: "You know, I hear these things. I don't know if they're true. But you know, like I need to know if they're true." The multimillionaire gave Axelrod a look that Axelrod would later describe as "glacial." Then Hull simply said: "There's no paper on that."

Recalled Axelrod: "So I'm thinking, Wow, that's really not the answer you really wanted to hear. . . . And I am thinking, He's a cold guy. I finally came to that conclusion. I also came to the conclusion that he could not get through the primary with that stuff. . . . I told him that he needed to think real long and hard about this because these kinds of things come out. It was a really awkward thing, though, because I knew stuff from a conversation that I felt was, in certain ways, privileged. I knew all this shit but I could not, and would not, use it against him."

Nevertheless, if Axelrod wanted some business from the 2004 Senate contest, after those talks, it became clear that he must find another Democrat to pay his bills.

LESS THAN TWO MONTHS BEFORE OBAMA'S SEASONAL HAWAII vacation in 2002, he delivered what he considers today to be his best speech—an address to a group of hundreds of war protestors in which he flatly stated his opposition to a U.S. invasion of Iraq. The American military action was still months away, but foment had been building against it, especially among the Democratic Party's left wing. Still, just a year after the 9/11 terrorist strike, President Bush's favorable rating among voters was very high. Polls showed that most Americans supported him if he ultimately deemed the invasion to be in the best interests of the country's foreign policy.

As a still unannounced Senate candidate, Obama for months had been quietly courting what he considered his two strongest bases of support—Chicago's so-called lakefront liberals and African Americans. The lead organizer of the downtown Chicago anti-war rally was Bettylu Saltzman, a liberal stalwart among the city's elite lakefront crowd whose admiration for Obama dated back more than a decade. Saltzman, a petite woman in her early seventies at this time, was the daughter of a late Chicago-area builder, Philip Klutznick, who left a fortune to Saltzman and her five brothers. Klutznick had held top posts in the administrations of Franklin Delano Roosevelt, John F. Kennedy and Jimmy Carter. Of Klutznick's children, Saltzman was the one drawn to political activism. She had been "enraptured" by the speeches of Adlai Stevenson, the senator from Illinois who lost the 1962 Democratic presidential nomination to Kennedy. At the Democratic National Convention in Chicago in 1968, she worked on behalf of Eugene McCarthy, the anti–Vietnam War candidate for president. But her biggest political role came when, for four years, she ran the Chicago-based office of Senator

Paul Simon. While working for Simon, Saltzman formed a close bond with one of Simon's chief political minds—Axelrod, who had comanaged Simon's first Senate campaign. The two would talk on the phone almost daily, each sharing a passion for political gossip and Chicago Bulls basketball. By 2002, Saltzman was a major Chicago fund-raiser who could not only tap into her own wealth but had big-money connections that could help raise substantial cash for any political candidate.

Saltzman first met Obama in 1992 after he graduated from Harvard Law and moved back to Chicago to practice civil rights law. Saltzman was volunteering for Bill Clinton's campaign and was assembling constituency groups when Obama stopped into Clinton's Chicago campaign office. Saltzman was immediately struck by Obama's understated presence. "He came in and not only did he speak well, but it was just the way he presented himself," Saltzman said. "He had never run for office, but he had all those qualities of a skilled politician." Saltzman and Obama formed a lasting political friendship and she was helpful to Obama when he ran for the Illinois senate. When I interviewed Saltzman about Obama, it was readily evident that his appeal to her went beyond the normal politician's. Having mentioned that she was "enraptured" by Adlai Stevenson, Saltzman said she felt the same way about Obama, only more so. She seemed almost spellbound by him, in fact. "When he speaks, it's like—it's like magic."

So when Saltzman was assembling speakers for the anti-war rally in late October 2002, Obama came to her mind. Saltzman knew from conversations with the lawmaker that he did not support an Iraq invasion, and she called him and asked him to participate in the rally. Obama, still an unannounced candidate for the Senate, did not immediately agree, but he told Saltzman that he would think it over. This was one of the biggest decisions of his potential Senate bid. Should he take a public stand on the likelihood of an Iraq war? He consulted with Shomon, still his main political adviser at the time, and Shomon told him that it was a no-brainer—if Saltzman was urging him to speak, he could not refuse. Moreover, Obama was trying to draw Axelrod onto his Senate campaign team. It would not be wise to disappoint Saltzman if he wanted her to continue lobbying Axelrod on his behalf. So Obama agreed to speak. But Shomon also warned Obama to be careful with his words because "there will be political ramifications to whatever you say." Shomon was indeed correct, and the importance of this decision to Obama's political future cannot be underestimated. It is particularly important when considering this: Obama made the decision to protest the impending war in part as a political calculation that he hoped would benefit him among Democrats. Around the same period, in the U.S. Senate, lawmakers were mulling over a similar political calculation—whether to give President Bush the authority to launch an Iraq invasion. Several of these senators with presidential ambitions—Hillary Clinton, John Kerry and John Edwards—would choose the opposite course from Obama's. As the next few years would unfold, Obama's calculation to oppose the war would fuel his rise in the Democratic Party beyond his imagination. Meanwhile, the decision of these three senators to give Bush invasion authority would prove to be horrendous for their careers.

Obama called Saltzman just a couple of days before the event and told her that he would participate. He then went home and wrote the speech longhand in a single evening. Up to this point in Obama's career, he had rarely used written text, relying on his talent for speaking extemporaneously. But this time he took Shomon's advice to heart about the importance of the precise language of his speech. "I knew that this was going to be an important statement on an important issue and I didn't want it subject to misinterpretation," Obama said. "I wanted to strike exactly what I truly felt on this thing. The nice thing is, because I thought the politics of [the invasion] were bad, it was liberating—because I said exactly what I truly believed." When Obama told me this in an interview, I found it fascinating. If one parses this statement, Obama is implying that he has given speeches with sentiments that he does not truly believe for the sake of politics. This is something that most politicians are loath to admit, at least on the record. Obama came close to conceding that point, but as usual, lacked real specificity.

Though the speech was delivered to a group that most likely included a good number of pacifists, Obama opened by announcing that he was "someone who is not opposed to war in all circumstances." This was something that I would see Obama do on many occasions during his Senate campaign—begin his talk with an opinion that did not necessarily coincide completely with the views of his audience. For an observer, these moments were affecting because they made it appear that he was not pandering to the crowd but authentically telling them what he believed.

Throughout the speech, Obama inserted the refrain "I don't oppose all wars" between his unusually blunt rationales for opposing the launch of military forces to Iraq:

> What I am opposed to is a dumb war. What I am opposed to is a rash war. What I am opposed to is the cynical attempt by Richard Perle, Paul Wolfowitz and other armchair, weekend warriors in this administration to shove their own ideological agendas down our throats, irrespective of the costs in lives lost and in hardships borne. What I am opposed to is the attempt by political hacks like Karl Rove to distract us from a rise in the uninsured, a rise in the poverty rate, a drop in the median income—to distract us from corporate scandals and a stock market that has just gone through the worst month since the Great Depression.
>
> That's what I'm opposed to. A dumb war. A rash war. A war based not on reason but on passion, not on principle but on politics.

By 2007, the speech looked positively prescient. Obama warned that a U.S. occupation would be of "undetermined length, at undetermined cost, with undetermined consequences." He said that he was certain that an invasion without strong international support would only "fan the flames of the Middle East and encourage the worst, rather than the best, impulses of the Arab world, and strengthen the recruitment arm of Al Qaeda."

Another aspect of this scenario that seemed prescient later: Shomon's prediction that there would be "political ramifications" to whatever Obama asserted. For Obama, those ramifications would be nothing but positive. Over time, Obama could tout the speech as his predictions evolved into reality, as the Iraq war devolved into a chaotic mess, killing several thousand Americans and tens of thousands of Iraqis. Indeed, the speech had placed Obama firmly and publicly against the Iraq venture before it happened. This was a position that most Americans would gradually move toward as the war dragged on and consequently sapped the political fortunes of the many Democrats and Republicans in Congress who voted to give the president the authority to invade. The speech would also make Obama the only candidate in the 2004 Senate race and the only top-tier Democrat in the 2008 presidential contest to flatly oppose the war before it was launched. In mid-2006, well after he delivered his keynote address in Boston and myriad other high-profile speeches as a senator, Obama told me that this had been his best-written and most courageous speech and was thus his favorite:

> That's the speech I'm most proud of. It was a hard speech to give, you know, because I was about to announce for the United States Senate and the politics were hard to read then. Bush is at sixty-five percent [approval]. You didn't know whether this thing was gonna play out like the first Gulf War, and you know, suddenly everybody's coming back to cheering. What I knew at the time, even as I gave the speech, was the

military aspects of it would be extraordinarily successful and very quick. I didn't have any doubt that the U.S. military would blow through there pretty quickly. And it was just, well, a well-constructed speech. I like it. I like the opening because it sets the tone. I mean, in some ways, it was not a typical anti-war speech.

Obama later in 2003 would correctly predict the sectarian violence between the Shiites and Sunnis that overtook Iraq in the aftermath of the U.S.-led invasion and subsequent occupation. He would also tell audiences that if he had been a senator at the time, he would have voted against the resolution giving President Bush the authority to invade Iraq because he feared the United States would stoke sectarian violence and be bogged down in the Middle East for the foreseeable future. After these pronouncements, when he finally entered the 2004 Senate contest, he could legitimately hold himself up as the candidate that spoke out the earliest and the loudest on the war. And by the time he entered the 2008 presidential contest, these on-the-mark forecasts would be one of his main selling points to an electorate that had long wearied of American bloodshed in Iraq.

Standing firmly against the Iraq war helped to define Obama, just as voting for the resolution defined the Democratic senators who ultimately would pay a steep political price for their choice. Perhaps there is a lesson in this story for politicians: Sometimes saying what you "truly believe," in Obama's words, can pay long-term dividends.

With Moseley Braun out of the Senate race, Obama was free and clear to announce his candidacy. He did so on January 21, 2003, at a downtown Chicago hotel, becoming the second announced candidate, along with Gery Chico. His event was not overwhelming, but it looked serious. He assembled a group of about a dozen established Democrats, including his Illinois Senate poker crew, Illinois senate president Emil Jones Jr., Congressmen Danny Davis, Jesse Jackson Jr. and various others. He focused his rhetoric on assailing the Republican incumbent, saying that Peter Fitzgerald had "bought himself a Senate seat and he's betrayed the people of Illinois ever since."

By now, Axelrod had all but formally committed to Obama. It appeared that the race was shaping up as something of a free-for-all—potentially a three- or even four-man contest by the end, with Obama, Hynes, Hull and Chico being the top contenders. Chico, by this time, had hired Adelstein. Axelrod was still weighing his options. Hull had the divorce issue and alcohol abuse problems in his past that Axelrod thought would ultimately fell his candidacy. Axelrod had reached out to Hynes, but Axelrod had never been particularly close to the Hynes's family-based political operation on the city's Southwest Side, while his and Obama's friendship went back years. And besides, the state comptroller wanted to keep his organization in the family. He hired his brother Matt as his campaign manager.

That essentially left Axelrod with Obama, whom, ironically, he had once counseled to pass on the Senate race and wait to run for Chicago mayor one day. But with coaxing from Obama fan Bettylu Saltzman, Axelrod had been warming to Obama's pitch on how to win the primary—marry the African-American and liberal vote as a base and then expand from there. Axelrod also knew that Obama and his credentials—community organizer, *Harvard Law Review* president, state lawmaker—might play well with an important constituency: newspaper editorial boards, which hand out endorsements. In Obama, he saw raw talent, tremendous drive and a decent opportunity at winning, especially with all of Hull's personal issues.

"My involvement was a leap of faith," Axelrod said. "Barack showed flashes of brilliance as a candidate during the early stages of the campaign, but there were times of absolute pure drudgery. . . . His speeches were very theoretical and intellectual and very long. But I thought that if I could help Barack Obama get to Washington, then I would have accomplished something great in my life."

To counter Obama's shortcomings as a candidate, Axelrod did two things: He urged Obama to think more in terms of people and their stories rather than pure policy, and he stressed the importance of this advice to Obama's other close adviser—Michelle Obama.

Axelrod told Obama to visualize the people he had met and would be meeting on the campaign trail, to try to bring their stories to life, to "invoke more humanity in his speeches." "In a classic way he grew under the tutelage of the people he was meeting out on the stump and realized that he was internalizing their everyday concerns and problems and realizing what the whole was about," Axelrod said. "It clicked in his head and he became a much better candidate over time. His learning curve is great. Once he realized that he was not taking orals at Harvard he became a better candidate."

This maturation process is not dissimilar to that of many great politicians. Again, John F. Kennedy was a poor speechmaker and even worse glad-hander at the beginning of his political career. But after coaching from aides and his own observations on the campaign trail, Kennedy developed into one of the great presidential orators and a politician who could work a room with the best of his kind. As for Michelle's influence, Obama listens to her advice as he does to that of no other individual, not even Axelrod. "She's my coconspirator," he said.

13
The Race Factor

Barack is a black man!
—MICHELLE OBAMA

The Democratic takeover of the Illinois senate in January 2003 could not have come at a more opportune time for Barack Obama and his candidacy for the U.S. Senate. With the close relationship between Obama and the senate's president, Emil Jones Jr., Obama was suddenly positioned perfectly to move from back-bench obscurity into a prominent lawmaking role. Obama cleverly approached Jones about supporting his Senate candidacy from the tack of increasing Jones's own political might. "You know, you have a lot of power," Obama told Jones. "You can make the next U.S. senator." Jones responded, "Wow, that sounds good! Got anybody in mind?" To which, Obama said, "Yes—me." Indeed, throughout 2003 and 2004, Jones offered Obama an array of high-profile bills to shepherd through the senate. This powerful kinship also allowed Obama to remain neutral, or fade to the background, on thorny issues that came before the legislative body. But perhaps more than anything, it provided Obama with crucial logistical assistance in running an effective statewide campaign. With Jones setting the senate timetable, Obama was no longer at the mercy of Republican leadership, which meant he was no longer in the dark about the senate voting and session schedule. If Obama needed to hit the campaign trail, attend a debate or throw a fund-raiser, he now could ask Jones not to schedule an important vote during that time period.

As a consequence of Jones's patronage, over the next two years in the legislature, Obama sponsored nearly eight hundred bills, with the new Democratic governor, Rod Blagojevich, signing more than two hundred and eighty into law. Many of those bills allowed Obama to curry favor with key Democratic constituencies— such as organized labor—that Obama would call upon in his campaign for the Senate. Obama, for example, sponsored legislation that blocked overtime restrictions mandated by the Bush administration, and he sponsored a law that extended the reach of the Earned Income Tax Credit for the working poor. Besides advancing Obama's political career, the bulk of this legislation reflected his commitment to the liberal doctrine of expanding government powers to help the most vulnerable in society: children, the elderly, the poor. "Being in the majority was critical. Accomplishing the things that I had wanted to accomplish, that pent-up demand of ideas that I had, was important," Obama said. "It gave me a lot of confidence. This is the kind of politics that I want to practice. This is why I am in this thing. It really bolstered my sense of why politics is important because I had seen what I could get done in a legislative context."

By early 2003, the field for the Democratic Party's race for the U.S. Senate seat from Illinois was nearly set, and surprisingly, it included no real powerhouse candidates. This was one of the reasons Obama felt he had a chance to win. Besides Obama, over the coming months, the contest to be decided in March 2004 would attract seven contenders, five of them seemingly having a shot at the nomination for the junior seat. Democrat Dick Durbin held the senior Senate seat.

The presumptive front-runner was Dan Hynes, the state's comptroller. Hynes was the only candidate to have competed in—and won—a statewide race, meaning that voters in every corner of Illinois had at least seen his name on the ballot, if not voted for him. He had been the youngest comptroller in the state's history when he was first elected in 1999, and most insiders perceived him as older and more experienced politically than he was. Though just in his late thirties, Hynes had graying temples and a serious face that gave him a mature look. He had seemingly the deepest political résumé in the field. His father, Thomas Hynes, had been president of the Illinois senate and still ran a powerful ward organization on the city's Southwest Side.

Daley was the foremost Irish political name in Chicago; Illinois house speaker Michael Madigan placed second in that category; and Hynes arguably came in third. The other top contenders for the Senate would be Blair Hull, the securities trader who had promised to dump as much as thirty million dollars into his campaign; Gery Chico, Mayor Daley's former chief of staff who had been Chicago school board president; and Maria Pappas, the Cook County treasurer. Also joining the contest were two others deemed more spoilers than anything: Joyce Washington, a health care consultant; and Nancy Skinner, a liberal bomb-throwing radio host.

Obama's campaign operation in early 2003 was being run by Dan Shomon from a small two-room office in Chicago's Loop business district. Shomon, however, did not want to spend the next year running a Senate campaign. Feeling burned out, he told Obama that he would help him launch an office, but he needed to find a long-term campaign manager by spring. Shomon explained it wasn't that he didn't believe Obama could win, but he didn't want to be involved in the grind of another campaign. This news infuriated Obama, who could not understand how Shomon could abandon him at this crucial juncture in his career. "This Senate thing," Shomon tried to explain to Obama, "that's your thing, Barack—it's not mine. My life is going a different way." Obama, however, still felt betrayed. While the two men spoke frequently over the next months and years, their tight-knit relationship was never fully repaired. Obama, who had been deemed a special person since his childhood, clearly was not accustomed to being left at the altar.

While Obama's campaign operation was stagnating under a lame-duck campaign manager, Bettylu Saltzman was working hard. She was pulling together fund-raisers and lobbying Axelrod hard on her belief that he indeed had a budding star on his hands in Obama. "I'm sure she used the word *magical*," Axelrod said with a smile. So as Axelrod moved fully on board, he reached out to Pete Giangreco to run the direct mail operation. Giangreco is a senior partner in the Strategy Group, based in suburban Evanston, and he is one of the leading political mail experts in the country. In that campaign season, he was handling targeted mail for John Edwards's presidential campaign. He was also fully versed in working a statewide campaign in Illinois, having been an integral adviser to Rod Blagojevich's successful governor's race in 2002. So with Axelrod and Giangreco behind him, Obama had the top media and direct mail operatives in Illinois. Once they pulled in top-notch Washington-based pollster Paul Harsted, Obama suddenly began looking like quite a formidable candidate. All he needed now was staff.

The first full-time hire was Nate Tamarin, the thirtyish son of a Chicago union organizer who hailed from Giangreco's Strategy Group. Tamarin joined the campaign as a deputy campaign manager in March, just a couple of days before Chicago's annual Saint Patrick's Day Parade. Politicians never miss the downtown Chicago parade. It's a ready-made opportunity to preen before a good-sized crowd and before the television cameras. The most telling aspect of a campaign's strength is often the placement of a candidate's float or car in the parade—the nearer to the front of the pack, the more powerful the candidate or officeholder. Tamarin might have wondered what kind of campaign he had just joined: "We were dead last," he said. Hynes, meanwhile, was among the first half dozen.

As the spring wound down, Obama's team still lacked a full-time manager, and Axelrod began beating the bushes. He wanted someone experienced in electing a black candidate in a predominantly white area. He had followed a race in which a new black mayor was elected in Jersey City, New Jersey, a community of diverse races and cultures. The campaign manager for that candidate was Jim Cauley, who surprisingly came from one of the least diverse places in the country—the Appalachian territory of Kentucky. Axelrod had worked with Cauley on the successful Baltimore mayoral campaign of Martin O'Malley, and it seemed to him that Cauley had a knack for urban political warfare. Cauley was in his mid-thirties at the time but was an experienced political hand. He had run O'Malley's campaign and a couple of congressional campaigns and had worked for the Democratic Congressional Campaign Committee. At the time, Cauley held a steady job at the DCCC, but he was chafing at the day-to-day office work, which lacked the adrenaline highs of a political campaign. So Axelrod called Cauley, and like most people at that time, Cauley reacted skeptically to the name "Barack Obama." But he did a bit of research and was impressed that Obama had raised more than half a million dollars in his first congressional effort. Besides, Cauley thought, Axelrod was a "big-time player," and if he was vouching for Obama, he must be worth looking at. "Just come to Chicago and meet Barack and see if you see what I see," Axelrod said. "If you don't see it, no sweat." So Cauley flew to Chicago and lunched with Obama near Chicago's famed Michigan Avenue. Cauley was impressed with Chicago in June, but he did not necessarily see what Axelrod saw in Obama. "They were working out of two small rooms, and frankly," Cauley said, "it looked like a campaign for the state senate, not the U.S. Senate." But, Cauley said, "a lot of influential people were asking me to do it and I respected them—so I said okay." Still, Cauley climbed aboard with some apprehension. Even Cauley's father, who was a certified public accountant and had been involved in Kentucky Democratic politics for decades, wondered if Cauley knew what he was doing by handling a candidate named Barack Obama. "Have you really thought this one through, Jim?" his father asked.

Cauley was something of the antithesis of whom one might expect to find as Obama's campaign manager. Slightly balding and barrel-chested, he exuded every bit of his Kentucky upbringing. His accent was thick and his phrasing southern, down-home and straightforward. His plainspoken nature, his lack of interest in policy matters and his attention to minor details all ran counter to Obama's larger, philosophical vision. Cauley did not claim to be a Rhodes Scholar, and he knew what his role was: to help Obama raise money, hire and organize staff and volunteers and then turn out Obama's voters on election day. "Somebody's got to keep the trains running on time and make sure the money is coming in—and that's me," Cauley said.

AS HE OFFICIALLY SET OUT ON THE CAMPAIGN TRAIL, OBAMA charted a course to shore up his two core constituencies, African Americans and liberals. Among key liberals, he won the support of Representative Jan Schakowsky, who represented a lakefront district on Chicago's North Side. She had considered a run for the Senate but decided against it and instead ardently threw her support behind Obama. Also officially coming on board was the Reverend Jesse Jackson Sr., the best-known black leader in the country. Jackson, whose Rainbow/PUSH organization was headquartered in Hyde Park, had been an informal adviser to Obama for several years. This was an example of Michelle helping to ingratiate her husband with Chicago's African-American network; she had been friends with Jackson's daughter, Jacqueline, while they grew up on Chicago's South Side. As a teenager, Michelle had even babysat young Jesse Jr.

Beyond the endorsements, Obama put his most vigorous on-the-stump efforts into winning votes in the black community. Blacks are a key voting bloc for any Democrat, with about 75 percent of black voters being Democrats compared with 45 percent of white voters. In Illinois, roughly one in five Democratic primary voters is black, and Obama would need a large percentage of the African-American

vote on election day to be successful. And judging by his uneven relationship with some members of the legislative black caucus and the
ill feelings from the Bobby Rush race, his black support was by no
means guaranteed. Indeed, Rush's grudge against Obama was as
intense as ever, and the congressman did everything he could to poi-
son Obama's image among blacks. Rush even signed on to be
cochairman of Blair Hull's campaign. Moreover, Obama's foes in
the Senate contest were more than willing to point out the race issues that percolated beneath the surface from the Rush race. This
left various observers to ask: Was Obama black enough to win widespread African-American votes?

"As a politician, the Obama character has a tragic flaw," worried
commentator Laura Washington in the *Chicago Sun-Times*. "He may
be too smart, too reserved, and perceived as too elitist for regular
black folk. It's the Uncle Leland problem. My uncle says that low-
income and working-class blacks don't think Obama is 'down'
enough. It's a cultural phenomenon, and it's rooted in an unfortunate strain of anti-intellectualism and distrust of those with close
associations with the white power structure. . . . Some of the black
nationalists are whispering that 'Barack is not black enough.' He's
of mixed race, he hangs out in Hyde Park, and is a darling of white
progressives; he's not to be trusted. And there are the black Machine
Democrats. They're all crabs in the barrel, trying to get to the top.
And they don't want Obama to get there first."

In these early days of his U.S. Senate campaign, aggressively
trying to establish a kinship with ordinary blacks, Obama still
seemed a tad uncomfortable trying to win acceptance, or at least
win votes. Sometimes he could work just a little too hard. When
addressing African Americans, he would drop into a southern drawl,
pepper his prose with a neatly placed "y'all" and call up various
black colloquialisms. The cadence of his sentences would change
dramatically, moving faster and then slower and then faster. He

would hit crescendos and then fall back into a slower rhythm again until the next crescendo, taking a page from the many African-American preachers he had first encountered in his days as a community organizer on Chicago's Far South Side. Sometimes this worked; other times it seemed forced.

Obama explained to me the change in his speech pattern this way: "There is going to be a certain rhythm you feel from the audience, any audience. An all-black audience is going to respond in a different way. They are not going to just sit there. I am not a minister and I can't pretend to be Dr. King, nor would I want to. He spoke poetry and I am prose. But it takes a different speaking pattern to connect with your audience." Aides would occasionally worry that his change of tone and diction from precise English could be viewed as talking down to his audience. "It's been mentioned, but most times, you know Barack, he just can't help himself," said David Katz, his campaign photographer. Sometimes Obama clearly would go overboard in all-black settings, yet no one could dispute that he did anything but campaign tirelessly in the black community. From early 2003, he filled nearly every Sunday morning with speaking appearances in Chicago's black churches. And when stepping before the congregation, always careful to be respectful and stand astride the preacher's pulpit, always certain to mention his own pastor, Jeremiah A. Wright, Obama would use inspiring oratory and his credentials as a state legislator to educate the audience about himself and his candidacy. By the end, he had clearly won over most of the crowd.

Obama's message to black audiences was similar to his proclamations to other Democratic crowds—a theme of hope in the goodness of the human spirit—although he wrapped that message in a language tailored to African Americans. He consistently mentioned his Christian faith and his home church on the city's South Side, Trinity United Church of Christ, and suggested that the teachings of the African-American church should be a primary guide for lawmakers and other decision makers. But in these speeches to blacks, Obama went one step further. He linked his own political ascendance to the forward momentum of the black cause in general. He spoke in terms of his own success being a step in the larger success of blacks nationwide. To a nearly all-black audience at Mars Hill Baptist Church in Chicago's Austin neighborhood in November 2003, Obama declared: "I am not running a race-based campaign. I am rooted in the African-American community, but not limited by it. I am campaigning everywhere. The way we are moving so far, if we can sustain it, and if our community can rise up and we recognize this opportunity, I am confident we are gonna win. This is a campaign based on truth and based on honesty and based on the values that I learned in the church."

Not enough emphasis can be placed on Obama's claim of not being "limited" to only winning black votes. This sentiment strikes at the heart of black America: a desire for full emancipation into the white world, not to be "limited" by skin color alone, to be given an equal shot. In making himself part of that cause, Obama was conveying that a vote for him was a vote for freedom of all blacks, a vote for blacks not to be held down.

In January 2004, at the opening of his south suburban campaign office, he implored an audience of largely black supporters to look beyond immediate disappointments, not to let painful thoughts of the black struggle cloud a vision of future possibilities—with the greatest possibility, of course, being his election as a U.S. senator:

> We just assume that young people in our communities won't aspire to higher education and we are not surprised when they drop out. We automatically assume that they are going to be two or three grades behind in national reading scores. That is just something that we have come to expect. We are not shocked that there are more African-American men in prison than there are in college. We are not shocked that fifty percent

of African-American men between twenty-one and twenty-four are not in school and are out of work. And when it comes to Washington, we just assume that the game is fixed for the powerful, for the special interests. . . . The essence of this campaign is for us to no longer accept the unacceptable, to raise the bar, to set a new set of standards, to start thinking differently about what is possible in our communities and in our politics.

This hopeful oratory aside, his most convincing argument to blacks was not his eloquent language but the basics of his political résumé. Obama had accumulated a list of concrete accomplishments during his tenure in the Illinois senate, many of which assisted blacks. In Obama's oft-repeated words, "I have not just talked the talk, I have walked the walk." Before black crowds, Obama ticked off all the laws aiding the black community that he had personally written and ushered through the legislature: a law that forced police to record the race of people pulled over in order to stem racial profiling; a law that forced authorities to videotape confessions in the wake of *Chicago Tribune* investigative reports that more than a dozen innocent men, all of them black, had been placed on Illinois's death row; a law that greatly expanded the number of poor children covered under the state's medical insurance. When blacks in churches like Mars Hill heard this, they would nod their heads approvingly, and any thoughts that they might have been harboring that Obama was a tool of whites soon evaporated.

Another factor that greatly inoculated him from these charges of lacking black authenticity was his wife, Michelle. Obama had married a black woman from the city's South Side. Cynthia Miller, his office manager in his congressional campaign, said when questions about Obama's blackness would arise among her African-American friends and acquaintances, she would consistently be asked if he married a black or a white woman. "It was the first question I would get— and I would get it a lot," Miller said. When she answered that he had married a black woman, the wariness would subside: His choice of a black wife seemed to give him legitimacy in the community.

Michelle, for her part, defended her husband vigorously against suggestions that he lacked black credentials. In a Chicago public television profile of her husband during the Senate race, she practically jumped out of her seat when the interviewer noted that Obama had been cast by a previous political opponent as a mercenary for white elites. "Barack *is* a black man!" she said emphatically, her eyes widening. In an interview with me, she expanded on that sentiment and explained why her response was so emotional. She said that blacks who achieve academic success and do not adhere to popular stereotypes within black culture can often be ostracized—and she encountered similar issues when growing up. So when her husband was accused of not being fully black, it struck a raw nerve. Said Michelle:

> The anger and frustration that comes with that whole issue has just as much to do with . . . the frustration of the challenge within the community, and I faced it too. Here I am, you know, South Side, doing just what I thought I was supposed to do. But Princeton and Harvard do certain things and, you know, it's like that growing up, you know. You talk a certain type of English and then you have to cover that up on your way to school so you don't get your butt kicked. You know, we grew up with that. My brother faced it, we all faced it where . . . there's frustration of feeling like you have to camouflage your intellect in order to survive in your own community. And fortunately I came from a family on both sides that didn't believe in that, that didn't foster it, that fostered any courage, intellectual discourse and, you know, all of that. So that just brings up all of [what] I experienced. So I didn't view that as something that was particular to Barack. . . . The point is . . . that he's a black man. . . . So that's not even the issue. You can't even focus on

that. But what it points to that is frustrating is there's still an ability of people to use intellect and race as a way to drive a wedge between certain people in their own community. And that's the frustration, I think, that I felt with that issue because it reminded me of just the kind of things that I had to deal with growing up.

The Real Deal

I have been chasing this same goal my entire adult career, and
that is creating an America that is fairer, more compassionate and
has greater understanding between its various peoples.
—BARACK OBAMA

My first encounter with Barack Obama in his U.S. Senate race came
on a cold fall morning in November 2003. I had been assigned to
cover the Democrats vying for the nomination and I began reporting profiles of the top-tier candidates: Obama, Dan Hynes, Gery
Chico and Blair Hull. By now, Obama had contracted with a communications director named Pam Smith. Good-natured and usually
smiling, Smith was a Chicago-based public relations consultant in
her forties who was rather inexperienced in politics at this level, and
she could be a bit overwhelmed with the daunting task of serving as
the lead spokeswoman for a high-profile Senate candidate. She was
aware of the importance of the campaign—and Obama understood
the significance of hiring a black woman as his public face to Chicago area voters—but she acted more as a conduit to reporters than
as a mouthpiece for Obama. Just as Obama had told Shomon years
before, Obama handled his own press for the most part. He wanted
control over his message, especially when it came to his image in
such a powerful local media outlet as the *Chicago Tribune*. I told
Smith that I wanted to attend one of Obama's constitutional law
classes at the University of Chicago and we arranged for me to sit
through a lecture later that week.

But when I arrived at 8:50 A.M. at the room number Smith had given me, I found no students and no Obama, even though the class was to begin in ten minutes. After a series of cell phone calls—to Smith, who called Obama, who called me—we straightened out the matter. His class was down the hall. Political campaigns, especially in the early stages, are rarely well-oiled machines. They are like a start-up company tossed together, usually populated by young people barely removed from college. So this disorganization was irritating, especially at 9 A.M., but not surprising. What surprised me was Obama's handling of the foul-up. He apologized and told me that he had given Smith the wrong room number. He emphasized that it was his fault. "Don't blame Pam," he said. "It was my mistake." I found this odd—and rather refreshing. Obama could easily have made his aide the culprit but chose to accept the blame himself. Politicians are not known for admitting to mistakes. Settling into a chair in his classroom, I thought, He is either honest, naive or endeavoring to change my first impression of him from several years ago—as one who would blame an ill child for his missing a key Illinois senate vote while he vacationed in Hawaii.

Dressed in blue shirt and tie, Obama strode into the classroom about five minutes late and did not acknowledge my presence, another act I found somewhat strange. How many times does a *Tribune* reporter sit through your class? He pulled off his winter coat and his navy suit coat, and I noted that his frame was even thinner than I had remembered from that press conference a couple of years earlier. He opened the discussion and, as he paced the room, he loosened his tie and reached down to roll up the sleeves of his shirt. He turned over each layer of light blue cotton so slowly and with such precision that it was impossible not to fix your eyes on his movements. Obama certainly knew how to call attention to himself, and in the most subtle manner. It was understated, but he definitely had a confident presence, if not overly confident.

To Obama's far right, I spotted two young African-American

female students gazing at him from above their laptop computer screens. One of the young women looked positively enraptured, while the other appeared just slightly entranced. Even if they did not have crushes on the instructor, they seemed more focused on his physical being than the subject of his lecture. Obama went through the material of the day, cases involving civil rights and voting rights, in a clear and methodical manner. He challenged some students but in no way seemed bent on embarrassing them. The University of Chicago is an elite institution, and the students were bright and alert. It was evident that they had read the assignments and were at least minimally prepared. As the class ended, the two young African-American women approached Obama. The woman with the intense gaze was fidgety and nervous in his presence, even as she asked a common question about an assignment. He folded his arms with a detached coolness that did little to put her at ease. I smiled. If Obama runs any kind of television ad campaign that gets him noticed by large numbers of voters, I thought, it's crystal clear where the vote of college-educated black women is going.

As students departed, Obama walked up and shook hands with me. His handshake was not as tight and firm as I had come to expect from politicians. He joked that he was going to call on me a couple of times in the class, but "wasn't sure you had read the assignment." I laughed mildly. I then explained that I would be covering the campaigns of the Democratic candidates for the *Tribune* over the next five months, and I suggested that we grab a cup of coffee and talk about the race. "Great idea," he said. So we walked out and headed down the hallway. A student with a blue-white "Obama, Democrat for Senate" button fastened to his backpack walked by. Obama broke into a wide grin and pointed, "Hey, look at that." The button didn't exactly impress me. Surely Obama had a good number of university students in his corner, considering that he was teaching there and lived just a few blocks away in Hyde Park. As we stepped down a corridor and neared some outside doors,

Obama turned and said, "Well, have a good day." Perplexed, I froze in my tracks. "I thought we were getting coffee?" I asked. "Oh, we will," he said. And he stepped away. This baffled me, but over time I would learn that Obama was a man accustomed to setting his own pace and his own schedule—and that day, a cup of coffee with the *Tribune* reporter was not in his plans. He had only learned the night before that I would be there, and he was a man who insisted on being fully prepared for these kinds of encounters.

I found Obama again when he was one of several speakers to a crowd of South Side black veterans. The event was outdoors and Obama was late arriving. He surprised me by driving up to the gathering by himself, with no staff accompaniment. Most politicians, and especially candidates for high office, are chauffeured from place to place with at least one aide at their side. Afterward, as I walked with Obama back to his Jeep Cherokee, he seemed to have only faint interest in setting up that cup of coffee with me. Yet when a black man slowed his car and waved encouragement to him in his Senate race, Obama was quick to point me out: "Hey, I'm doing good," he told the man. "Look, I've got the *Chicago Tribune* here with me right now." I guess I came in handy for show-and-tell, I thought.

OBAMA'S CANDIDACY INTRIGUED ME, BUT NOT NEARLY AS MUCH AS the newest face on the political scene: multimillionaire Blair Hull. His résumé of making hundreds of millions as a securities trader was interesting enough, but as the story went, he had moved into trading after parlaying twenty-five thousand dollars in blackjack winnings from Las Vegas into a successful trading company. Besides that, rumors swirled about those personal issues—alcohol abuse and ex-wife problems—that had scared David Axelrod away from working for his campaign. Hull had long ago put together a campaign apparatus, and it was so well assembled that it resembled a minicorporation. Professional-sounding secretaries answered the phone and there were up-to-date magazines in the waiting area. The campaign was housed in a historic-looking building in the bustling Near North neighborhood of downtown Chicago, within walking distance of the "Magnificent Mile" of North Michigan Avenue. I counted more than two dozen signs or buttons or stickers with the name "HULL" emblazoned on them in the waiting area. Hull entered the race with zero name recognition, and the first order of business was building his name into a brand. He had already been running television commercials across the state. The poll-driven theme was that, as an outsider to Washington not beholden to any special interests, Hull would cut through the muck of the Beltway cesspool to solve the nation's health care crisis. His health care plan, in fact, would ensure that all Americans had medical coverage. The ads were slick and highly professional—and Hull had an unlimited bank account to flood the airwaves with them.

I had called the Hull campaign to establish contact and meet with his press secretary, a good-humored, buttoned-down spokesman named Jim O'Connor. I wanted to meet Hull, and O'Connor said that would be "fantastic." But after weeks of chatting, the meeting still had not come together. O'Connor seemed to find one excuse after another for why we could not sit down. I would learn later that Hull's aides had been working for months to prepare the political neophyte for talking with reporters. Finally, in late November, O'Connor, Hull and I had lunch, and Hull seemed amiable enough. I told him that his ads were sharply produced and appeared to be having some effect on an electorate that, at the moment, was barely aware that a Senate contest was under way. But I also took notice of something that a colleague had pointed out: "Those Blair Hull ads are good, but you know, I can never remember what he looks like afterward, even though I have seen that commercial a dozen times." Indeed, Hull had such an undistinctive face—none of his characteristics were particularly defining—that his physical presence could go largely unnoticed and be soon forgotten. Bespectacled, with gray hair, Hull smiled throughout the lunch and appeared to be reciting lines that had been written for him, but I expected this from a first-time candidate for office.

What I did not expect was the call from his office that came about 6 P.M. several days later—on Monday, December 8. Someone named Jason Erkes said there would be an important announcement from the Hull campaign at any moment. When Erkes called back, he explained that he was a second campaign spokesman and he had information to release: A young woman had been found dead in the garage of Hull's home in Chicago's tony Gold Coast neighborhood on Saturday evening. Hull had not been living in the home and he did not know the woman—she apparently was the close friend of a former young girlfriend of Hull's. The two women had been sharing the home and Hull had moved elsewhere. The police were investigating but did not believe foul play was involved.

Later, the authorities ruled the woman had died from carbon monoxide that had emanated from a malfunctioning swimming pool heater located inside the garage. The pool was atop the garage. The young woman had stepped into the garage and was felled by the poisonous fumes before she reached her car.

The story was buried in the newspapers and only lasted a day. But four months before the primary, Blair Hull's bizarre, scrambled personal life was already taking center stage in the Senate race.

THE WEEKEND BEFORE THE HULL STORY BROKE, I WROTE A curtain-raising story about the Senate race for the *Tribune*. The piece called Dan Hynes the front-runner and mentioned Blair Hull, Gery Chico and another candidate who had now entered the race, Cook County treasurer Maria Pappas. She was perhaps the hardest candidate to explain. Pappas was an eccentric who had high name recognition, but she had jumped into the campaign so late that observers wondered if she was really serious or was a stalking horse for another candidate. I had spoken with her on the telephone and the conversation left me more puzzled than anything. She was known for her quirky nature, and in the chat with me, she stressed that she might even hop on a bicycle and go through neighborhoods on two wheels. Then she invited me to bike with her. "Um, it's December and this whole campaign is running through the winter," I said. "I think I'll take a pass on the Sunday bike rides." First there was Blair Hull and the death of a young woman in his home. Now here came this odd woman insisting that I bicycle with her in December in Chicago—this race had more characters to it than I had ever anticipated.

In the *Tribune* story, I mentioned that Obama was trying to pull together the key Chicago Democratic voting blocs of North Shore liberals and African Americans. In doing so, I wrote, Obama had campaigned "almost exclusively" in the black community, at least so far. I wrote this because I had limited information about Obama's campaigning, since he had not yet sat down to speak with me. Smith had told me that he was spending every Sunday morning in black churches—and I had spent one Sunday observing him in several of them. But once the story appeared, Smith called to say that he had been campaigning in far more settings than the African-American community, and Obama wanted to talk to me in person about my misrepresentation of his campaign. And could I do it that afternoon? she asked.

Obama's campaign office was located high inside a beautiful white terra-cotta building along South Michigan Avenue across from downtown Chicago's Grant Park. When I stepped into his private corner office, my eyes were drawn to two things: the room's clutter and a huge framed poster of heavyweight boxer Muhammad Ali. The striking photograph caught the fighter with a

"Barry" Obama in his 1979 senior class portrait at the Punahou Academy in Honolulu. *(Courtesy of Punahou Academy)*

Obama, fourth from the right in the front row, with his ninth-grade graduation class at Punahou Academy. In Honolulu, the private school was known as the school for the "haole," or the school for whites. *(Courtesy of Punahou Academy)*

Obama with the Ka Wai Ola Club at Punahou Academy in 1976. *(Courtesy of Punahou Academy)*

Obama is making a layup here, but he was largely relegated to warming the bench at Punahou after arguing with his coach about the lack of playing time. The experience taught Obama the pitfalls of challenging authority. *(Courtesy of Punahou Academy)*

David Axelrod, the premier Democratic political consultant in Chicago, is Obama's chief media strategist and was the lead architect of Obama's Senate campaign. A Republican strategist once put Axelrod at the top of a list of "Guys I Never Want to See Lobbing Grenades at Me Again." *(Courtesy of Paul D'Amato)*

Obama files petitions with the Illinois Board of Elections in Springfield, Illinois, to be placed on the Democratic primary ballot for the U.S. Senate on December 8, 2003. At the right is his first long-term political adviser, Dan Shomon. *(Courtesy of Associated Press/Seth Perlman)*

Obama confers with Illinois Senate president Emil Jones Jr. on the floor of the state senate on July 24, 2004. Jones was Obama's political patron in Springfield, helping Obama mold his legislative record in a fashion that would help Obama win his U.S. Senate race that year. *(Courtesy of Associated Press/ Randy Squires)*

Obama delivers his now-famous keynote address to the Democratic National Convention in Boston on July 27, 2004. The speech propelled him onto the national stage. *(Courtesy of Associated Press/Ron Edmonds)*

Jack Ryan was Obama's first Republican opponent in the 2004 U.S. Senate race until the former investment banker was felled by allegations from his ex-wife, a Hollywood actress, that he pressured her to have sex in public. *(Courtesy of Associated Press/ Nam Y. Huh)*

Alan Keyes was Obama's second Republican foe in the 2004 Senate race. The bombastic Keyes got under Obama's skin when he continually challenged Obama's Christianity, even charging that "Jesus would not vote for Barack Obama." *(Courtesy of Associated Press/Nam Y. Huh)*

Obama leaves a Chicago polling station with Michelle and his daughters, Sasha, front left, and Malia, after voting in November 2004. *(Courtesy of Associated Press/Nam Y. Huh)*

Obama and his oldest daughter, Malia, then 6, along with his wife, Michelle, and their second daughter, Sasha, then 3, celebrate Obama's Senate victory in November 2004. Obama became the third African American since Reconstruction to hold a U.S. Senate seat. *(Courtesy of Associated Press/M. Spencer Green)*

Obama talks with his staff in his office in the Hart Senate Office Building on Capitol Hill in February 2006. *(Courtesy of Associated Press/Manuel Balce Ceneta)*

Craig Robinson, Obama's brother-in-law, initially worried that his sister Michelle would "jettison" Obama if he failed to meet her expectations. *(Courtesy of Associated Press/Stew Milne)*

As media chaos engulfs them, Obama greets his paternal grand-mother, Sarah Hussein Obama, at his father's farming compound in the village of Kolego in western Kenya in August 2006. *(Courtesy of Associated Press/Sayyid Azim)*

Obama and his traveling entourage are swarmed by Kenyans as he visits the Nairobi slum of Kibera in August 2006. "I love all of you, my brothers, all of you, my sisters!" Obama told the crowd. *(Courtesy of Associated Press/Gary Knight VII)*

Obama comforts Antoinette Sitole as they view the historic photo of her and her slain brother in the Hector Pieterson Museum in Soweto, South Africa, in August 2006. *(Courtesy of Associated Press/Themba Hadebe)*

In Nairobi, Kenya, Gregory Ochieng held aloft a portrait of Barack Obama that he had painted and then delivered to the visiting U.S. senator in August 2006. "He is my tribesman," Ochieng said of Obama. *(Courtesy of David Mendell)*

Obama's traveling media entourage snuggles up close in August 2006 as the U.S. senator tours the Robben Island prison in South Africa, where Nelson Mandela and other anti-apartheid activists were imprisoned. *(Courtesy of David Mendell)*

Obama strikes a pose as he peers through the bars of the jail cell where Nelson Mandela was imprisoned on Robben Island just off the coast of Cape Town, South Africa. *(Courtesy of Associated Press/Obed Zilwa)*

Joined by his half sister, Auma Obama, and his paternal grandmother, Sarah Hussein Obama, Barack Obama answers media questions on his father's farming compound in western Kenya in August 2006. Auma said she worries that her brother is driven by the same perfectionism and ambition that overtook their father's life. *(Courtesy of David Mendell)*

Obama chats with top aide Robert Gibbs, one of the key architects of his ascension, while they await Obama's appearance on CBS's *Face the Nation* in January 2007. *(Courtesy of Associated Press/Aynsley Floyd)*

Anthony Direnzo of Norfolk, Massachusetts, and Lauren McGill of Blacksburg, Virginia, appear to be enraptured as Obama speaks at a rally at George Mason University in Fairfax, Virginia, in February 2007. Obama's support is especially strong on college campuses. *(Courtesy of Associated Press/Susan Walsh)*

Ten-year-old Donavan Dodds seems thrilled to shake hands with Obama after a campaign rally at Georgia Tech University in Atlanta in April 2007. *(Courtesy of Associated Press/Gregory Smith)*

Now a full-fledged presidential candidate, Obama speaks at a campaign stop in Sioux City, Iowa, on April 1, 2007. *(Courtesy of Associated Press/ Nati Harnik)*

Maya Soetoro-Ng, Obama's half sister, said her ambitious, wandering brother was compelled to leave Hawaii, in part because of the isolated atmosphere of the Pacific islands. *(Courtesy of Associated Press/Lucy Pemoni)*

Michelle Obama, whom her husband describes as "my co-conspirator," has become a partner on the campaign trail as he seeks the presidency. She speaks here in Windham, New Hampshire, in May 2007. *(Courtesy of Associated Press/Jim Cole)*

clenched right arm and a bloodthirsty, celebratory gaze in his eyes just moments after blasting opponent Sonny Liston to the canvas. The famous Ali image hovered above Obama's head at his desk. The candidate explained that he recently grabbed it on the spot from a street vendor. Was this a metaphor for his underdog campaign pulverizing the opposition with glee? I asked with a grin. He returned a smile.

I sat in front of Obama's desk, with Smith taking her place in the chair beside me, and was prepared for Obama to launch into a diatribe about what he believed I had written inaccurately about his campaigning. Instead, Obama said, "Well, go ahead, fire away. Ask whatever you want." I realized that this was our profile interview, not a dressing-down session. Obama was now obviously ready to speak to the *Tribune* and had summoned me for that purpose, and somehow the communication wires were crossed again. I hadn't fully prepared for such an extensive discussion, but the candidate was right here in front of me, so away we went. After about a half hour, I had managed to ask only three questions. Once he was fully prepared, Obama could talk endlessly about a subject dear to him: himself. He spoke quietly and slowly, measuring each of his statements. Indeed, his low-key, conversational delivery in a one-on-one interview can be the complete opposite of his stage persona. His rich baritone is very clear in personal settings, but sometimes it is almost hushed. As Obama would do with so many reporters as the months wore on, he impressed me that evening with his mixture of intelligence, eloquence and a committed idealism. He framed his entrance into politics as part of an ongoing quest for social change. He was truly an activist at heart, not a politician, he claimed. I would see each of his life experiences shape the course of his discussion—the happy-go-lucky prep school teen from Hawaii, the adventurous community organizer with trained listening skills, the Harvard Law School graduate who presided over combative intellectuals on the *Law Review,* the state lawmaker who had endeavored to remain out of Chicago's nasty ward politics, the devoted husband and father:

> I think politics was really an extension or progression from a broader set of goals and concerns. When I was in college, I decided I wanted to be part of bringing about social change in this country, and some of that is based on the values my family gave me. Some of it is based on, I think, my status as an African American in this country. And some of it is informed by my having lived abroad and having family in underdeveloped countries where the contrast between rich and poor is so sharp that it is hard to ignore injustice. But I didn't know in college how that would take shape. And I was actually pretty cynical in college about electoral politics. That's when I decided to get involved in community organizing as opposed to signing up with someone's campaign. I took a lot of inspiration from the civil rights movement and the way the movement brought ordinary people into extraordinary positions of leadership. It struck me that lasting change came from the bottom up and not from the top down. I have been chasing this same goal my entire adult career, and that is creating an America that is fairer, more compassionate and has greater understanding between its various peoples.

I asked Obama something of a reporter's question—designed in an impossible fashion to elicit a certain response. At the time, the city was building a grandiose park area, now called Millennium Park, just outside Obama's window on the other side of South Michigan Avenue. The project was well beyond its target date for completion and far over budget—into the hundreds of millions of dollars. I asked him about his relationship with Mayor Daley, and he responded that it was "cordial, not close." So I queried Obama about what he thought of the park, if it wouldn't have been wiser to spend those hundreds of millions on the beleaguered city school system or to spur economic development in the poor neighborhoods of his district. Spending those kinds of public resources on a park largely for tourists and the elites—how was that advancing social change? And why hadn't he spoken out on this issue? Obama winced at the question, and I readied myself for a politic answer, perhaps something about how the center city needed to be developed, how Chicago was a tourist metropolis and needed that kind of urban investment. Instead, Obama leaned forward and said, "How do you really expect me to answer that? If I told you how I really felt, I'd be committing political suicide right here in front of you." I found his candor refreshing. But it also told me something: Even if he was driven by an activist heart, he was no radical. Rather, he was a polished professional politician who knew that, despite being a community organizer and staunch do-gooder, as a U.S. Senate candidate, he was now working from within the established political order. And when I pressed further for a real answer, he indeed found that politically correct response, saying that the park would advance the city's national reputation and he understood its importance to overall economic development, but he would also like to see more resources devoted to the problems in the city's economically depressed neighborhoods.

In the interview, Obama said the three men he most admired were Mahatma Gandhi, Abraham Lincoln and Martin Luther King Jr.—"men who were able to bring about extraordinary changes and place themselves in a difficult historical moment and be a moral center." (It's worth noting that Obama might fall into that category if he were to win the presidency in a difficult period fraught with a foreign war. As a presidential candidate, he has portrayed himself as a moral leader who wants to change not only policy but the combative nature of politics.) As we spoke, Obama also emphasized various themes that I would hear over the coming months and years in his stump speeches as a Senate candidate, in his town hall meetings as a senator and in his trips across the country in his White House quest. Even though he spent his teen years ridden with feelings of parental abandonment—again, he had often deemed himself "an orphan"—he said that he had been blessed with so much good fortune in his life that it was incumbent upon him to give something back to society. His devoted mother and grandparents, as well as college professors and adult friends, were aligned to steer him from a path of self-destruction, he said. And too many black teenagers are not that fortunate:

When I see young African-American men out there and the struggles that they go through, then I connect with that. I know what that means. I know that, in my book, I mention that I dabbled in drugs or that I was acting tough. I put that in there explicitly because what I wanted to communicate was the degree to which many young men, particularly young African-American men, engage in self-destructive behavior because they don't have a clear sense of direction. But I also wanted to point out that there is way to pull out of that and refocus, and in my case, it was tying myself to something much larger than myself. In my case, that was trying to promote a fair and just society. That is the reason I work on ex-offender legislation. I say to myself that if I had been growing up in low-income neighborhoods in Chicago, there is no reason to think that I wouldn't be in jail today, that I could have easily taken that same wrong turn. That is something that I am very mindful of and it is something that motivates me. Thinking about how you provide hope and opportunity to every kid is my biggest motivator. When I see my five-year-old and my two-year-old, it makes me weep because I see children who are just as smart and just as beautiful as they are, who just don't get a shot. It's unacceptable in a country as wealthy as ours that children every bit as special as my own children are not getting a decent shot at life.

Obama's notion of attaching oneself to a larger ideal is a persistent theme in his public rhetoric. Through the first months of his presidential campaign, he would challenge audiences of all political stripes, educations and financial levels to work on behalf of a greater good and not to spend their lives strictly in pursuit of material possessions. This fell right in line with his mother's post–World War II brand of liberalism and humanism. His message could be preachy, especially coming from a Harvard Law graduate who eventually would become a millionaire presidential candidate, but even conservatives would be hard-pressed to argue against the premise of helping those less fortunate. Throughout our first encounter, Obama also exuded a certain authenticity and great ease with himself, the polar opposite of, say, Blair Hull, who could seem uncomfortable just saying "Hello."

IN SUMMER AND FALL 2003, DAVID AXELROD HAD BEGUN TO PULL together facets of Obama's résumé to sell to his Washington media contacts and Illinois-based reporters. He managed to get some pieces in Washington-based political journals that touted Obama as a potential star in the making. These were helpful not only to build up his candidate in the media but to present to potential fund-raising contacts to show Obama's viability as a strong contender in the race. This helped Obama raise three million dollars overall and still have two million on hand heading into the final months of the campaign. This was nothing compared with Hull's tens of millions, but it would be enough to run a two- to three-week television campaign and present Obama as a serious candidate to influential political insiders.

Also in the fall of 2003, as most professional campaigns would do, Obama's advisers hosted a series of focus groups to determine their candidate's strengths and weaknesses. Focus groups are a far cry from scientific experiments, but they can open eyes to what voters might be thinking. In these groups, a cross section of people were assembled to review television footage of the candidates. Obama's consultants watched from the next room, munching on junk food and swilling canned diet soda. The sessions revealed that Obama's self-confidence as a political candidate and as the purveyor of a desirous message was not misplaced. There was even an epiphany or two among Obama's campaign team.

The aides learned that various parts of Obama's unique résumé appealed to different demographic groups. In explaining Obama's experience to a white voter, all they had to do was mention "first black president of the *Harvard Law Review*" and the voter suddenly had an image drawn—a positive portrait of a black man. "It worked on two levels," Axelrod said. "There were those who thought that breaking barriers is very important. And for others, the *Harvard Law Review* was a major credential in itself." With black voters, however, it was not Harvard Law that evoked positive responses but Obama's community-organizing experience and the legislation he had successfully sponsored, such as the racial profiling law and the expansion of health care coverage to poor children.

But Obama's campaign tacticians also learned that he would probably play fabulously on television. Indeed, their candidate had an amazing telegenic quality. Jim Cauley, the campaign manager, had not necessarily swallowed Axelrod's predictions of Obama's potential star power. Perhaps it was because Axelrod had to be sold on it himself by Bettylu Saltzman. But as these two and others would learn, Obama could be viewed very differently by women than by men. And when it came to suburban white women, he could be viewed in "magical" terms, to quote Saltzman. Obama not only had good looks, but his charm with women apparently emanated through the television screen. His face had a certain honesty and handsome warmth that was not offending, much like his personality. In addition, many strong women had shaped Obama's character, from his mother and his grandmother as a child, to his sister and his wife as an adult.

"My moment was a focus group," Cauley recalled in his Kentucky twang. "The moderator was talking to [liberal, North Shore] women voters, thirty-five to fifty-five and fifty-five plus. He asked the older group, 'Who do each of these guys remind you of?' For Dan Hynes, a woman said, 'Dan Quayle.' For Hull, she said, 'Embalmed.' And she looked at Barack, and the lady said, 'Sidney Poitier.' At that moment, I was like, 'Shit, this is real!'"

Hull on Wheels

Don't you think it would be cool to be a senator?
—BLAIR HULL, DEMOCRATIC CANDIDATE FOR THE SENATE

The Senate contest of 2004 taught Dan Hynes how difficult it can be to be anointed front-runner in a high-profile political race. When something does not go your way, your spot at the top and the strength of your candidacy are instantly questioned. And few things went Hynes's way in this campaign.

With his powerful father's longtime ties to organized labor, a key Democratic Party constituency, Hynes was expected to lock up labor support far and wide. In fact, he had won the endorsement of nearly every trade union in the state, groups representing seven hundred thousand workers overall. But Hynes was the state comptroller, whose only real responsibility was cutting payments for the state's bills. Throughout 2003, Barack Obama had been working in the legislature on behalf of various labor groups. And thanks to his tight friendship with senate president Emil Jones Jr., he was successful at pushing pieces of legislation that benefited labor interests. Jones had also given Obama the chairmanship of the senate's Health Committee, which gave him a close working relationship with the Service Employees International Union. The SEIU had more than one hundred thousand members in Illinois, representing tens of thousands of nursing and other medical workers. So when Obama plucked the

SEIU's endorsement away from Hynes, it was a major coup. The SEIU was a younger, more racially diverse union than the mostly white trade unions. In the 1990s, the SEIU began building an effective grassroots political mechanism, running phone banks and amassing armies of volunteers to work on behalf of endorsed political candidates. On a cold, dreary Saturday morning in December 2003, Obama addressed the annual state convention of enthusiastic SEIU members at Chicago's McCormick Place convention center. He donned a purple SEIU jacket and spoke passionately about the union movement in America. Impressively, thousands had turned out early that morning to hear Obama preach about how he could win the Senate race. But, of course, he needed their help, he told them. "I can't do this alone," he said.

"Barack has taken the lead on issues of significant importance to our members," said Tom Balanoff, the SEIU president, in explaining the Obama endorsement. "He's also been out there for us when we have been in trouble, during strikes and things like that." In Springfield, Obama indeed had carried SEIU's water. He was instrumental in expanding child-care benefits for workers and had been an ardent proponent of universal health care coverage. He was also a leader on the so-called hospital report card act, which, among other things, required hospitals to post staffing levels and mortality rates on the Internet. After the SEIU's blessing, endorsements followed from the Chicago teachers' union and the American Federation of State, County and Municipal Employees. This shifted some momentum toward Obama and away from Hynes.

Hynes struck back with the hard-fought endorsement of the labor umbrella group, the AFL-CIO, as well as with John Stroger, the African-American president of the Cook County Board of Commissioners. But in the end, the Stroger endorsement was an empty gesture. Stroger even conceded that he was endorsing Hynes strictly as a favor to Hynes's father, Thomas, Stroger's longtime political friend. Hynes's aides contended that this showed Obama's weakness among black voters. But when I scanned the room during the Hynes event, it was clear from the lack of enthusiasm in their physical reactions that the two dozen blacks in attendance had come more because they were part of Stroger's large political operation than because of any good feelings toward Hynes. I surmised that at least half of these blacks would ultimately vote for Obama if he were in contention on election day. Indeed, history shows that if there is a viable black candidate on the ballot, African Americans will over-whelmingly support that person in the privacy of the voting booth, no matter how many verbal endorsements the white candidates in the race have sewn up in the black community.

Hynes had other problems too. For one, like the various Demo-cratic presidential candidates, Hynes supported giving President Bush the authority to invade Iraq. In the first debate in the race, Obama sought to exploit what he believed was an Achilles' heel for a Demo-crat in a primary contest. He pressured Hynes about his support of the impending war in Iraq and tweaked him for having no legislative ex-perience when Hynes seemed to waver on that support. "The legisla-ture is full of tough calls," Obama said in a lecturing tone. "It's not like an administrative job; it requires tough calls." Hynes was also dogged by questions from the panel about a story I had written in the *Tribune* about Hynes's questionable bundling of campaign contribu-tions from a donor who did business with his office. The donor had his employees give thousands of dollars to Hynes's campaign fund and then reimbursed the workers from the company pot, something that federal elections officials ultimately found was illegal.

With all this controversy swirling around Hynes, I looked around for him when the radio debate was concluded, but he had ducked out of the postdebate Q&A with reporters. It's never a good sign when a candidate for public office feels compelled to run from the press.

By December 2003, opinion polling showed that a great num-ber of Illinois Democrats had no idea whom they would vote for in the contest, with "undecided" being the overwhelming favorite.

Among voters with a preference, Hynes was leading, with about 20 percent of respondents saying they would cast a ballot for him. Blair Hull was coming up on Hynes's heels on the strength of his massive television campaign. Gery Chico had raised a lot of cash and had spent almost as much as he had collected, but he was languishing in the single digits and looking more like an also-ran. Maria Pappas, who had been in politics for years and possessed high name recognition in Chicago, polled in the double digits, but had no discernible campaign operation. The still obscure Obama was around 10 percent, but he had two bases that he was working diligently, and Axelrod was singing his praises to people of influence.

So the burgeoning story was Hull. His steady movement in the polls from nowhere to just behind Hynes greatly concerned Hynes, as well as the other candidates. Most campaign strategists were aiming to get their candidate to 30 percent. With so many contenders, this thinking went, the first candidate to reach 30 would be hard to stop. It was increasingly looking as if only three candidates had a shot at getting to this point: Hynes, Hull and Obama.

But Hull's rapid ascent had put Hynes and his staff into a mild panic. The problem for Hynes: Hull was grabbing voters downstate and in other rural corners of Illinois, where life was slower and his television advertising was seeping into the public consciousness. In Chicago, his ads were more likely to get lost amid the urban frenzy. But Hull's name and message were gaining notice in these small towns even though he had never set foot in them. These were voters that Hynes was counting on. Obama would draw blacks in and around Chicago, lakefront liberals and perhaps college students. But if Hynes was to win, he needed rural voters on his side. Believing he had to blunt Hull's early movement, Hynes dropped several hundred thousand dollars in television commercials in downstate markets in late 2003, months before the March primary election. Unfortunately, with voters still not engaged in the race, the brief ad campaign by Hynes had little penetration; in fact, it served only to take a chunk of money from his campaign fund that he would need down the stretch.

I turned my attention to Hull. I had heard from *Tribune* political writer Rick Pearson, among others, that Hull's sketchy past deserved looking into at some depth. Hull had the oddest of political résumés, although it was becoming more commonplace for the excessively wealthy to enter electoral politics using their personal fortunes to bankroll a campaign. A federal campaign law, called the "Millionaires' Amendment," had been enacted to try to even the playing field for less well-heeled candidates in races with candidates of extreme personal wealth. The amendment allowed other candidates to surpass federal donation limits when raising money for such races. In the Illinois contest, this was quite advantageous to Obama, who relied heavily on wealthy lakefront donors like the Pritzkers. It seemed that every member of the Pritzker clan had given the new maximum of twelve thousand dollars to Obama. Hynes, meanwhile, was funded by labor unions and political action committees, which could not spread out their maximum contributions among friends and relatives.

Hull, who possessed a scarily keen mathematical mind, had been a professional blackjack gambler who turned his winnings into Wall Street success. At the urging of his partners, he had sold his securities firm for more than half a billion dollars and then, looking for a new professional interest, turned to Illinois politics. But what intrigued me was not his past but his current campaign. When I tracked him on the campaign trail for a few days, I was stunned by the extreme artificiality of both the candidate and his message. I had lunched with a longtime political source in the city who told me that, like Axelrod, he had interviewed with Hull but declined a job offer. Just as one line from Hull had frightened off Axelrod, this source offered a similar story. When he had asked Hull why he was seeking office, Hull responded: "Don't you think it would be cool to be a senator?" The source was stunned: "How do you work for a guy whose sole purpose for running seems to be that it would be a cool job?" Moreover, in briefings with his many aides, Hull even expressed skepticism about the effectiveness of a representative democracy. "For goodness sake, don't say that in public," he was warned. "You are running for the U.S. Senate, after all."

I didn't find much more depth on the campaign trail. Hull had tapped his vast finances to construct one of the most sophisticated political operations anywhere in the country that campaign season. His staff and payroll were larger than those of any Democratic presidential contenders, and he had hired some of the most savvy consultants in the business at top-dollar wages. But this was also part of his undoing. His aides joked that they were working on the "Noah's Ark" of campaigns because there were two of each of them: two pollsters, two communications directors, two campaign chairmen. This could often mean far too many conflicting voices around the strategy table. His campaign manager earned twenty thousand dollars a month and his policy director fifteen thousand a month. He had as many as twenty-eight consultants on the payroll in the last three months of 2003. He tooled around Chicago in a huge recreational vehicle that had cost the campaign forty thousand dollars. At a joint appearance of the candidates in suburban DuPage County, I was following Obama for the day when he spotted the RV with its huge red-white-and-blue lettering on the side: HULL FOR SENATE. "Gee, what's that?" Obama asked with a sense of wonderment. "That's Hull on Wheels," I explained, using the Hull campaign's moniker for the vehicle. "How do I compete with that?" he asked rhetorically.

It's true that all political campaigns have something of an illusory quality, with candidates hiring consultants to craft and sell a certain image of the client. Hull's skilled image makers had been paid handsomely to conjure a message and a vision of Hull as an independent fighter in Washington for common Illinoisans. But the media blitz was so dizzying that it was more reminiscent of a company building a brand, or of Hollywood marketing a blockbuster film. Hull's campaign had shelled out hundreds of thousands of dollars a week to run television commercials that touted his thick and detailed economic and health care plans. Hull billboards appeared in virtually every corner of the state. Even Internet surfers could not escape Hull's bespectacled mug, with ads gracing websites as varied as the *Washington Post* and Yahoo e-mail pages. The media barrage was so relentless that even Hull wondered about the value of running a campaign this way. "Don't you think people kind of, you know, get sick of you after a while?" he asked me. Because he had no prior grassroots operation in place and no political base, he paid supporters fifty dollars a day to act as volunteers and a cheering section for his public appearances. (In a weird irony, one of Michelle Obama's distant relatives took a job in this capacity.) As Hull would step into an event, they would line up and yell, "Give 'em Hull!" But Hull was perhaps one of the most uncomfortable stump speakers in the history of U.S. Senate races. He would trip over his lines and rarely seemed to convey an extemporaneous thought. His supporters, however, would cheer him madly, even after a verbal flub. To me, all these staged theatrics gave the campaign a feeling of utter artificiality— and as a voter, I was aghast that his type of campaign seemed to be resonating in a democracy. As I wrote in the *Tribune,* it was "sort of like *The Truman Show* meets *The Candidate*."

Nevertheless, Hull's ads were working. And when Hynes's quick hit of television had no effect, the Hynes brain trust began worrying even more about Hull. Hynes's campaign spokesperson, Chris Mather, stepped up her phone calls to me and other reporters in hopes of slowing the Hull momentum. However, the intense lobbying effort actually had the opposite effect with me. Hynes's obvious fear gave Hull even more credibility. At about this time, I met with a Hynes operative for lunch. When I had gone to meet Mather earlier in the campaign season, we convened near Hynes's office. But this operative wanted to come to me, so we gathered at a North Michigan

Avenue restaurant just a couple of doors from the Tribune Tower. Before I had taken a bite of my grilled chicken sandwich, I was handed a folder of opposition research on Hull. Among the papers was a copy of the outside sheet of the filing of one of Hull's two divorces in Illinois. Hull, in fact, had been divorced three times. He was married to his first wife for nearly thirty years, raising three children with her. After moving to Chicago, he then twice married and divorced the same woman. The rest of the divorce file had been sealed, and this vague court order was the only document publicly available. The order contained only one salient fact: Hull's second wife, Brenda Sexton, had once been granted an order of protection against him.

As this was occurring behind the scenes, Hull continued ascending in the polls, cruising past Hynes and the rest of the field. Hull was nearing the 30 percent mark when I interviewed him for my Sunday profile of him and his candidacy. Like many encounters with Hull, it was an uncomfortable experience for both subject and interviewer. In his campaign office, several aides and Hull sat at one end of a long table and I sat at the other. When I brought up the divorces, Hull squirmed in his chair, nervously shifting from leaning on an elbow to folding his arms to various other poses. He steadfastly refused to discuss the circumstances of his marriages, divorces or the court order, saying they were private matters. Because he had been reluctant to explain these issues, particularly the court order, I felt compelled to include this in my profile. I placed this nugget fairly deep inside the story, but it served the purpose of the other candidates—the behind-the-scenes gossip had now slipped into the largest circulation newspaper in the state. Other political reporters and pundits jumped at the tasty morsel. *Tribune* columnist Eric Zorn was the first on board, penning a column asking what Hull was hiding and maintaining that he owed it to the voters to release the divorce files. Mike Flannery, a political reporter at CBS-affiliated

Channel 2 in Chicago, pressed Hull incessantly about the divorce files as Hull opened a campaign office on the city's West Side. Flannery and Zorn were among Chicago's more skilled political reporters, and yet I could not help but notice that both were also guests of Axelrod at his annual holiday party. For his part, Hull continued to stonewall, citing privacy concerns.

It was not long before the Hulls' divorce story assumed a life of its own, dominating headlines, leading newscasts and consuming public debates. At a televised candidates' forum on public television, Hull was peppered with questions about the sealed divorce files—and he stammered no-comments when prompted to talk about the issue. This was not pretty to watch, but Hull faced a legitimate question of a candidate for such an important public office: Had Sexton accused him of something untoward, and didn't voters have a right to know if she had? His staunch resistance to answering questions seemed to indicate so. Yet Hull was a magnificently wealthy man, and his aides were suggesting off the record that she had made accusations against him to wrest more money from him in the divorce settlement. Nevertheless, I had to admit: witnessing this spectacle was not something that I enjoyed, even though my story had instigated the feeding frenzy.

At the TV debate, various reporters and campaign advisers watched from a nearby room. Axelrod and Giangreco, who had been on the road with the John Edwards campaign, had returned to Chicago to prepare and guide Obama through the forum. A grimacing Axelrod paced the floor as a bemused Giangreco watched more patiently from a chair. Axelrod had a habit of pacing whenever Obama began speaking in a public place. If he disliked what Obama was saying, his pacing would quicken. Even though he had been dealing with the issue for days, Hull looked like a man dying from a thousand cuts. Axelrod stopped in midpace and pulled up beside me. "You know, you're responsible for this," Axelrod told me, apparently trying to stroke my journalist's ego. The comment did nothing of the sort. Seeing a man's reputation unravel in slow motion in the glare of the public eye gave me little sense of accomplishment. "David," I told Axelrod, "if it wasn't through me, you folks would have figured out another way to get this mess out there. I just fired the first bullet loaded in the chamber."

IN A DESPERATE ATTEMPT TO CALM THE STORY, JASON ERKES, HULL'S campaign spokesman, offered to let me review the divorce records on an off-the-record, nonprintable basis. I declined the offer, saying that the information had become too vital to the Senate campaign to be kept from the public if a reporter had seen it. Soon, the *Tribune* and WLS-TV sued for the unsealing of the divorce records. When it became likely that a judge would have ruled in their favor before election day, Hull and Sexton jointly asked the judge to release the records.

After I spent thirty seconds with the documents, it was apparent that Hull's chances of winning the race were over. The files showed that Sexton had accused Hull of becoming violent, profane and verbally abusive in the waning days of their second marriage. She accused him of calling her a "cunt." During one incident, Sexton alleged that he "hung on the canopy bar of my bed, leered at me and stated, 'Do you want to die? I am going to kill you. . . .'" Only once, however, did she accuse him of striking her, which led to Hull's brief arrest. But authorities declined to press charges against Hull because they determined that "mutual combat" had occurred. Hull said he struck Sexton's shin in retaliation for Sexton allegedly kicking him.

If this weren't enough, in the post–debate press conference at a later TV forum, each candidate addressing the press was asked by Channel 2's Flannery if he or she had ever used drugs or sought counseling for drug or alcohol abuse. Once again, Hull became the focus of the story when he answered that he had used cocaine and sought counseling in the 1980s for alcohol use. Hynes and Chico admitted to the minor infractions of having smoked marijuana in college, and Obama had already conceded in his memoir his drug activity as a youth. So Hull's drug use as a Wall Street trader was the news of the night. Hull only made matters worse for himself when he sensed the news conference turning sour on him and abruptly stopped taking questions. Television cameras chased him from the TV studio like a defendant fleeing the courthouse. Reporters shouted questions at him. His adult daughter, who had come to Chicago to blunt criticism of her father as a violent man, scurried alongside her father and exclaimed, "Don't answer that question! Don't answer that question!" The bizarre scene was chaotic and every bit as surreal as Hull's campaign. When I called the *Tribune* city desk to report what had transpired after the debate was over, my editors were incredulous. "Forget what you saw on TV during the debate," I said. "Blair Hull just admitted that he used cocaine." At the *Tribune,* my colleagues and I began referring to Hull as the "Velcro" candidate—everything stuck to him. "Each day we come in here determined not to write another Blair Hull story, and each night, here we are, writing a Blair Hull story!" said an exasperated *Tribune* colleague, John Chase, as he furiously recast the debate story into a drug story under deadline pressure.

As Hull fell into ignominy, voters were left looking for a candidate to support in the contest. Obama's campaign had been following the simple strategy long advanced by Axelrod and Giangreco: *Hold on to your money and TV advertising until the final weeks when voters finally are paying attention—and then blast the airwaves with as much force as you can.* This was not as easy as it seemed. As Hull was rising and Hynes was still hovering around 20 percent in the polls, Obama was making only minor advancements. Just a month from the election, Obama was still an unknown commodity to most blacks and most Democrats in general. This made some supporters extremely anxious, and they began advising him to jettison Axelrod's earlier strategy and start running TV ads immediately. "Barack was concerned that we needed to be out there," Axelrod said. Obama talked to his advisers about these concerns but ultimately chose to follow their plan and "hold our powder," he said.

Axelrod was privy to Hull's messy divorce from their earlier interview, after all, and knew that his candidacy most likely would end at any moment—as soon as the divorce details went public. There was one concern, however. As Hull was falling fast, he seriously considered dropping out of the race altogether. One morning on my way in to the office, I received a frantic cell phone call from a panicked Giangreco concerned about just that. Had I heard if Hull was dropping? This could throw a monkey wrench into the racial dynamics of the contest. Obama needed for Hull to siphon white votes and downstate votes away from Hynes. "I sure hope we haven't overshot the runway!" Giangreco worried.

MANAGING OBAMA'S HEALTHY EGO HAS BEEN ONE OF THE MORE trying tasks for his staff and paid consultants. As Obama himself will acknowledge, his mother went to great lengths to shore up her son's confidence. She worried that because his father was absent and he was biracial he might fall prey to a lack of self-worth. "As a consequence, there was no shortage of self-esteem," Obama told me with a wry smile.

In a politician, a show of grandiose ego can be off-putting and cost support from all quarters—media, colleagues and, in particular, constituents—and there were moments in the Senate campaign when I found myself in the midst of the effort to rein in Obama's ego. In January 2004, for instance, the *Chicago Sun-Times* ran a story about prominent Chicago-area politicians who had what the newspaper referred to as the "IT Factor." Said the newspaper: "Some politicians acquire it. Some hire it. Others earn it. But for this coiffed crew . . . no spin doctor is required. Call it 'charisma,' if you prefer. Or 'packaging,' if you're going to be cynical about it. But there's one thing we look for in our political candidates, whether we admit it or not: Sex appeal."

Bill Clinton headed the list of politicians who stirred the libido, according to the newspaper. But in Chicago, Obama was listed among the dozen or so politicos who had "IT." Beneath a flattering photo of a smiling, confident-looking Obama, the *Sun-Times* breathed heavily: "The first African-American president of the *Harvard Law Review* has a movie-star smile and more than a little mystique. Also, we just like to say his name. We are considering taking it as a mantra."

One can only imagine the reaction to this designation from Michelle, who considered it her personal mission to keep her husband's ego from inflating beyond all proportion. Obama's take on the story certainly drew rolled eyes from some staff members. Obama walked through the campaign office with a copy of the tabloid newspaper folded under his arm and open to the story. He proclaimed with glee, "See, told you I've got 'IT'!"

When I brought up the story to an aide that day, the aide told me, "For god's sakes, don't mention that story to him. He's walking around here with a huge grin on his face and saying 'I've got IT, I've got IT.' He sure doesn't need any reinforcement in the IT department."

The Small Screen

Can you believe it? He is just as fine in person as on TV.
—A YOUNG FEMALE ADMIRER

Barack Obama's "IT" factor came to the fore at the strangest moments. And just as his natural talents often bred jealousy among his colleagues in the Illinois General Assembly, his charismatic appeal drew resentment from other campaigns, particularly supporters of the stoic Dan Hynes and the artless Blair Hull. Case in point: the morning of January 24, 2004.

On this Saturday, sleep was still in my eyes as my wife set breakfast on the table. Before I could take a bite, the phone rang. It was a few minutes before nine, the time of the day and week when custom allows only family members or close friends into your world. So when my wife came marching toward me with the phone in hand and a sour look on her face, my keen reportorial instincts told me that good news was not afoot. "It's somebody from the Hynes campaign," she grumbled, with a half-quizzical look on her face that clearly said: "Why would they be calling at this hour on a weekend? Can't you ever get some peace from these politicians?"

My gut told me what to expect—my stomach fluttered with that little uneasy feeling common to reporters when a source is calling to complain about something in a story, possibly an error. The *Tribune*'s first Sunday edition, called the "bulldog," had hit the newsstands that Saturday morning, and it featured my campaign profile of Obama. The article concluded with a short anecdote in which I hoped to convey what I considered key aspects of Obama's persona: his political brazenness and his infectious personal charm, especially with women. The anecdote read:

> There's no doubt Obama can draw attention. Shortly after sign-ing autographs at the recent forum, Obama grabbed the hand of Christina Hynes, the wife of one of his opponents, Illinois Comptroller Dan Hynes, and then kissed her cheek, prompting her to flush and smile broadly. "He has a smooth personality, sometimes a little too smooth," said his campaign manager, Jim Cauley. "He's still young, and we have a ways to go, but he has the potential to be something very special in this business."

Sure enough, I grabbed the phone to find an irate Matt Hynes on the other end, the younger brother of Dan Hynes and the manager of his older brother's Senate campaign. In a sharp tone that ranged from incredulity to anger to outright paranoia, he launched into a several-minute diatribe. The thrust of his speech was that this anec-dote suggested that his brother's wife had reacted sexually to Obama's smoochy greeting—which was not only unfair but intolerable.

The short version of Matt Hynes's emotional tirade went like this: "You have Christina falling into Barack's arms like she can't control herself around his magnetism. This is a cheap shot, a total cheap shot. Something has to be done about this in Sunday's paper or Dan is never going to speak to you again." I mulled over his asser-tion. Was it a cheap shot? The moment certainly had occurred, since I had witnessed it firsthand only days before. I believed I had accu-rately described it—although, if anything, I had toned down Mrs. Hynes's overt response to Obama. A somewhat shy woman, she ac-tually giggled and teetered backward like an awkward high school freshman given attention by the school's star senior quarterback. But was I being unfair to Hynes by bringing his wife into the profile of Obama at all? Had I unintentionally caused a schism in his marriage? "This is a cheap shot, and you know it," Matt Hynes repeated again and again.

Though I meant only to make an observation about Obama, the Hynes people clearly believed deeply in this ulterior motive. Their emotional response was far over the top and suggested the paranoia about the media and political opponents ingrained in many successful political families in Chicago, such as the Hyneses and the Daleys. Weeks later, a writer for the alternative paper the *Chicago Reader* learned of the matter and wrote a short piece about it. In the *Reader* story, Chris Mather, Hynes's communications director, said that my piece "talked about how women were wowed by [Obama]." The weekly dubbed the less-than-sordid matter "The Kiss and the Cover-Up." "I think it was inappropriate to bring Dan's wife into the story by saying that—by implying that she was reacting for the same reason the other women were reacting," Mather said.

On the phone, I had attempted in vain to calm Matt Hynes, to persuade him that I was not taking a cheap shot at his brother, that this anecdote simply explained much about Obama's appeal. Yet despite making that argument and opening a moral dialogue with myself, I have to admit that the words that echoed in my head were "Dan is never going to speak to you again." If that were true, it would make my assignment of covering the Senate race more than a little difficult and perhaps impossible. Hynes, after all, was the frontrunner in the race at that time. In my thinking, it seemed ridiculous to allow such a seemingly innocuous thing to end my professional relationship with a key political candidate just as the race was heating up. So I hung up with Hynes's brother and consulted with the *Tribune*'s political editor, Bob Secter. After some discussion, we both agreed that the anecdote was effective at conveying Obama's charm and charisma and agreed to leave it in subsequent editions of the Sunday paper. But we decided to edit out the reference to Hynes's wife, changing her identifier to "a supporter of an opponent."

For me, it was a matter of placating a source and keeping my journalism integrity intact, while still serving the *Tribune* readers. Secter seemed more worried about extricating our reportage from the Hynes's marital relationship. Secter later told the *Reader,* "It seemed gratuitous, and it didn't seem to be necessary. I think they thought it made an unnecessarily demeaning impression about Hynes' relationship with his wife. We're dealing with a difficult dimension here—a guy's personal relationship with his wife. And I don't know anything about their relationship, and it wasn't our place to try to. If they inferred something out of it that we didn't originally see, I didn't think it was worth the hurt feelings to Hynes, since it was really a total side issue to the main point we were writing about."

Near the end of the campaign, while on a plane during Obama's tour of the state, I mentioned the much-ado-about-nothing matter to Obama, who, appropriately, appeared less than interested. One of Obama's strengths is that he rarely busies his mind with matters that seem trivial to him, and this certainly was, at its essence, a trivial matter—but a trivial matter that said much about its participants. Indeed, as Obama stepped away from Mrs. Hynes that morning months before, I suggested that it "might not be wise" to greet your opponent's wife with a kiss. Even then, he just sloughed it off, saying he had seen her many times on the campaign trail and they had gotten to know each other. Hynes's brother, however, said much the opposite. He said that Christina Hynes was taken aback by the physical nature of the greeting.

In any case, on the campaign plane, Obama himself summed up the matter succinctly: "Sounds like [the Hynes campaign] spent too much time arguing with the refs."

Surely that was true. But by design or happenstance, here was another instance in which Obama's charming and bold nature had again subtly unnerved a political opponent and helped to further Obama's own political cause.

AS THE PRIMARY ELECTION DREW CLOSER, THE CAMPAIGNS OF ALL the candidates naturally took on more urgency and more vitriol. Volunteers, consultants and the candidates themselves were all a bit more on edge and more aggressive in their tactics. Hull and Hynes, in particular, were growing more venomous toward each other. Hynes's campaign was anemic from start to finish, except when it came to attacking Hull behind the scenes. As the front-runner with a powerful political father, Hynes had chosen a Rose Garden strategy—lie low publicly and the vote will be there on election day when the unions and ward groups churn it out. This proved a fatal error. Hynes was running in a U.S. Senate race, not a local alder-manic contest or even a race for a state executive office. He had been adept at hitting his lines at editorial boards and in public de-bates, but he was offering nothing particularly special to the voters. For their senator in Washington, voters wanted something more than "the favorite son of the Democratic Party," in Obama's words. In most polling, Hynes hovered around 20 percent and, as the stretch run began, had moved only a few percentage points, if that. Hynes's play-it-safe strategy was taken to the extreme in nearly ev-ery aspect of his campaign. For example, in contrast to the couple of hours I spent talking with Michelle Obama, Hynes declined to make his wife available for an interview, even before the Obama kiss incident. When I shadowed Hynes, he seemed reluctant to talk about anything in his personal life, for fear of alienating a constitu-ency. "I knew Dan Hynes was not going to win that race," a high-ranking Illinois Democrat told me. "I've played basketball with both him and Obama—and Hynes played soft."

Nevertheless, the one aspect of the Hynes campaign that showed no timidity was its criticism of Hull. To a large extent, Hull and

Hynes were targeting the same Democratic constituencies—suburban and downstate Democrats. This made the two candidates natural enemies. Chris Mather, Hynes's spokeswoman, would bend my ear daily about Hull's many political weaknesses and liabilities. And I would get the same treatment of Hynes from Hull staffers. As these two duked it out, Obama was putting one foot in front of the other, moving forward slowly but surely. "They're running a very smart campaign over there," allowed Anita Dunn, Hull's media strategist. Catching up with Obama one afternoon after he spoke to a progressive crowd in suburban Oak Park, I mentioned the fierce battle between Hull and Hynes. Obama laughed and ducked his head as he emphasized this point: "I'm just trying to keep my head down while they fling arrows at each other."

As Obama largely ignored Hull and Hynes, there was a candidate who could irritate him—Gery Chico. The former school board president was the first to announce his candidacy and he was well funded, but his money began drying up when his campaign had trouble getting off the ground. Chico had overestimated the strength of the Latino vote and, most damaging, became distracted by the controversial disintegration of the venerable law firm that he had co-led. He did not project a positive television presence—his gruff manner could be a turnoff to viewers—but he was especially skilled at turning a complicated policy matter into a pithy, two-sentence sound bite. This was a skill that Obama did not possess. Obama flourished in extended interviews and longer speeches, where he impressed his audience with elongated, eloquent, thoughtful passages. But in the many candidate forums on the stump, Chico overshadowed Obama with snappy phrasing that cut to the heart of an issue.

After attending one forum, Obama's campaign manager, Jim Cauley, advised his candidate: "You know, Chico is good out there. You're going to have to raise your game." Obama was the intellectual Harvard Law graduate and was expected to excel in a debate setting. But that wasn't necessarily the case in this Senate race. At one forum in suburban Evanston, a liberal lakefront bastion surrounding Northwestern University that Obama needed to carry in overwhelming fashion, Chico tartly challenged Obama about arcane specifics of his education policy. Most of the Democrats in the race agreed on policy matters, and substantial argument over policy was uncharacteristic. Obama is a policy wonk in his soul. But on education, he generally used a standard Democratic applause line about President Bush's No Child Left Behind law—"George Bush left the money behind"—and he was taken aback by Chico's spicy criticism. Chico, having served as president of the Chicago school board, was expert in the subject area and got the better of the exchange. When I asked Obama afterward what he thought about Chico's attack, Obama responded by beginning to relitigate the policy argument. I stopped him. "Not that," I said. "Why do you think he went after you? Because you are the liberal favorite up here in Evanston?" Obama stammered and said, "Oh, you mean you think it was a political thing?" He was so caught up in the intellectual policy aspect of the incident that he completely missed the politics of it. Chico clearly wanted to score a point on Obama on Obama's home turf, not hash out education policy.

THE FOCAL POINT OF OBAMA'S SHOE-LEATHER CAMPAIGN WAS unquestionably his stump speech. His main address throughout his Senate race was never formally assigned to paper or computer file. He started with a basic framework and spoke extemporaneously from that, usually with similar themes carried throughout. He typically opened with the same joke about his odd name by saying that people invariably call him something else—"Yo mama" or "Alabama." This line always drew a laugh, and Obama delivered it with such frequency that, as the primary wound down, he had to remind himself to smile in response to the audience laughter. The quip served to begin a short explanation of his unique biography. ("My father was from Kenya, in Africa, which is where I got my name. *Barack* means 'blessed by God' in Swahili. My mother was from Kansas, which is where I got my accent from.") He then launched into the meat of the speech, which usually involved the gap he perceived between his own values and the course of the country as set by Republican leadership.

These primary speeches were fairly consistent addresses that he would tweak on the spot, depending on feedback from that particular audience. One weekend, as the day wore on and he delivered his speech at least half-a-dozen times, he told me, "It's getting harder and harder to change this speech each time for you, Mendell." The comment surprised me. First, I hadn't noticed many substantial changes, just a bit different language to emphasize the same themes. Second, having shadowed political candidates previously, I was accustomed to the identical speech being delivered day in, day out. One candidate I had covered had given the same speech, almost word for word, at each stop, and I told Obama there was no reason to alter his addresses on my account. But after some thought, I realized that he was using my nearly permanent presence as an intellectual device. These speeches were mental exercises for him—to hone his message and advance his already polished oratory skills. Varying his addresses, no matter how slightly, helped him mature as a public speaker. "My general attitude is practice, practice, practice," Obama said. "I was just getting more experienced and seeing what is working and what isn't, when I am going too long and when it is going flat. Besides campaigning, I have always said that one of the best places for me to learn public speaking was actually teaching—standing in a room full of thirty or forty kids and keeping them engaged, interested and challenged. I also think that David [Axelrod] was always very helpful in identifying what worked and what didn't in my speeches." Axelrod described their chats about message and speech delivery as akin to "musicians riffing together."

Obama's ability to connect with a black audience, once questionable, grew tremendously during the Senate contest. As a community organizer and as a candidate against Bobby Rush, Obama spent countless hours in Chicago's African-American churches digesting the cadence of a preacher's rhythm and the themes that stoke an African-American crowd. So when he came before black audiences as a Senate candidate, particularly in a church, Obama spoke in start-and-stop passages, imbuing his delivery with a touch of soulfulness and building complex thoughts about social justice and economic inequity into bold emotional crescendos. His message remained remarkably consistent: Despite many superficial differences, Americans are linked by a common bond of humanity, and the country's government must reflect that benevolent core. "I am my brother's keeper! I am my sister's keeper!" he proclaimed, his rich voice booming as he reached the height of the speech. Such addresses from Obama were mostly secular and political in nature, but he made sure to pepper them with hints of the Bible, Christian orthodoxy and borrowed phrases from the nation's African-American civil rights icon, the Reverend Martin Luther King Jr. He would end many of these speeches by intoning his favorite quote from King: "'The arc of the moral universe is long, but it bends toward justice.' But it doesn't bend on its own," he told his audiences. "It bends because you put your hand on that arc and you bend it in the direction of justice."

ONLY ONE REAL MISFORTUNE BEFELL OBAMA'S PRIMARY CAMPAIGN. During the final week of November 2003, Paul Simon, the bespectacled, gracious former two-term senator from Illinois who had run for president, stopped by the *Tribune*. I caught him briefly in the hall and asked if he had a preference in the U.S. Senate race. Simon said he would make a formal endorsement in a couple of weeks, but would not divulge at that moment which candidate he supported. "But," he added with a grin, "I'm going to be very proud of this one." Knowing that Obama was an acolyte of Simon's, I simply said,

"Obama." Simon responded, "Just wait two weeks," and he hopped in the elevator. The next week, Simon suffered serious complications during surgery to repair a valve in his heart and died the following day. The down-to-earth Simon, who hailed from central Illinois, was a figure beloved by Democrats throughout the state for his honesty, integrity and unabashed progressive politics. Axelrod had planned to shoot a television commercial with Simon's personal endorsement of Obama. By linking Obama with Simon, Axelrod hoped to create the image for both progressives and downstate residents that Obama had the same characteristics as Simon: passionately liberal, disarmingly honest and definitely his own man. The ad would run primarily in downstate markets and it was considered crucial to help Obama pick up some votes outside the Chicago metro area. So Simon's death was a blow to the Obama camp.

Initially, Axelrod was unsure how to proceed. Then he took a risk. He cut a Simon-themed ad featuring Simon's adult daughter comparing Obama with her father. In the commercial, Sheila Simon talked about the values her father brought to politics and proclaimed that Obama was the person who would best carry on his legacy. "For half a century," she said, as images of her father flashed on the screen, "Paul Simon stood for something very special—integrity, principle and a commitment to fight for those who most needed a voice. Barack Obama is cut from that same cloth." The ad was risky, to some extent, because it could have been perceived as exploiting Simon's death for the sake of Obama's political career. In fact, while test-marketing the ad in focus groups, consultants found that one woman reacted negatively for that very reason. But Axelrod felt strongly that the commercial would resonate with Democratic voters and devoted liberals. Going with his gut over the consultants' findings, Axelrod ran the ad anyway. It turned out to be a huge success in downstate Illinois, projecting a positive image of Obama to voters who had most likely never heard of him. Indeed, their introduction to Obama was framed in these terms: Here comes the next Paul Simon.

In those final three weeks, Obama's campaign ran several more of Axelrod's handsome television ads. The first was a biographical piece with footage of Obama that introduced his candidacy and his life to voters. One of Axelrod's foremost talents was identifying the key selling points of a candidate's biography and creating an attractive video package to highlight those points. Axelrod had learned in his focus groups that whites were drawn to Obama's Harvard résumé and blacks to his community-organizing experience—so both were accentuated in the advertising. In Obama's case, it was rather easy to produce an appealing video commercial because of his television charisma and legislative accomplishments.

The theme was "Yes, we can," which implied many things depending on who was interpreting its meaning: Yes, a politician with the ideals and track record of Obama could make a difference and change lives for the better. Yes, a black man could win a U.S. Senate seat. Yes, "we"—meaning all people—could make a difference too. Axelrod framed this message primarily in terms of Obama's barrier-breaking *Harvard Law Review* presidency (which whites had reacted to favorably in focus groups) and the landmark legislation that he passed in Springfield. The legislation aided key Democratic constituencies that Obama was courting: for women, a law that forced insurance companies to cover routine mammograms; for liberals, blacks and the poor, laws that expanded health care coverage to twenty thousand more impoverished children, provided tax relief to the working poor and reformed the death penalty. "Now they say we can't change Washington?" Obama asked in an earnest voice while stepping forward to fill the camera frame. "I'm Barack Obama and I am running for the United States Senate to say, 'Yes, we can.'"

When Obama first saw the "Yes, we can" theme, he was far from impressed. In fact, he did not care for the idea. Obama understood the implication of this sound bite, but intellectually he found the simple refrain "Yes, we can" rather trite considering the seriousness of his cause. He was inclined toward something with more depth. But

Axelrod felt so strongly about this message that he stood fast behind it. So Obama went to his most trusted adviser—Michelle—and asked what she thought. She told him that it was a good idea, that it would penetrate the African-American community and that he should use it. Obama knew that his wife understood the culture and psyche of South Side blacks, and he deferred to her judgment. This indicated the maturity of Obama as a political candidate. Earlier in his career, he might have fought Axelrod on the concept. "Barack is extremely intelligent, and one of the pitfalls of extreme intelligence is you are so accustomed to being right that you believe you are always right," Axelrod observed. But after the Bobby Rush debacle, Obama discovered that, when it came to politics, there were professionals in the field who could offer wisdom beyond his. It was probably wise to heed the guidance of a political professional like Axelrod, especially if his wife agreed with the plan.

Other commercials used the same "Yes, we can" mantra to appeal to different constituencies. Pollsters have consistently found that urban voters lean toward candidates who are change agents, while voters in rural areas are more conservative, perhaps a bit less jaded, and tend to look for political experience in their candidates. So Axelrod's downstate ad touted Obama's legislative experience and his proclivity to work hand in hand with Republicans. Over narration about Obama's bipartisan nature, there was an image of Obama walking next to a blue-jean-clad farmer, with silos and green fields in the background. Wearing a less formal beige suit, Obama stood in front of a small-town courthouse and asked: "What if folks in office spent their time attacking problems instead of each other?" Another ad targeted for rural areas featured a protectionist theme. Obama was seen shaking hands with union members as he promised to work to enforce trade laws and slash tax breaks for corporations that moved jobs overseas. "Give [those tax breaks] to companies that create jobs here—in America," he said firmly.

In perhaps Axelrod's most effective ad, he harked back to the days of both Simon and the iconic Harold Washington—and then morphed into Obama as the modern-day ambassador of their causes. "There have been moments in our history when hope defeated cynicism, when the power of people triumphed over money and machines," a deep-throated narrator intones as images of Simon and Washington wash over the screen. The ad then quoted various newspaper endorsements of Obama that called him "the man for this time and place" (*Chicago Sun-Times*) who has a "proven record of spirited, principled and effective leadership" (*Chicago Tribune*).

As much hard work as Obama put into his on-the-stump campaigning, it was this television campaign produced by Axelrod that pushed him over the top in the final three weeks of the race. As Hull's divorce files became part of the public consciousness around Chicago, Obama's ads were hitting the small screen. It was difficult to know specifically if they were resonating with the voting public, but internal polling from the campaigns indicated that Obama was surging and filling up the void opened by Hull's collapse. "Obama is on fire!" Jason Erkes, the Hull spokesman, told me.

As usual in the final two weeks of any major election, the *Tribune* beefed up coverage of the Senate race, assigning a single reporter to cover each major candidate. Hull had been the story of the campaign so far, and my editor, Bob Secter, told me that he wanted me to shadow Hull. Since the *Tribune* had sued for the Hull records, Secter explained to me, "We can't get beat on the Blair Hull story." But I told Secter that I should be assigned to Obama. I could see how the race was playing out—Obama was about to cruise into the lead and probably win. If I were to continue on and cover the Democratic nominee in the general election over the next six months, these last two weeks could be crucial in building a better working relationship with that candidate. After Obama was nominated, there would probably be more staff hired and more obstacles between the candidate and me. It would be easier to cut through that small bureaucracy if I got to him now, when there was no real protection day to day. Then, in August, he would be much more comfortable with me tagging along and my access would be improved. "Besides, the Hull story is over," I told Secter. "The story on election night will be the Barack Obama story." My editor was not necessarily in agreement. "What makes you so sure Obama is going to win?" Secter asked. I told him that the polling was clearly headed in that direction, and suddenly you could feel something in the air when Obama appeared in public—a buzz in the crowd, a certain look of sheer devotion on the faces of his followers. "It's like when you're watching a ball game and you can feel the momentum shift toward one team," I said. Secter relented. I would now enter Obama's orbit full-time, hanging on his coattails for the next couple of weeks and beyond. And what a strange trip it would prove to be.

ON THE WEEKEND BEFORE THE ELECTION, OBAMA'S EASE IN DIFFERENT ethnic and racial settings was on vivid display as he bounced from black church to white event to Latino neighborhood.

After a morning breakfast in the black community on the South Side, Obama's small entourage—just Obama, his driver and me—was headed downtown for Chicago's annual Saint Patrick's Day Parade, the predominantly white celebration not to be missed by prominent elected officials, and especially not to be overlooked by those hoping to become elected officials. This was the parade in which he was dead last the year before. Indeed, just a month earlier, Obama was still little more than an obscure state lawmaker campaigning tirelessly to spread his unusual name and his political message to a largely disengaged Illinois electorate. Now, he was seventy-two hours from the biggest election in his career, perhaps the biggest single moment of his life. And suddenly it appeared as if his long-held lofty vision for himself might be on the verge of realization—he might just be rising from obscurity. Nothing was certain yet, but after one of the most bizarre campaign seasons in

Illinois history, newspaper and campaign polls all indicated that Obama had surged from third place to the top of the crowded field of Democratic candidates.

His driver parked Obama's campaign-leased black sport utility vehicle beneath a row of leafless trees lining downtown Chicago's Grant Park, a sprawling and scenic public green space situated between the Loop commercial district and Lake Michigan. Chicago's interminable gray winter cloud cover had parted in recent days, and Obama stepped into bright noon sunshine. He pulled a well-worn charcoal gray wool overcoat onto his even thinner than usual frame, reduced by months of fourteen-hour days on the campaign trail. He reached into a frayed coat pocket and secured an always present pair of sleek black Ray Ban sunglasses, ever so slowly raised them to his face, assuredly cocked his head a quarter turn sideways and with chin pointed upward, coolly slipped on the glasses.

Just then, three young white women, none of them beyond their early twenties, spotted the political candidate and darted from a nearby throng of parade-goers. Judging by their accents, the women hailed from one of the mostly white ethnic sections of the town, where heavy Chicago accents are inescapable.

"Mr. Obaaaama! Mr. Obaaaama!" one of the young women said, rushing to the candidate's side and pronouncing the first *a* in Obama's name with a nasal Chicago inflection, as a sheep says "baaaaa." "Can we pleeeeze get a picture with you, Mr. Obaaaaama?" she pleaded. "Pleeeeeze?! We have seen you on TV and we are all going to vote for you!" With no visible reaction to the flattering attention, Obama turned to me.

A photo was snapped of Obama smiling broadly, his arms loosely fastened around the waists of two of the women, and I was immediately struck by how remarkable this moment was. In all likelihood, these women came from a place in Chicago that has never produced a substantial number of votes for a black political candidate. In one of the most racially and ethnically segregated metropolitan regions in the country, they most likely came from a part of town where, even at the dawn of the twenty-first century, blacks still rarely felt comfortable enough to venture.

Besides the racial disconnect, I wondered how these three women even knew who Obama was. Immersed in establishing a career, building a romantic relationship and following pop culture, the typical twenty-something is the least reliable vote on election day. At that point, the vast majority of white Democrats had never even heard of this state lawmaker with the strange-sounding name. These women were not in Obama's target groups, being neither black nor lakefront liberals, and just two months earlier, Obama had the backing of fewer than three in ten *black* Democratic voters. Moreover, he was running for the U.S. Senate, not Illinois governor or Chicago mayor, the political jobs in this midwestern state with real public cachet. Had Axelrod's ads really sunk into the public consciousness to this extent, to produce these three young white women gushing over him as if he were some rising rock star and assuring him of their electoral support three days hence?

After the photo was taken and all arms unlocked, one young lady looked at another and said, "Can you believe it? He is just as *fine* in person as on TV."

Obama's SUV driver, a barrel-chested black man named Mike Signator, glanced at me with eyebrows raised and a wry smile on his face. Obama smiled to himself. At long last, here was firsthand evidence that his television splash had transformed his campaign, and at the perfect political moment. "You know, if you have the votes of those three young white women," I said, "you really are going to run away with this thing."

Obama flashed a confident and knowing smile. "You just wait to see where all our votes are coming from on Tuesday," he said. "You just wait."

A Victory Lap

Truthfully, it feels like a movement. I think for people of our generation, we haven't been a part of something like this before.
—A THIRTY-SOMETHING OBAMA VOTER

David Axelrod's television advertising defined Barack Obama in the public consciousness in those final three weeks of the primary campaign. Axelrod's long-held strategy of keeping a lock on the money and spending it on TV at the end proved to be positively genius, especially in light of the ugly demise of Blair Hull. Axelrod, of course, had full knowledge of Hull's unseemly personal baggage and he was confident that, in such a high-profile contest, it would not remain under wraps forever. Dan Hynes, meanwhile, ran a series of commercials, but they had a cutesy quality and got lost in the seasonal blitz of political ads. In one commercial intended to court suburban women, Hynes stood at a kitchen counter attired in an apron and cradling an egg—"a nest egg," he called it—and then cracked the egg over a skillet. This was a reference to Republican indifference to safeguarding a retirement nest egg for everyday people. (Note to campaign commercial scriptwriters: Never put your candidate in an apron.) Another ad featured Hynes and his wife, but again, it seemed lightweight for a U.S. Senate contest.

With Obama's ads resonating, momentum had swung his way and it seemed unstoppable. He was soaring in the polls. This was the very beginning of the Obama phenomenon that would sweep through Illinois and then spread nationwide, carrying him into the U.S. Senate and ultimately into the 2008 presidential contest.

Obama's Democratic primary opponents huddled with their staffs to brainstorm about how to stop Obama in those final weeks, but they came up with nothing. Hynes had been concentrating almost solely on derailing Hull and seemed clueless about how to respond to Obama's surge. The Hull campaign had assembled opposition research on Obama, but nothing was of much substance. Obama had been inoculated from harsh criticism of his teenage and college cocaine snorting because he had divulged the matter himself in a book published ten years earlier. Hoping to turn Democratic women away from Obama, operatives for one opponent feverishly lobbied me behind the scenes to write a story about Obama's many "present" votes in the legislature on abortion bills. But when I interviewed abortion rights activists in Springfield about Obama's posture on abortion, they stood firmly behind him as a staunch pro-choice advocate. "There was nowhere to go, nothing to make an issue of," said Mark Blumenthal, one of Hull's pollsters. "Obama had the best of all worlds. He said, 'I embody change, but I have experience.' The Obama persona born the last week of the campaign—we couldn't look at that with any depth. He was just on the right side of Democratic primary voters that season. And nobody was able to create a news story to take him down."

As a black man, Obama, in some ways, wore a suit of armor. No candidate wanted to alienate the significant black voting bloc by sharply attacking him as he was fast becoming a symbol of pride in the African-American community. When Obama stepped into public spaces, he was recognized as never before. "Hey, that's Barack Obama," a black man whispered to a friend with a beaming smile on his face as Obama marched through Chicago's McCormick Place Convention Center to an event. "Let's face it—he's black. And we don't want to look racist," explained Jason Erkes, the Hull communications director. There was also more to Hull's reluctance—Hull's campaign cochairman, Bobby Rush, had assured the Hull team early in the race that Obama could not draw votes from the black community. So he had never developed a strategy for attacking Obama. Indeed, the bad blood was with Hynes. Said Erkes, "If we don't win, and it looks like we can't, we are going to do everything we can to make sure Hynes doesn't win."

The only other event that could have damaged Obama's pristine image came in the final televised debate. Obama, now the clear front-runner, was visibly nervous from the outset and did not deliver his opening statement with his usual cocksure attitude. To offer support and prepare Obama for the debate, Axelrod and Pete Giangreco had traveled back to Chicago from their duties with the John Edwards presidential campaign and other clients. They assumed that the other candidates would come out swinging at Obama, and that was true. And one could sense that Obama was thinking about what might soon be coming his way. Hynes, for one, assailed Obama for doing little to curb state spending under the previous Republican gubernatorial administration and thus asserted that Obama was partially responsible for creating the state's fiscal mess. "He stayed silent," Hynes said of Obama. "He did nothing." But, Giangreco noted, while Obama's even temperament can make him poorly equipped to take the first swing at an opponent, "Barack is a great counterpuncher when he's attacked." Sure enough, Obama swatted back effectively by saying that Hynes was the state comptroller at the time and "he signed off on every one of these budgets."

Obama came out of the forum largely unscathed, with only days to the election. For all intents and purposes, the nomination was his.

THE TYPICALLY GRAY WINTER SKY OF CHICAGO HUNG BEFORE ME on March 10, 2004, framing the impressive city skyline in dreariness. I was juggling my morning coffee, a manual-shift Saturn and speeding traffic along the Eisenhower Expressway en route to the Tribune Tower in downtown Chicago when my cell phone showed that Obama was calling. It is a rare occurrence when a candidate calls a reporter himself, but to this point, Obama had pretty much run his campaign on his own terms, and pretty cheaply as well. Instantly I sensed why he was dialing me. For the first time, Obama had led the *Tribune*'s daily wrap-up story on the Senate race. The *Tribune* had run weekly stories on the policy issues in the race, but the coverage that had absorbed most of the media's attention concerned Hull's marital troubles. And in this case, not only was Obama the lead of the story, but the article featured a negative finding about him. Talk about rare.

A rival campaign had passed along to me an Obama flyer dubbed a "Legislative Update" that looked suspiciously like campaign advertising, although its cost was borne by state taxpayers. The flyer had been mailed in early February to every household in his state senate district on Chicago's South Side under even more suspicious circumstances. It arrived in mailboxes just days before an ethics law prohibited elected officials running for office from dispersing such taxpayer-funded literature. And there was irony to the story. Obama had written the ethics law himself, touting it as an example of both his probity and his legislative accomplishments. Thus, it seemed he was caught in an impropriety: By mailing the state-funded positive piece about himself just before the deadline, he was clearly doing an end run around the very law that he had sponsored with pride.

The *Tribune* story, which I cowrote with colleague John Chase, began like this: "State Sen. Barack Obama claims the mantle of a reformer, but early last month the Democratic U.S. Senate candidate spent $17,191 in state taxpayer money on a mailer that had the look and feel of a campaign flier. The mailing went out just days before a new ban on the pre-election dissemination of such state-paid constituent newsletters went into effect, part of a package of ethics reforms that Obama takes credit for getting passed."

Considering that his main opponents in the Senate race had been caught up in improperly bundling campaign contributions (Hynes) and allegations of spousal abuse (Hull), this story hardly merited a misdemeanor. But it did put a slight chink in Obama's armor, which at that point hadn't endured so much as a minor surface scratch.

Like most reporters who receive such phone calls, I quickly realized I was going to have to defend my story. In this case, that didn't seem like a difficult chore. It was hardly the crime of the century, but the story was solid and legitimate, far from a cheap journalistic shot. "Hey, this story today," Obama began, before pausing a moment. "Uh, I guess with me at the top, I guess this means I'm the front-runner, huh?" His hesitation and his less-than-assured tone told me he didn't seem comfortable making the call, but felt he must defend himself. "You know, Dave," he continued, "this story today, we didn't do anything illegal here. The implication is, we did something illegal." I explained that the story never implied illegality, but said that he appeared to have breached "the spirit of the law." I then waited for his defense of that argument.

But, to my shock, Obama did something that politicians rarely do—he backed down and concurred with me, showing a rare glimpse of both humility and candor. "Okay, I'll give you that," he said. "And between you and me, I chewed out my staff for mailing that out when they did. It should have gone out a long time ago."

And with that, the matter seemed to be laid to rest. That is, until I chatted with Obama's lead campaign consultant later in the day. Axelrod called to inform me that my story was bogus—his candidate had done nothing wrong. When I used the same response as with Obama—I never implied illegality, only that he had violated the spirit of his own ethics law—Axelrod disagreed entirely. He argued vigorously that the mailing in no way resembled campaign literature and it was strictly distributed by Obama's senate office to inform his constituents about his work in Springfield. No laws were broken— just an unfair shot from the newspaper at his man, Axelrod insisted.

When I told Axelrod that his candidate had conceded to me earlier in the day that he had mistakenly sent out the flyer when he did, that it should have gone out earlier, and that Obama had further admitted to breaking the spirit of the law, Axelrod responded with utter amazement.

"He did?" Axelrod said. "He said that?"

"Yes," I said. "How about that? It would appear that you have an honest man on your hands here."

"Yeah, I know. And you know what?" Axelrod said. "That can be a real problem."

AS THE PRIMARY CAMPAIGN WOUND DOWN AND OBAMA'S VICTORY seemed in the bag, he naturally started thinking about the general election and how to strengthen his campaign organization. He would now be the Democratic Party nominee, and resources from the national party would be forthcoming. The question was, how much did he need to avail himself of these resources. As someone who endeavored to keep his career as unencumbered as possible by the organized political structure, he wanted to maintain autonomy over his message, his media and his policy. In the primary, Obama had hired savvy professional consultants and staff with ties to the mainstream political establishment in Axelrod, Jim Cauley and Pete Giangreco. But for the most part, Obama's message and his core beliefs were the main thrust. He also had a flock of volunteers who believed in him. Many of them were students at the University of Chicago and other Chicago-area colleges who had heard of Obama by word of mouth or had seen him speak. "We have all these save-the-world types showing up at the door," campaign manager Cauley mused one day. "Sometimes, I don't know what to do with them all." I almost laughed aloud at this comment. Wasn't Obama a save-the-world type himself?

In any case, Obama could still look at his primary campaign and say that victory pretty much came organically—not just because the powers-that-be wished it so. Obama, to be sure, had worked within the established order, currying favor with insiders, raising millions in campaign money and hiring aides with powerful connections. He had secured Illinois senate president Emil Jones's backing. He had tapped many of the same financial donors who also backed Mayor Richard Daley. Unions, trial lawyers and other so-called special interests had backed him financially and with grassroots help. And Michelle, of course, had worked in city hall and helped to introduce her husband to a network of important African-American business leaders. But Obama operated mostly on the fringes of Chicago's legendary machine politics. He was not considered a vital cog in the wheel of any political operation but his own.

So Obama wanted to keep his freedom and his independence intact through the fall election. "I don't want this campaign to be taken over by Washington," he said, while riding in the campaign SUV between appearances at African-American churches. That will be easier said than done, I thought to myself. Perhaps Obama didn't realize the power of his candidacy. Illinois was trending heavily Democratic, and he had an excellent chance to be only the third black elected to the Senate since Reconstruction and most likely the only sitting black in the upper chamber. This in itself would catapult his profile above all the black representatives who had been in Washington for years.

THE BUZZ AROUND OBAMA IN THOSE FINAL WEEKS BECAME INtense. African Americans, especially, jumped aboard his candidacy with fervor. Private polling showed that Obama shot from less than 15 percent of the black vote to nearly 50 percent just a week after his television ads hit the airwaves. "It was a straight arrow up," said Blumenthal, the Hull pollster. "And it just kept going up and up." This electricity among blacks was palpable in nearly every African-American setting Obama walked into. Rather than sitting anonymously in a church pew on Sunday morning and then having to introduce himself from the pulpit, he now turned heads wherever he went. A fund-raiser thrown by a young black professional at a trendy downtown bar was packed to capacity. It took Obama half an hour to push himself through the crowd to the back of the huge night-club, where he was to speak. When the host introduced him to a cheering crowd as "*the* best and *the* brightest we have to offer the world," even the ambitious, self-assured Obama raised an eyebrow at this obsequious treatment. He needed a burly escort to help him back through the crowd and into the waiting SUV. Finally hopping back into the vehicle, he seemed stunned himself at the outpouring of affection.

Wherever Obama went in public as primary election day neared, he put on his game face. "I am fired up!" he would exclaim in joy-ous rallies filled with exuberant union members and teachers and progressives and blacks. Privately, however, Obama's driving ambi-tion, and now the seeming fait accompli of becoming a U.S. sena-tor, was having a different effect—it was eating him up.

What is worse: Getting a job that you desperately want but that will dramatically alter your relatively happy life, whisking you away from your beloved wife and children and curtailing your time with close friends, or not getting the job and living out your life quite comfortably? Obama dearly loved his wife and his two young girls, and it was dawning on him that being a U.S. senator—especially one with star power—was going to pull him away from them more than he might have foreseen. Obama's abiding belief in his own per-sonal destiny created this paradoxical effect—he would strive for something passionately and then rebel against its deleterious effects on his life. "Ambition has always been both Barack's downfall and his greatest attribute," his former aide Dan Shomon said.

This internal conflict came to the attention of his close friend Valerie Jarrett at a picnic she threw for the campaign's volunteers and staff. Throughout the event, Obama's face was tight and he was on edge—almost the antithesis of his easygoing public persona. Jarrett noticed this and asked him about it when the two had a private lunch not long after. "Well, you are on your way now," she told him. "It looks like you are going to be a U.S. senator and who knows what's next? So what's the matter, Barack?" Obama's eyes had been downcast since he greeted her, and he now hung his head low as Jarrett spoke. When he lifted his head to answer, a tear rolled down his cheek. "I'm really going to miss those little girls," he said.

ELECTION NIGHT WAS A RAPTUROUS EXPERIENCE FOR MOST OF Obama's supporters, except Obama. Highly disciplined and focused, he displayed few signs that he was about to be the Democratic nominee to the U.S. Senate. The victory party was held at the Chicago Hyatt Regency, owned by the Pritzker family. As a roomful of two hundred guests watched election results pour in across the television screen, Obama paced about endlessly, checking notes for the speech he had cobbled together and greeting the many smiling well-wishers who hugged him and shook his hand. Obama had stepped into his Hawaii calm, cool exterior. "He's really pretty excited," Michelle told a quizzical-looking Eric Zorn, the *Tribune* columnist. "He's basically a calm guy. It takes a lot to push his buttons. He has incredibly low blood pressure."

Axelrod was stationed at the Cook County Board of Elections, monitoring the results of the race. On election night, Axelrod's crusader instincts took over from those of the Machiavellian consultant bent on winning a race at all costs. As he looked over the numbers, he began thinking about the historic nature of what was occurring. A black man was running away with a statewide race for the U.S. Senate, and he was not squeaking through. He was winning predominantly white wards all over the city and precincts in the suburbs that most blacks would have considered far out of reach. The breadth of Obama's support shocked even his chief strategists. "The most surprising and gratifying thing was when those numbers rolled in on primary night," Axelrod said. "And you saw numbers from the

Northwest Side of Chicago, and you saw the numbers from the collar counties, and you realized that, you know, I mean, I was covering Chicago politics when the issue of race was at a jagged edge here. And I was around when Harold Washington went to Saint Pasquale's church on the Northwest Side and was roundly booed and the hatred was, you know, palpable. And that night, that primary night, I was moved, if you could be moved by watching numbers come across a computer screen. What those numbers meant was that we had passed a Rubicon in the politics of this state, where a guy could come along who was an African-American candidate, but who had universal appeal and people were willing to look beyond race."

I had run back to the *Tribune* from Obama's fete in order to assemble the story on his victory for the next day's paper. And I was surprised by the numbers as well, even if my roots in Chicago and its racial schisms were more recent. The final tally: Obama had a whopping 53 percent of the vote; Hynes had 24 percent; and Hull finished with about 10 percent. These numbers were beyond anything that Cauley, Giangreco or Axelrod had fathomed was possible. Giangreco initially figured that if Obama pulled in 80 percent of blacks and half of college-educated liberals—which was the initial goal—he could win 35 percent of Democratic voters and prevail in such a crowded field. Giangreco's analysis of voting patterns had Obama's high-end threshold in the upper 30s or, maybe, if all went perfectly, in the low 40s. Cracking 50 percent was beyond hope. But in the end, Obama raked in more than 95 percent of the black vote and even won some city wards with a heavy percentage of non-college-educated whites.

As I sat at my desk to write the "Obama Wins" story, I looked through the various feeds arriving from reporters in the field who were interviewing voters at polling stations. There was one theme throughout: Barack Obama was a "breath of fresh air," in the words of one voter in a white Northwest Side ward that would have seemed safely in Hynes's pocket. But Obama displayed a unique quality that she could not quite put her finger on. In the end, she summed it up by saying: "He knows his stuff. I'm tired of the machine."

This was the kind of Obama voter that shocked pollsters and strategists. Obama destroyed the stereotypical appeal of a black candidate. Illinois surely had a history of electing blacks to statewide offices, but usually this was done purely on the strength of liberals, college-age voters and African Americans themselves. A postprimary analysis by Blumenthal showed that even when he subtracted college-educated whites from his sample, Obama still garnered nearly 30 percent of the rest of the white vote, which was extremely unusual for a black candidate. He decisively won all the heavily white Chicago collar counties and even captured nearly one in four Democratic votes outside the Chicago region, though he barely campaigned there.

Those who were inclined toward Obama in the first place were swept away in the heady moment. They talked about Obama representing a new generation of politician whose language was not shrill and who eloquently conveyed a feeling of hope and honesty. "Truthfully, it feels like a movement," said Leslie Corbett, who had met Obama the year before at a meeting of the National Poverty Law Center and instantly became a supporter. "I think for people of our generation, we haven't been a part of something like this before." Twenty-six-year-old Deborah Landis of Chicago said she first saw Obama a year before and was overwhelmed by his aura. "When I first met him, I registered to vote that evening just so I could vote for him," Landis said.

Back at his victory party, Obama ambled onstage surrounded by his family—Michelle, Sasha and Malia; his brother-in-law, Craig Robinson; and his half sister Maya Soetoro-Ng, from Hawaii. A proud-looking Reverend Jesse Jackson, who had endorsed Obama, proclaimed: "Tonight, surely Dr. King and the martyrs smiled upon us." Obama himself told the crowd that they, not he, were responsible for this victory. He reiterated the Democratic ideals that he had cited throughout the campaign, saying that he was on a mission to enact broad social change to better the condition of society's most vulnerable citizens. "At its best, the idea of this party has been that we are going to expand opportunity and include people that have not been included, that we are going to give voice to the voiceless, and power to the powerless, and embrace people from the outside and bring them inside, and give them a piece of the American dream," Obama said.

Throughout his speech, a chant rose from the jubilant and attentive crowd, and Obama used the chant to launch a call-and-response. The words of this chant could be found on the huge white-and-blue banners hanging throughout the hotel ballroom: "Yes, we can! Yes, we can!"

CHAPTER

A Dash to the Center

*I don't think you are going to see me tacking to the center,
because I never feel like I left what I consider to be the
mainstream of American thinking. . . .*
—BARACK OBAMA

Whether he wanted to admit it or not, Barack Obama and his "Yes,
we can" campaign would soon be headed straight into the belly of
the Washington establishment—and that establishment would embrace him. Because of Obama's race and David Axelrod's promotion
of his client as a star in the making, Obama's general election contest
would soon gain national attention. The Republican incumbent,
Peter Fitzgerald, decided against seeking reelection, and GOP voters
in Illinois nominated a new youthful face of its own to run against
Obama—Jack Ryan, who was Hollywood handsome and independently wealthy. In a state trending so heavily Democratic, early polling showed Obama with a definite advantage in the race, but this
would be a matchup between two attractive, highly telegenic candidates anchored in distant political ideologies. Ryan was a fervent
capitalist; Obama was a fervent big-government liberal.

Word of Obama's rising star was now extending beyond Illinois,
spreading especially fast through influential Washington political
circles like blue-chip law firms, party insiders, lobbying houses.
They were all hearing about this rare, exciting, charismatic, up-
and-coming African-American Democrat who unbelievably could
win votes across color lines. *The New Yorker* sent a writer to Illinois

to do an extended profile of Obama, a piece that was largely laudatory and served to introduce him in more depth to the liberal
cognoscenti across the country. *The New Republic* put him on the
cover and ran a piece that dissected the racial reasons behind his
ascendance and the ramifications of it. These were major literary
stamps of approval, and a sign of things to come.

Obama took advantage of this groundswell of modest political
celebrity and within weeks of his primary victory was on his way
to Washington to raise campaign cash. Axelrod, Jim Cauley and
Obama's influential Chicago supporters and fund-raisers all vigorously worked their D.C. contacts to help Obama make the rounds
with the Democrats' set of power brokers. Even though Obama had
spent that disappointing weekend at the Congressional Black Caucus in 2002, this would be his grand introduction to the major players inside the Beltway. He spent a couple of days and nights shaking
hands, making small talk and delivering speeches before liberal
groups, national union leaders, lobbyists, fund-raisers and well-
heeled money donors. In setting after setting, Obama's Harvard
Law résumé and his reasonable tone impressed this elite crowd.
"Barack was nervous a couple times, but he wowed them," Cauley
said. Obama gained the attention of liberal billionaire George So-
ros, who hosted a fund-raiser for him in New York. Senator Hillary
Clinton opened her home in Washington to him.

As he had so often before, Obama sold his message to both liberals and centrists, as well as to some who tilted toward the right. His
message, after all, was both liberal and conservative. His policy positions were decidedly to the left, but he offered them in such a pas-
sive, two-pronged way that it made him sound almost conservative.
He talked at length about the importance of committed parents and
communities in raising children. But, depending on his audience, he
was liable to follow that with the responsibility of government to
assist parents and communities struggling to stay committed. In the
era of George Bush's running up huge federal deficits, Obama advo-

cated fiscal restraint, calling for pay-as-you-go government. He extolled the merits of free trade and charter schools, but he also pushed for tax incentives to keep businesses from moving abroad and for more money for ailing school systems to help the less fortunate. He waxed on about the power of the free market to create wealth and change lives. But he also had an afterthought on a market-based economy straight from liberal economist Paul Krugman: "Sometimes markets fail, and that's when labor laws and government regulation are necessary correctives." In other words, he was saying that capitalism is magnificent, but it does have its drawbacks. It would be hard for anyone to argue with such a balanced statement. "Obama figures out ways to present himself like a conservative to conservatives," said David Wilhelm, a former campaign manager for Bill Clinton who informally advised Obama in his Senate race. "He has the whole venture capital industry here in Chicago, nothing but Republicans, thinking he is their champion. He has supported entrepreneurship. It is a pro-growth message and he is brilliant at delivering it." Indeed, when Vernon Jordan, the ultimate Washington and corporate player, who was a close adviser to Bill Clinton, hosted a fund-raiser for Obama at his home, Obama had securely moved beyond being an obscure good-government reformer to being a candidate more than palatable to the moneyed and political establishment.

This moderate manner was in direct contrast to some of the language Obama used in the primary, where he often sounded like a fiery liberal. He would give eloquent speeches, but among crowds of angry, out-of-power Democrats, he would always be sure to toss them a juicy applause line. That exhortation typically involved excoriating the Bush administration, which, of course, was the bane of nearly all Democrats in that election cycle. This could be frustrating to Obama. Crowds listened attentively to his professorial prose, responding to his thoughtful oratory with polite applause. But "all I have to say is George Bush is a bad person and they all go wild," Obama said with a shrug. He seemed genuinely distressed at this pattern, probably because bashing the other side for a jolt of audience electricity or a media sound bite is not in his true nature. He would much rather listen to all sides of an issue and offer a constructive solution than fan the flames of partisanship. Yet in a primary election the masses salivate for red meat, and Obama was aware of that and complied.

Now that he was a general election candidate, however, his overtly left-leaning lines would be reserved only for tried-and-true Democratic audiences. He now took a more centered, softer approach—and he held firmly to that mild manner all the way into his presidential bid. This darling of Illinois liberals was now engaged in the timeless dance of politics. Once the party nomination was in hand, Obama was gingerly stepping toward the center—Bill Clinton famously called it the "vital center"—in an effort to court independent and swing voters in the fall general election. When I posed this shift-toward-the-middle scenario to Obama, he insisted that he would remain true to his core beliefs. "I think you will see consistency in my message from the primary through the general election," he told me. "I don't think you are going to see me tacking to the center, because I never feel like I left what I consider to be the mainstream of American thinking and the mainstream of Illinois views." However, when I wrote a story for the *Chicago Tribune* in late April 2004 that flatly stated that Obama was moderating his message and dashing toward the center, I heard nothing from him or his campaign disputing this assertion. My article delved into the unattractive political motivations behind this move, but in truth, the story most likely did more to alleviate concerns among moderates that Obama might be a liberal firebrand than it did to anger true believers on the left.

As an example of Obama's movement toward the middle, I mentioned a recent vote on a bill in the Illinois senate that allowed retired law enforcement officers to carry concealed weapons. If there was any issue on which Obama rarely deviated, it was gun control.

His district housed many economically depressed neighborhoods savaged by gangs and crime. In his answers to primary campaign questionnaires for the *Tribune,* he was the most strident candidate when it came to enforcing and expanding gun control laws. So this vote jumped out as inconsistent.

When I queried him about the vote in an interview in his campaign office, he grew defensive. I told him that I had taken a day-long class on firearms in which the instructor said that the people carrying guns who often concerned him were aging law enforcement officers. The firearms instructor worried that some in this group had an arrogance about how well they handled their deadly weapon because they had been carrying it for so long. And that arrogance sometimes bred a sloppiness that could cause a tragic accident. In addition, many failed to keep up with target-shooting and regular practice with their weapon because they were no longer assigned to the streets. When I ran this by Obama, his facial expression tightened. "Look," he said, "I didn't find that [vote] surprising. I mean, I am consistently on record and will continue to be on record as opposing concealed carry. This was a narrow exception in an exceptional circumstance where a retired police officer might find himself vulnerable as a consequence of the work he had previously done—and had been trained extensively in the proper use of firearms."

It wasn't until a few weeks later that another theory came forward about the uncharacteristic vote. Obama was battling with his GOP opponent to win the endorsement of the Fraternal Order of Police. Obama ultimately won the fight, and during the news conference on the endorsement, Obama stood beside police officers who noted that they had some qualms about his past legislative record. For example, Obama had voted against stiffer sentences for gang-related crimes, saying that these laws 'unfairly targeted minorities. But the union president told reporters that Obama stood with them on other issues, and he specifically cited the concealed carry vote for retired police officers. "It's impossible to tell you how important it is for a black Democrat from Chicago to get the FOP endorsement," an Obama aide told me. "Downstate, that endorsement can mean a lot. Obama might not be a big fan of guns, but he is a big fan of the FOP endorsing him."

NOT ONLY DID OBAMA TAKE HIS FIRST BABY STEPS INSIDE THE Beltway during this period, but the Washington system also arrived on Obama's doorstep in Illinois. With more resources at his disposal and a high-profile contest before him, he now attracted seasoned staff personnel—and quality résumés dropped on his desk like never before. He hired three thirty-something Washington-bred campaign veterans: Amanda Fuchs as his policy director, Darrel Thompson as his chief of staff and Robert Gibbs as his communications director. Fuchs had worked in organized labor and for various Democratic candidates, most notably as an issues and opposition researcher for Geraldine Ferraro in her ill-fated 1998 campaign for senator in New York. Thompson had been an aide for five years to then House minority leader Richard Gephardt. Gibbs had worked on more than a handful of Senate campaigns and had recently quit as chief spokesman for John Kerry's presidential campaign amid an internal turf war.

Of this trio of new aides, Gibbs quickly established himself as the most influential. Through hard work, personal charm and keen political instincts, Gibbs would evolve into Obama's most powerful staff adviser, projecting a voice among his inner circle eclipsed only by Axelrod. "Robert stepped into this incredible void—and he filled it up completely," said Valerie Jarrett, Obama's close friend and finance chairwoman. I first met Gibbs at a hastily organized campaign event in Chicago's Chinatown neighborhood, an event seemingly assembled at the last moment to give a CNN reporting team some on-the-stump footage for a profile on the Illinois race. For the past couple of months, I had been hanging on to Obama's coattails, shadowing him almost daily. This meant there was little buffer zone between him and the *Chicago Tribune* on a daily basis. It was fine to have me this close in the primary, when Obama was looking to spread his name and build relationships with campaign reporters. Now he was in a general election in which he was the favorite, and the strategy became less about introducing him to the populace and more about avoiding mistakes. Having a reporter at his side hour by hour only increased the chances that a verbal miscue would get a full airing in the media.

Even more than keeping my wary ears at bay, Obama needed more structure to his schedule in a general election campaign. An efficient daily schedule and structure were not among the strengths of his primary campaign, and this could have been Obama's fault. He did not care for being chained to a schedule, and he and his driver would sometimes slip off on their own, many times so Obama could squeeze in one of his daily exercise workouts. During a campaign trip in the primary to Metro East, the Illinois section of greater St. Louis, Obama himself had to direct his small caravan into a pizza parlor because lunch had not been built into the schedule. There, Obama pulled out twenty-five dollars from his wallet to pay for the pizza, and then collected five-dollar donations from his hungry entourage. Needless to say, the candidate should not be worrying about feeding his staff or the media. That role should be delegated. In addition, on one of the primary debate days, Dan Shomon scheduled Obama to speak at a prison in southern Illinois in the morning and Obama had to be whisked back to Chicago in the evening for the debate. (Shomon was still assisting with downstate campaigning.) Obama scurried into the television studio and dropped into his chair just minutes before airtime. And he was so fatigued that at one point in the debate he nodded off for a moment.

When I saw Gibbs and Fuchs trailing their new boss through Chinatown that morning, I vividly recalled Obama's anti-Washington words. I could only think: Good thing his campaign is not going

Washington! Obama introduced me to Gibbs by putting his hand on Gibbs's shoulder and casually pulling his new staff member to stand between us. "This is someone I'd like you to meet," Obama said. The symbolism could not have been more apparent. *Here comes the blocker between me and Obama.* Gone were the days of Obama directly calling my cell phone when he had an issue. And vice versa. I'd now be calling Gibbs, who would be calling Obama for a response that would probably be drafted by Gibbs.

In his mid-thirties, Gibbs had cut his teeth in a number of campaigns, dating back to his first political job in college when he interned with a congressman. Wearing fashionable thin-framed glasses below his receding reddish-blond hair, Gibbs at first glance appeared to be every bit the son of two librarians from Alabama. But his scholarly appearance belied an undergirding of competitive intensity. He exuded a southern charm that was immediately apparent, typically greeting professional acquaintances with a hearty smile and friendly squeeze of the arm. "What's neeeew?" he'd ask through a light south-of-Dixie accent. But in many ways, Gibbs was the anti-Obama, adding the tougher, rougher edges to Obama's softer, more tranquil demeanor. Gibbs was an indispensable aide for a politician with the lofty long-term ambitions that Obama harbored. He was Obama's hired gun, skillfully trained to shoot at reporters whose coverage was deemed unfair, as well as a cutthroat pragmatist who could brainstorm on message and tactical strategy. Gibbs was a ruthless political operative who relished personal confrontation as much as Obama fled from it. "Robert is a bully," said a former Obama aide. "Stuff landed on his desk that should never be on the desk of a communications director. But nine out of ten times, his gut instincts are right. That tenth time could be ugly, though." Cunningly smart, Gibbs understood the importance of a pithy sound bite and he thrived on manipulating reporters to the benefit of his candidate. But Obama and Gibbs did have two things in common: raw ambition and a burning competitive nature. A former college soccer player at North Carolina State University, Gibbs was a sports enthusiast with a particular fondness for fantasy sports leagues, the kind of guy who treated late-evening video golf games in a bar along the campaign trail as if they were life-and-death sport. "Plain and simple, Robert wants to be the communications director of the White House," Cauley told me.

Gibbs was one of the few aides in Obama's orbit who was fearless when it came to pushing back on the boss. When Obama's oratory meandered toward the wonkish or his news conference answers drifted off message, Gibbs was not shy about schooling his boss in the vital importance of verbal restraint. After Obama had been elected to the Senate, he approached Gibbs in his Senate office one day and asked, "Gibbs, who is the president of Tanzania?" Many aides would wilt at this question, mostly because they would not know the answer. Gibbs's response: "Who the fuck cares?" That answer got a laugh from Obama. Gibbs was fond of recounting his worst day in politics, a now infamous moment in the presidential campaign of John Kerry. While campaigning in Philadelphia for the Democratic nomination, Kerry had been handed a Philly cheesesteak sandwich. He asked if he could have Swiss cheese on it instead of the normal Cheese Whiz coating. Altering the ingredients of a traditional blue-collar hometown delicacy fed into the very blue blood, elitist image of Kerry that turned off voters, and Gibbs spent the rest of the day working the phones to temper the media fallout. When Kerry campaigned through a state fair, Gibbs yelled at traveling aides on their cell phones: "Get a fucking hot dog in his hand—now!"

Gibbs sold himself to Obama during a fairly brief interview in a bland conference room in Washington. Gibbs said what struck him about Obama in that first meeting was "Barack's total ease with himself"—which is the first thing almost everyone notices about him. What helped to sell Obama on Gibbs was Gibbs's experience in the campaign of Ron Kirk, an African American who made a valiant but unsuccessful Senate run in Texas. Obama assumed he would have little difficulty in the fall election with urban Chicago voters, and Gibbs seemed to understand the potential perils of a black man trying to win votes in conservative southern rural areas. "I told Barack there was no way Texas was going to elect a black Democrat like Kirk," Gibbs said. "And that experience was unique, because the focus in that campaign was largely on race."

KEEPING ME AT ARM'S LENGTH WAS NOT A DIFFICULT TRANSITION for Obama. It was not as if he had ever delighted in my presence. We had pleasant off-the-record conversations along the campaign trail about common interests such as raising young children, sports and jazz music, passions for both of us. But I always sensed that he would much rather be on his own. Perhaps realizing that his personal freedom would eventually fall victim to celebrity, he fought to hold on to it as long as possible. In fact, the first time I trailed him for a full day, back in January, I could feel him chafing at my proximity. Until then, Obama had been driving himself to campaign events. Using me as an excuse, Cauley finally forced him to be transported by a driver in a leased SUV like the typical candidate for high office like the Senate. "You know, it would slow me down looking for parking all the time," Obama acknowledged. But it wasn't as if we had a big entourage—there were just the three of us. Stopping that first afternoon at a suburban forum with the other Democratic candidates, Obama stepped from the jet-black SUV and spied Hull's massive RV, the "Hull on Wheels." Hull walked from the vehicle surrounded by at least half a dozen people. "I don't know how he handles that," Obama said about Hull and his flock of supporters. "I'm just not an entourage kind of guy. I'm more of a solo act."

But if Obama was to be a senator, he needed to get accustomed to traveling in entourages and to have reporters trailing after him. His friend Valerie Jarrett, who holds degrees in both law and psy-

chology, constantly psychoanalyzed Obama. She and Michelle
would endlessly discuss what they believed made him tick, each
adding an observation that would paint another layer on their overall portrait. In an interview with Jarrett, I asked her if she thought
his father's abandonment of Obama as a child contributed to his
desire to seek public attention. "Absolutely," she said. "Having a
parent who leaves you makes you particularly energized for approval. I think that's a real part of it. Rejection is a tough thing for
a kid to accept. That's a hard thing and you spend your life trying to
get approval." But often, getting that attention seems stifling to
him, I said to her. He feeds on the energy of a crowd connecting
with his words, but he can lose patience with hangers-on or people
who he perceives are invading his personal space. She agreed, and
said this was perhaps the most frustrating aspect of his personality.
"He is so complicated," Jarrett said. "He has this whole mercurial
side to him. It's like, no one made you run for U.S. senator. So stop
complaining about things."

This interview occurred months later. Earlier in the campaign,
Obama's occasional unease with my presence baffled me. Editorially,
the *Tribune* was an ardent supporter of his candidacy, and he told me
on various occasions that he believed my coverage of him had been
fair. So what was the problem? This came to a head in the final
weekend of the campaign during a scheduled fly-around of the state.
Arriving at Midway Airport after a couple of morning events in
Chicago, Obama shook my hand and said, "Nice having you around,
David. See you later." I was perplexed. An aide had assured me that
I would be on the charter flight. Didn't Obama realize this? I rushed
over to Peter Coffey, one of his scheduling aides, and asked what was
going on. My job, I explained, was to cover Obama every step of the
way. Coffey conferred with Obama, and after a few minutes Obama
shook his head and shrugged. Coffey walked over and told me that
I was free to go along. As we nestled into our seats on the plane,
Obama asked somewhat caustically, "Haven't you had enough of this

yet?" I explained that just as his job was to campaign, my job was to watch him campaign. "Okay," he said, nodding. He now understood this reality but was not happy about it. He tried his best to hide it behind his normally gracious exterior, but the mercurial nature of Obama had seeped out.

It wasn't until a few weeks later that I learned there was another reason besides personal privacy why Obama had been so resistant to my presence: Obama was a secret smoker—and he did not want to light up a Marlboro in front of a reporter. Some politicians are comfortable smoking in front of the media or in public, while others believe the habit will reflect poorly on their public image. Obama was in the latter group, almost to an obsessive degree. The public portrait of Obama now bordered on saintly, especially for a politician. Learning that he smoked might tarnish this picture. So Obama went to great lengths to conceal the habit. "He was fine with you personally," Coffey told me weeks later. "But he wanted to have that cigarette, and he either couldn't have it when you were with him or he had to sneak it."

It really came as no surprise to me that Obama smoked. His wife mentioned in our interview that Obama had a cigarette dangling from his lips on their first lunch together. He had written in *Dreams from My Father* about smoking in the college dorms. But most telling, like most smokers, he occasionally smelled of tobacco. One morning, cigarette smoke still hung in the air of the campaign SUV as I boarded. When I asked his driver if he smoked, the driver replied that he did not, and so I surmised that Obama still did. But what was I to do with this information? At the time, it was pretty insignificant. I figured I would mention it in the lengthy personality profile I would be writing about him in the general election. Obama's sense of abandonment as a child made him seek universal affection as an adult, but he was slowly learning that being in the public eye has a sharp downside.

Maybe the sharpest downside involved the pitfalls of celebrity.

Obama would now consistently get autograph requests in public rather than just at political events. And his "IT" factor became something of a problem with his marital life. Michelle Obama had been accustomed to women finding her husband attractive, but she was always confident in his fidelity. In social settings, Dan Shomon, who was divorced, said it could be difficult being a friend of Obama's because Obama would swallow up all the female attention, albeit quite unintentionally. "You really didn't want him around anyone you wanted to date. He was the worst wingman in the world," Shomon said, referring to the male role of helping your friend attain female companionship. "All the women would fall in love with him." That equation would only expand to the general populace in Illinois now that he was gaining ever more recognition. Michelle, knowing of Obama's devotion to her, could usually brush aside any fawning that women might do over her husband. But his growing status as a local celebrity was making her start to question things. A friend mentioned to her that she overheard two women at the health club discussing Obama. "Let's go down and watch Barack Obama work out," one woman excitedly said to the other. With women going out of their way just to watch your husband run on a treadmill, it would be hard not to feel some discomfort. I talked to Jarrett about this development and how Michelle was coping. Jarrett was blunt. "He knows that if he messes up, she'll leave him. You know, she'll kill him first—and then she'll leave him," Jarrett said with a laugh. "And I think there is a subtle element of fear on his part, which is good."

The Ryan Files

I don't think I am going to have a hard time convincing anyone that Barack is the mainstream candidate in this race.

—ROBERT GIBBS, AFTER OBAMA'S OPPONENT WAS
CAUGHT UP IN A SEX SCANDAL

Barack Obama's campaign caught a huge break in the primary with the sordid divorce files of Blair Hull taking center stage. But the real break was the improbable entrance of Hull into the race in the first place. Hull had a brilliant mathematical mind that helped him build a mountain of wealth on Wall Street; but if he'd had a keener mind for human nature, he would probably never have entered politics and suffered such public disgrace. David Axelrod had warned him privately about the potentially devastating effect of his divorce files, but Hull went ahead anyway. Without Hull, it is impossible to know how events would have played out. Would Axelrod have signed with Obama? Would Dan Hynes have run a better campaign and raised more money without having to constantly worry about Hull? Would Obama have raised and spent a couple of million dollars for a big TV splash without the Millionaire's Amendment that Hull triggered?

Hull's influence did not stop at the end of the primary. And for Obama, Hull was the political gift that just kept on giving. Even though Hull had now disappeared from the race—and largely from Chicago politics—his divorce mess created aftershocks on the Republican side that would tilt the general election heavily in Obama's favor.

Jack Ryan, Obama's GOP opponent, had had portions of his divorce files sealed, just as Hull had. And when the *Chicago Tribune* and WLS-TV decided it was in the public interest to sue for the opening of Hull's file, they looked around at all the other candidates in the Senate race and discovered that Ryan had sealed a portion of the public records in his split from his ex-wife, Hollywood actress Jeri Ryan. So in the interest of fairness, the news outlets sued to open the entirety of Ryan's file as well. Ryan's lawyers had managed to keep the records private through the primary, but they had no compelling legal case to keep them closed forever. Ryan maintained that the files contained delicate information involving his nine-year-old son and held nothing embarrassing to him personally. He simply wanted to keep the files closed to protect his son, he said. Republican and political insiders were leery about this defense. Ryan had been married to an attractive TV actress, and Ryan himself looked as if he had been ordered from central casting. He was tall, lean, square-jawed, Ivy League–educated and well-spoken. After becoming rich in investment banking, he spent a few years teaching in a private high school in Chicago's inner city and articulated a Jack Kemp–esque, pro-business brand of compassionate conservatism. In some ways, he looked more mainstream than the liberal Obama from the South Side, and his campaign strategy was to paint Obama as an extreme leftist. A Republican in the Illinois legislature who had cosponsored bills with Obama dubbed Obama "to the left of Mao Tse-tung." In an interview with me, Kirk Dillard, another Republican in the Illinois senate who was close to Obama, awkwardly walked a tightrope between his loyalties to the GOP and his friendship with Obama. Dillard complained that Obama was "soft on crime and borderline socialistic," but ended the interview by emphatically saying that Obama was "truly a wonderful human being." As for Ryan, the rumor mills in Hollywood and Illinois political circles swirled about what caused the divorce between him and Jeri Ryan. Those rumors intensified in Illinois after Ryan confided to some Republicans that his sexual predilections could be a bit outside the norm. Illinois GOP chairwoman Judy Baar Topinka questioned Ryan sternly about the divorce records, and he assured her that they contained nothing that could derail his candidacy.

Besides being haunted by the closed divorce files, Ryan's general election campaign got off to an inauspicious start in another way. Ryan had sent a field tracker with a video camera to follow Obama in the hope of catching him in some sort of public gaffe. With the advent of the video-posting website YouTube.com, this strategy has become commonplace in campaigns today. A candidate can utter something unfortunate in the morning, and the video is up on the Internet that afternoon. But there is a certain physical space that a tracker generally affords his subject, and Ryan's young staffer stepped over that line. Instead of simply taking video of Obama speaking publicly, the man decided to follow Obama everywhere he went, down hallways to the bathroom and along sidewalks to his vehicle— all with a video camera trained on the subject. Obama was less than thrilled with a reporter from the mainstream media covering him every day, so one can imagine how this overly aggressive young Republican went over with him. The issue came to a head in Springfield during a senate session when Obama's patience ran out. With the tracker in tow, Obama walked into the press room in the state capitol and announced to a roomful of astonished journalists, "Meet my stalker." The man never stopped recording, and the next day's papers were filled with stories of this "stalker" from the Ryan campaign who wouldn't give Obama any room. Whether it was calculated or visceral, Obama's move proved ingenious. Public opinion fell heavily against Ryan. Hounded by questions, a Ryan spokesman finally succumbed and issued an apology to Obama. The tracker was withdrawn, but the story had a certain unique and easy-to-understand quality that drew the public's attention for several days. So the first major news story of the general election campaign was a blow to Ryan.

Assuming that he had urban Chicago well in hand, Obama turned his attention to wooing voters in moderate and Republican territories. In mid-June, he headed downstate for a campaign swing. While Obama was on this tour, Ryan's lawyers finally ran out of appeals and a California judge ordered the divorce records released. Ryan and his staff spent the day trying to figure out the best method of releasing them, a decision they ultimately bungled as badly as they did the tracker. Ryan called reporters to a news conference at six o'clock in the evening to see the files. But someone must have realized that this would allow television stations to go live on the evening news with the records release. So the media was kept waiting until after eight o'clock. This only set up an even bigger feeding frenzy, with news outlets now having hungered for a couple of hours to see details of the records.

The media had good reason to salivate. The files were even more voyeuristic than the Hull records. They showed that Jeri Ryan, who had gained fame primarily through her role in TV's *Star Trek: Voyager,* had accused her husband of taking her against her will to public sex clubs in New York and Paris, where he tried to coerce her into having sex in front of strangers. At the club in Paris, the file said, his wife alleged that "people were having sex everywhere. I cried. I was physically ill. Respondent became very upset with me and said it was not a 'turn-on' for me to cry. I could not get over the incident and my loss of any attraction to him as a result. Respondent knew this was a serious problem. I told him I did not know if we could work it out." Jeri Ryan described one of the New York spots as "a bizarre club with cages, whips and other apparatus hanging from the ceiling. . . . Respondent wanted me to have sex with him there with another couple watching. I refused. Respondent asked me to perform a sexual activity upon him and he specifically asked other people to watch. I was very upset."

As these embarrassing details appeared in full view, Obama made sure to be as far away from Chicago as possible. The strategy was to just let the Ryan saga play out all on its own. In the midst of a hectic two-day downstate campaign swing, Obama was giving a speech that evening at a fifty-dollar-a-plate fund-raiser at Southern Illinois University in far southern Carbondale. The first Associated Press story about the Ryan files flashed across Robert Gibbs's Blackberry mobile device about nine o'clock. Reading the story, one of Obama's young aides couldn't help but laugh at his own candidate's good fortune. "Awesome," he said, breaking into a big grin. When Obama finished speaking, he was swallowed up by autograph seekers and well-wishers. Gibbs waded in and pulled Obama to the side to relay the breaking news. Obama listened attentively and made sure that his body language gave away nothing. A few minutes later, I asked Obama for a response. His consultation with Gibbs had elicited this: "I've tried to make it clear throughout the campaign that my focus is on what I can do to help the families of Illinois and I'm not considering this something appropriate for me to comment on."

Outside, walking up to the campaign caravan, Obama flipped on his cell phone to find it filled with messages from reporters across the state seeking his comment. At campaign appearances all day long, he had declined to discuss Ryan's potential problems, saying that whatever was in Ryan's divorce files was not his concern. Gibbs advised Obama not to return any of the media calls, since the AP had already picked up these no-comments from throughout the day. "I wanna call back [Dave] McKinney with the *Sun-Times*," Obama said earnestly. "He's a nice guy and treated me well when I was a nobody." Gibbs gave him the go-ahead but reiterated that he was not to comment on the substance of the Ryan problems. As we stood around in the parking lot readying to leave, Obama shook his head at this bizarre turn of events. "This sounds worse than the Hull stuff, at least more embarrassing," he said. "I mean, what was Ryan thinking to get in this race?"

For his part, Jack Ryan denied the allegations, which came as the couple battled for custody over their son in 2000. But the political fallout was intense. The *Tribune* blared the story across its front page the next morning with the headline "Ryan File a Bombshell: Ex-wife Alleges GOP Candidate Took Her to Sex Clubs." Chicago television stations fed on the story for days. A Republican congressman from Illinois, Ray LaHood, asked Ryan to withdraw from the race. Gibbs privately joked, "I don't think I am going to have a hard time convincing anyone that Barack is the mainstream candidate in this race." Even late-night comedians got in their shots. Jay Leno cracked, "Jack Ryan, I've heard of going after the 'swing vote,' but this is ridiculous!"

As for Obama, his campaign caravan roared through Illinois again the next day, which he spent carefully keeping himself out of the Ryan story. Incredibly, the Illinois race had again turned into the dissection of a man's sordid marriage. Reporters tossed question after question at Obama about the unfolding Ryan saga, and time and again Obama did not take the bait, dutifully projecting a sense of morality in the process. In Peoria, he told a gaggle of reporters: "I don't take any pleasure from watching this kind of circus atmosphere. I want the outcome to be this: that the voters of Illinois have sent me to Washington to improve their job prospects, to improve the prospects of their kids going to college, to make sure that health care is affordable." As Obama endeavored to remain above the muck, some reporters waded right in. After Obama broke from the Peoria media gaggle in which he declined over and over to discuss the Ryan matter, a radio reporter ran after him through the parking lot and followed him to his awaiting SUV, yelling at full throat to Obama: "Do you think a sexual fetish defines a person's character?! Senator, do you think a sexual fetish defines a person's character?! Well, Senator, Senator?!" Despite the fact that the woman was thrusting her microphone over his shoulder and into his face, Obama miraculously kept his composure and kept walking.

The long two-day campaign trip through Illinois ended that evening with a fund-raiser at the Children's Museum on Chicago's

Navy Pier, an event attended by a score of Obama's longtime Chicago supporters. With the Chicago media waiting to ask Obama about the Ryan matter, David Axelrod stopped by to make sure his candidate did not fumble the ball during this tumult in his opponent's campaign.

But at Navy Pier, a fatigued Obama came closest to his breaking point. The night before, the campaign caravan had not arrived at the hotel until nearly two o'clock in the morning, and the first event was just seven hours later at nine o'clock. It had been a tiring couple of days, and when Obama addressed the Chicago press, he strayed from his no-comment theme a couple of times, once going so far as to say that "all of us have to take responsibility for ourselves and how we conduct ourselves in our campaigns." This unwise statement sent Axelrod into his mode of worried pacing. Obama then plunged back into his response, that voters ultimately would decide this election on the issues that mattered to them. But asked if he worried that a tougher opponent like former Illinois governor Jim Edgar might emerge if Ryan were to drop from the race, Obama was surprisingly candid about his upcoming campaign strategy. "When I started twenty months ago, everybody looked like a formidable opponent," he said. "And when people suggested to me that we shouldn't bother, because people with more money or higher name recognition, a higher profile or more organization were in the race, then we would have folded up our tent back in February. I think what we have shown in this campaign and what we will continue to show is, if we stay on message and talk about the issues that matter to voters in Illinois, that we are going to do just fine."

Axelrod paced faster and faster. Did his candidate really just say "stay on message"? Indeed, when Obama reiterated several more times this plan to stay "on message," Axelrod halted midstride, winced and shot a glance downward, as if somebody had punched him hard in the gut. His star client was juggling the ball all over the place. After all, candidates are supposed to stay on message, but they aren't supposed to announce to the world that they are staying on message. Finally, after getting lost in another answer, a weary Obama realized he had drifted into no-man's-land and abruptly ended the press conference himself. Gibbs saw my surprised reaction and sauntered over. "He's tired," Gibbs explained. "It's been a long couple of days." It certainly had been an exhausting campaign trip, and Obama had to be close to brain-dead after two straight days of talking, posing and politicking. But mental and physical stamina are vital components of a successful federal candidate, and fatigue would be a consistent part of Obama's life as his career sped forward. In his presidential run in May 2007, a sleep-deprived Obama would accidentally say that ten thousand people died in a tornado in Kansas when the actual number of fatalities was just twelve. That launched a series of news stories questioning his discipline and stamina. Few humans are as disciplined as Obama, but his stamina can be questionable. In any event, the poor performance at Navy Pier revealed not only that he was susceptible to fatigue, but that he was still learning the game. There would be many more long, tedious campaign swings in his future, and Obama would have to carry himself at a level higher than this if he were to realize his own ambitious dreams.

EVEN WITH THE PUBLIC JOKES AND PUBLIC EMBARRASSMENT, JACK Ryan's team insisted that he would remain in the race. But that was close to delusional. Ryan had lost the confidence of not only many Republican voters, but the GOP establishment in Illinois. Most important, Topinka, the state Republican Party chairwoman, felt betrayed because Ryan had assured her that the files contained nothing unflattering. Within weeks, as pressure built from top Republicans in Illinois, Ryan ended his candidacy, leaving the reeling GOP lost about what to do next.

Several months later, in an interview with the college newspaper at Dartmouth, his alma mater, Ryan explained that he believed the divorce files would never become public and he cast himself as a victim of an overzealous media. He made a vigorous defense that the records should have been kept sealed. He noted that Democratic presidential candidate John Kerry had sealed divorce files that no media venture had sued to open. "There's no other example of someone having to turn over their sealed divorce records. So this is obviously a higher standard than anyone else in the history of the United States has been held to," Ryan told the paper. "But I don't know whether it's because of my background they held me to a higher standard or not; I just don't know. I really don't know what motivated the *Chicago Tribune* to do that. They said it was a matter of principle. But if it were, it wouldn't have only applied to me."

As Illinois Republicans waded into the wilderness without a candidate, Obama was headed the other direction—straight into orbit. His campaign fund-raising was now moving at a furious pace and the buzz around him as a rising star was increasing by the hour. And there was more at work: Axelrod, Cauley, Gibbs and Obama's fast-growing network of influential fans were lobbying high party officials to win their candidate a key speaking slot at the Democratic National Convention in Boston the next month. The party was looking for a fresh new minority voice—and Obama fit the role perfectly. The main thrust of the Obama team's effort was to get him "in the window." That meant solidifying a speaking slot in the evening when the national TV networks would be carrying the convention live. This kind of widespread exposure could raise Obama's profile nationally and shift his fund-raising into fifth gear.

Kerry, now the presumptive Democratic presidential nominee, had met Obama twice at political events in Chicago, and Obama had impressed him each time. In particular, a joint appearance on Chicago's Near West Side at a small industrial plant gave Kerry a dose of the smooth grace of Obama's public style. Surrounded by employees of the company, the two men sat on stools next to each other and answered questions from the workers about health care, free trade and other matters. Obama wore a tie and a light blue dress shirt similar to the one he'd worn that day I met him in his constitutional law classroom, and he began answering his first question while rolling up his sleeves in the same cool, methodical, look-at-me manner. He loosely held a cordless microphone and walked about the room answering questions with complete ease. Kerry, by contrast, appeared stiff, uneasy and his answers sounded prescripted compared with Obama's engaging, conversational tone. The presidential candidate, in a reversal of protocol, looked to be picking up cues from the Senate candidate. "Barack just killed at that event," observed Nate Tamarin, Obama's deputy campaign manager. "You could see Kerry was looking at him with some awe."

Ultimately, Kerry and Democratic National Committee officials offered Obama the Tuesday-night keynote address to the convention. At first glance, this appeared to be a big deal. Obama, after all, was still a state senator from Illinois. His election as the third black U.S. Senator since Reconstruction looked very promising, considering he was without an opponent and Illinois Republicans seemed lost. Yet past keynote speeches had been delivered by the likes of Barbara Jordan, Mario Cuomo, Jesse Jackson and Bill Clinton. Cuomo was a speaker whom Obama had long admired. While Obama's advisers were honored to follow in this history, they were not pleased overall, because the major networks were not carrying that night's events live. "We wanted Monday night between nine and eleven and we were not thrilled with our time slot in the least," Cauley said. "We felt that if they really wanted to show us off, they would have put us in the window. At the end of the day, I think Barack fit the flower-power picture they wanted to paint. And as it has all turned out, we all look like geniuses. But back then, we were totally pissed."

Being denied the window, however, provided Obama with leverage to negotiate control over the content of his speech. Former congressman Harold Ford of Tennessee had delivered the 2000 keynote speech, and most reviewers called it flat. Like Obama, Ford was a young, charismatic, up-and-coming mixed race (and self-identified) African American. Obama had heard that the writing of Ford's speech had largely been co-opted by the DNC, and Obama feared that same result—he wanted to give a speech from his own heart. Obama was also surprisingly ineffective at delivering a speech written by another author. He simply didn't read well from a prepared script, sometimes sounding wooden and bored with his material. He spoke with much more authenticity and clarity when the words flowed directly from his own pen or straight from his thoughts at the moment. For this reason, Obama's advisers left it up to him to write the first draft of the speech. Gibbs forwarded along past keynote addresses from Ann Richards, Cuomo, Jordan and others.

But the speech itself came rather quickly to Obama. He wrote it longhand in hotel rooms along the campaign trail and in his hotel room in Springfield while the General Assembly was in session. Then he transferred it to computer at home in Hyde Park. Just as when he penned *Dreams from My Father,* most of his writing was done between nine o'clock in the evening and one o'clock in the morning. Obama is a night owl, and this has long been his prized time alone, to catch some sports highlights on TV, to return e-mails from friends, to write, to think. Having grown up largely as an only child, Obama has an introverted side that makes him savor this time to himself. And on the busy campaign trail, these moments were fewer and fewer. In their Hyde Park town house, his wife called his writing and thinking spot "The Hole," because her husband would disappear into a small room off the kitchen after they put their children to bed at night. The Hole was about the size of two typical workplace cubicles, down a long corridor away from the bedrooms and family room. A dark wooden desk sat in the left corner, with a dingy white chair and ottoman beside it. When Obama showed the room to me, it was rather tidy. Having heard of his sloppy personal habits at home, I asked, "Where's the clutter?" He replied, "Well, just imagine it with piles of papers and books everywhere."

The words of the speech came fast. "This was not laborious, writing this speech," Obama said. "It came out fairly easy. I had been thinking about these things for two years at that point. I had the opportunity to reflect on what had moved me the most during the course of the campaign and to distill those things. It was more a distillation process than it was a composition process." Obama e-mailed early drafts of the speech to Gibbs, Axelrod, Giangreco and others during the wee hours, with one draft arriving about four in the morning. By nine o'clock that morning, he would call and ask what they thought. Several hadn't even seen the new draft yet.

This was the moment Obama had been waiting for—his opportunity to convey his message of hope and unity to the greater society.

The Speech

We must make the American people hear our tale of two cities. We must convince them that we can have one city, indivisible, shining for all its people.

—MARIO CUOMO IN HIS 1984 KEYNOTE ADDRESS TO THE
DEMOCRATIC NATIONAL CONVENTION

We've been told that the interests of the South and the Southwest are not the same interests as the North and the Northeast. They pit one group against the other. They've divided this country, and in our isolation we think government isn't gonna help us, and we're alone in our feelings. We feel forgotten. Well, the fact is that we are not an isolated piece of their puzzle. We are one nation. We are the United States of America.

—ANN RICHARDS IN HER 1988 KEYNOTE ADDRESS
TO THE CONVENTION

The pundits like to slice-and-dice our country into Red States and Blue States; Red States for Republicans, Blue States for Democrats. But I've got news for them, too. We worship an awesome God in the Blue States, and we don't like federal agents poking around in our libraries in the Red States. We coach Little League in the Blue States and yes, we've got some gay friends in the Red States. There are patriots who opposed the war in Iraq

and there are patriots who supported the war in Iraq. We are
one people, all of us pledging allegiance to the Stars and Stripes,
all of us defending the United States *of America.*

—BARACK OBAMA IN HIS 2004 KEYNOTE ADDRESS
TO THE CONVENTION

Barack Obama's advisers might not have been successful at getting
their candidate into the network television lineup at the Democratic
National Convention in July 2004, but winning him the keynote
address placed historical importance on Obama's speech. Mostly, it
alerted the media elites that he was someone the party wanted to
showcase for the future; the speech would be broadcast nationally
on various cable networks. So when Obama arrived in Boston that
week, he was greeted as a celebrity-in-the-making. Political pundits and national journalists pontificated that by week's end Obama
would either be a major star or a major dud.

The convention was held in Boston's FleetCenter arena, with fifteen thousand media personnel credentialed to cover the events of
the week. Political conventions used to be where the presidential
candidates waged their final battles with party bosses for the nomination. But over the past generation, those battles were shifted to the
electorate in key primary states. Thus, the convention had evolved
primarily into a week of pageantry in which the party sells its candidates and its ideas to the American electorate, largely through the
prism of those fifteen thousand media members. Senator John Kerry
of Massachusetts had won the nomination, and his convention theme
expanded on the former Vietnam War veteran's pledge to make
America "stronger at home and more respected abroad."

The specifics of Obama's attendance at the convention were
fraught with uncertainty from the beginning. Illinois legislative
leaders and Governor Rod Blagojevich were locked in an intense

budget battle that threatened to spill the senate session into convention week and force Obama to fly back and forth between Boston and Springfield. But after a late-night session in the legislature, the budget was finally resolved, and Obama arrived at his Boston hotel after midnight on Sunday morning. Running on sheer adrenaline, Obama and his aides found themselves bumping into each other in the hotel overnight as they walked about, too hyped to sleep and trying to work off nervous energy. Obama's frenetic convention week was primed to start off with a bang—he was scheduled to appear on NBC's highly influential *Meet the Press* on Sunday morning, just a few hours later. David Axelrod, Robert Gibbs and other advisers had spent more time prepping him for this appearance than for his keynote speech. Tim Russert, the show's host, had become known as one of the sternest questioners in Washington. He had a dogged research staff and he was fond of pulling up unfortunate past quotations from his guests and making them squirm trying to explain the statements. Gibbs had negotiated with a Russert producer over what types of questions would be fair game in the interview. Gibbs stressed that Obama was still only a state senator and he tried to blunt any complex foreign policy questions that might trip up his boss, who at this point was more schooled on domestic and state issues than foreign affairs.

A few minutes before *Meet the Press,* I caught up with Obama and his team in a waiting room at the arena. Despite the huge week ahead of him, Obama emanated his typically calm, confident demeanor. Asked how he was feeling, he responded only, "Tired." Indeed, he had a now routinely fatigued look about him—heavy eyelids, puffy bags beneath his eyes. As he disappeared to do the show, Axelrod and Gibbs settled into two chairs near a television monitor. Russert first asked Obama what he hoped to achieve in his speech. Obama told Russert, "If we can project an optimistic vision that says we can be stronger at home, more respected abroad, and that John Kerry has the message and the strength to lead us in that fashion, then I think we'll be successful." Gibbs and Axelrod turned to each other and smiled. This was exactly the message that party officials had scripted for their candidates and speakers. Their man had hit his marks perfectly—and on the very first question. It wasn't long, however, before Obama found some nasty curve balls thrown his way. The upcoming issue of the *Atlantic Monthly* magazine, Russert said, quoted Obama as saying that sometimes Kerry lacked the necessary "oomph" as a candidate. "What does that mean?" Russert asked. At this, Axelrod got up from his chair and began pacing. The magazine had not yet hit the newsstands, and Obama and his team had no knowledge that he had been quoted disparaging the party nominee. But Obama handled the tough pitch magnificently. "Well," Obama said. "I think that, you know, early on in the campaign, and this was an interview that took place several months ago, you hadn't gotten a sense of John Kerry as the man, and I think this convention is going to be consolidating the impression that we've been getting over several months that this is somebody who's going to be fighting for working families, somebody who has the strength to lead internationally. This is somebody who has the life experience as a soldier, as a prosecutor, as a lieutenant governor, and for two decades as a U.S. senator, who is as well prepared as any candidate has ever been to lead our country to the kinds of promise that I think all of us hope for." Not only did Obama deftly handle the question, but again, he hit his talking points about Kerry's résumé. Axelrod and Gibbs both exhaled, and Axelrod took his seat.

The questions did not get much easier. Later, Russert flashed a quote from a *Cleveland Plain Dealer* story way back in 1996, when the Democratic convention was held in Chicago. In the article, Obama complained that people with money gained undue access in politics. Obama told the newspaper that "Chicagoans have grown especially jaded watching the Democrats raise cash for this month's national convention in Chicago. The convention's for sale, right? You got these ten-thousand-dollar-a-plate dinners, Golden Circle Clubs. I think when the average voter looks at that, they rightly feel they've been locked out of the process. They can't attend a ten-thousand-dollar breakfast. They know that those who can are going to get the kind of access they can't imagine." Gibbs and Axelrod again looked at each other. "Where did that come from?" Gibbs asked, obviously blindsided again. Axelrod lifted his shoulders and shook his head. Russert then said to Obama, "A hundred and fifty donors gave forty million dollars to this convention. It's worse than Chicago, using your standards. Are you offended by that, and what message does that send the average voter?" Gibbs looked down. "Oh lord," he said. But again Obama handled the difficult query with aplomb. He responded, "You know, I think that politics and money are a problem in this country for both parties. And I don't think there's any doubt about that. One of the things I'm proud about, though, is that when you look at John Kerry's record, what you know is here's a person who is consistently voting on behalf of what he thinks is best for America and the country. I don't think a convention changes that. I do think that the more we as Democrats can encourage participation from people who at this point feel locked out of the process, the stronger we are. One of the strengths of our party has always been the fact that we are closer to the average Joe, the guy who is trying to make a living, the guy who's trying to send his kids to college and pay his bills." Another winning answer to a difficult question.

At the conclusion of the interview, Axelrod and Gibbs hopped up from their chairs. Gibbs donned a grand smile. Axelrod simply looked relieved. Their prize pupil was unflappable in his first major hazing by a Washington journalist. "This was his 'Welcome to the NBA' moment," Axelrod said, heading out the door to find Obama. The next day, in commentary with Russert, *NBC News* anchor

Tom Brokaw called Obama's interview "a very strong appearance." Obama had passed his first major test in Boston, and he had passed with flying colors.

THE NEXT DAY, HOWEVER, HE WOULD LEARN HOW FICKLE THE PRESS can be with celebrities. Despite his shining performance amid the media circus, the *Chicago Sun-Times* chose to focus on his Kerry comment in the *Atlantic Monthly*. The tabloid splashed a headline across its cover: "Oomph!" This move perplexed Obama, who was just beginning to bask in a gauzy haze of media adulation. "Is that good reporting, good journalism—to take something I said months ago and print it now?" he asked me. Obama was learning that everything he said, past and present, now blared through a megaphone.

Over the next couple of days, Obama hopscotched through a gauntlet of media interviews, fund-raisers and breakfasts, lunches and dinners with various people of influence. Celebrity at a political convention can be determined by the size of the media gaggle that naturally surrounds an individual. Walking through arena corridors or near the huge media tents outside, one would come across an occasional huge huddle of humanity, with the inhabitants thrusting cameras and voice recorders toward a famous subject in the middle. The biggest of these huddles, at least that I saw, engulfed the left-wing bomb-throwing filmmaker Michael Moore. But senators and congressmen and big-city mayors drew their own gaggles. And of these, Obama's was among the largest. He could rarely take a few steps before being stopped by an autograph hound, well-wisher or reporter. He consistently had two or three staff members around him, guiding him from one appearance to the next and trying to manage the free-flowing entourage of reporters trailing him. Obama was now under an intense media glare unlike anything he had ever endured. On the arena floor, as Obama stepped up to the set of CBS's *Face the Nation,* host Bob Schieffer offered him a hearty handshake. "You're the rock star now!" the smiling veteran newsman said in his mild southern drawl. Obama demurred. "Talk to my wife and she'll tell you that isn't so," Obama said in a self-deprecating manner, using Michelle as a foil to tamp down his ego. (This would become standard routine as his fame grew—using his wife's taskmistress side as a way to display public humility and keep his feet on the ground.) As Obama gave interviews to a dozen reporters at a time, and appeared on one national media show after another, I could not help but think back to a chilly Chicago evening in January when it was just me making him uncomfortable as I shadowed him day by day. And I recalled a specific moment heading into a Chicago fund-raiser thrown by black professionals when Obama was notified that he had to do a radio interview before going into the event. Taking the call from the radio station, Obama asked that I step out of the SUV while he did the interview because having a second reporter eavesdropping gave him a sense of "being in a hall of mirrors." Now, with this scene in Boston, with a handful of journalists listening to his every utterance, what a hall of mirrors this was!

Tuesday, the day of the keynote speech, was even more maddening. Throughout the day, Obama was clearly the convention's hottest commodity, with more than a dozen reporters and photographers keeping pace with his every step. His day began at six o'clock in the morning with a green-pepper omelet that aides had fetched from an all-night diner because the hotel restaurant had not yet opened. Obama then headed to the FleetCenter, where he appeared on the morning shows of all three TV networks before sitting down with Ted Koppel of ABC's *Nightline*. Next, Obama had breakfast at the Sheraton Hotel with the Illinois delegation. Here, he allowed that his toughest critic—Michelle—had given a modest thumbs-up to his speech. "We brought her into the practice room," he told reporters. "Her assessment was that I wasn't going to embarrass the Obama family." Like an athlete warming up for the big game, he allowed his extraordinary self-confidence to flow freely. "I have high expectations of myself," he said. "And I usually meet them." As Obama's entourage rushed out of the delegation to the next event on his list, Dick Kay, a boisterous, barrel-chested television reporter from Chicago, hounded Obama's recently hired campaign press secretary, Julian Green. "When am I going to get my time with him, Julian? When? When?" Green, an attentive and always nattily dressed African-American man in his mid-thirties, could only mutter that they were running late and he would do the best he could. This foreshadowed a growing dilemma for Green—how to keep the hometown media happy and still satiate the hungry national press corps. After all, Obama still had an election to win in Illinois.

Shortly after noon, Obama delivered a short address at a rally sponsored by the League of Conservation Voters, which had indirectly contributed several hundred thousand dollars to his primary campaign in the form of television advertising on his behalf, run mostly in the Chicago suburbs. "I can't give you a long stemwinder," he told the group, apologizing. "I can't throw out my throat for tonight or I've had it." As he came off the stage, a corps of reporters mobbed him and insisted on an interview session. They interrupted each other with questions on topics ranging from reparations for black Americans to Obama's African-American heritage to his proposals for boosting the economy. Trying to manage the chaos, Green appeared near the end of his rope—and it was still early afternoon. "I need five Baracks today," Green said in frustration. "Everyone wants a piece of him. This is crazy, man."

Racing back to the FleetCenter in an SUV, Obama inhaled a turkey and cheese sandwich with spicy mustard as he tried to field questions from the half-dozen reporters traveling along. When one reporter rambled on about major political figures who had given keynote speeches before him, Obama answered through a mouthful of turkey and bread, "Are you asking me how I suffer in comparison?" After interviews with Illinois television stations, Obama made the rookie celebrity mistake of slipping away from his handlers to grab a cup of green tea at a Dunkin' Donuts counter in the arena. Immediately descending on him were reporters from BET, NBC News, ABC News and various publications. Obama gave up on the tea, which he wanted to help soothe an overworked throat. Maintaining his composure, he answered questions for about five minutes before announcing that he had to use the restroom. He confided to Green that he wanted to use the portable restrooms outside because, "You know, when I go into the regular restroom, all these people want to shake my hand, and that's not the place I want to be shaking hands." Yet when Obama neared the portable toilets with the media horde still at his heels, he turned plaintively to his pursuers: "Can y'all just give me one moment to use the Port-O-Let?" But the group kept moving apace, until Green threw out his arms, at last stopping the entourage. "Guys, guys, guys!" Green shouted. "Can you let him use the Porta Potti? Please! Thank you!" Soon, Gibbs appeared with Obama's tea and whisked his boss away to practice the speech in private.

Obama had never used a teleprompter or spoken before an audience of that size before. Five thousand delegates were in attendance. So Obama practiced the speech several times during the week. Jim Cauley became convinced during these practice sessions that the speech would be a hit. The final time Obama rehearsed it, Cauley noticed a DNC staff member with tears in her eyes. "Even that last try, though, Obama was only about eighty percent there," another observer said. "He didn't really nail it completely until he gave it before the crowd."

Gibbs had advised Obama that keynote addresses generally fell into two categories—thematic and programmatic. Thematic speeches generally involved broad, sweeping ideas about how to strengthen the country. Programmatic speeches homed in on specific policy details and offered solutions to major problems. Obama knew immediately that he was shooting for a thematic approach. He had been thinking in broad terms about his overall message and how he believed the country was swerving down the wrong path. He loosely based the speech on two previous well-received keynote addresses from oratorically gifted Democrats: Mario Cuomo's 1984 address in San Francisco titled "A Tale of Two Cities" and Ann Richards's 1988 speech in Atlanta. Cuomo described his vision of a country led by Democrats who want to spread wealth to people of all socioeconomic classes, races and ethnicities. He compared that vision with the way he perceived America evolving under Republican president Ronald Reagan—a society dividing into haves and have-nots based on wealth and education, a society being restructured by a social Darwinism in which the strong prevail over the weak. "We must make the American people hear our tale of two cities," Cuomo said. "We must convince them that we can have one city, indivisible, shining for all its people." Richards did much the same thing four years later, excoriating the Reagan administration and Republicans in general for a "divide and conquer" strategy, pitting different interests and different geographic regions of the country against one another for political gain.

Obama, whose career path forced him to chase his fiction-writing muse into political composition, leaned on storytelling in his speeches. In the keynote address, he took his campaign rhetoric about a common humanity and blended it with the biography of Kerry, weaving all of this into a tight seventeen-minute speech. His campaign speeches in 2004 were a stew of general political prose, the sermonizing poetry he had experienced in the African-American church and his past readings, primarily of Martin Luther King Jr. He launched the keynote in the same fashion as his stump speeches, by introducing himself and his unique family ancestry—mother from Kansas, father from Kenya. He wrapped it up by returning to that biography, saying that America's greatness lay in its unique ability to instill hope in "a skinny kid with a funny name" like him. In between, he concentrated on this basic notion of a unifying force in America, a hope in the American Dream; "the audacity of hope," he called it. Obama took that phrase straight from a sermon by his pastor, Jeremiah A. Wright, who himself had plucked it from King. (Said King: "I have the audacity to believe that peoples everywhere can have three meals a day for their bodies, education and culture for their minds, and dignity, quality and freedom for their spirit. I believe that what self-centered men have torn down, other-centered men can build up.") These were Obama's best-received lines from his earlier campaigning. For example, this section is a direct pickup from his stump speeches: "If there is a child on the South Side of Chicago who can't read, that matters to me, even if it's not my child. If there is a senior citizen somewhere who can't pay for their prescription drugs, and having to choose between medicine and the rent, that makes my life poorer, even if it's not my grandparent. If there's an Arab-American family being rounded up without benefit of an attorney or due process, that threatens my civil liberties. It is that fundamental belief—it is that fundamental belief: I am my brother's keeper. I am my sister's keeper. . . ." The final statement was a biblical reference that was generally a crescendo line with African-American crowds. Kerry's people, in fact, had edited that line out; but Gibbs, knowing its sure-fire popularity on the trail, made sure to restore it.

Obama's first versions of his speech included much more of his own biography, but that eventually fell to the cutting-room floor to keep the speech less than twenty minutes. He also wrote sections on his definition of the American Dream and his belief in the exceptional nature of America that were edited out. In the end, the essence of his speech consisted of the lines and themes that worked on the campaign trail.

Obama and his team exercised nearly full editorial control over the speech. Kerry's staff did make one substantial change, however. After Obama's riff about carving up the country into red states and blue states, he tied this color mosaic together by saying that all the country was "pledging allegiance to the red, white *and* blue." But Kerry's people said they might want to use that line in Kerry's speech at the end of the week. Obama was "incredulous" at this request. "Of all the lines in the speech, Barack was the proudest of that one," Axelrod said. "Barack said, 'They are taking my line.' Literally, throughout the week he was saying that he did not know but that he may say it anyway. But then, when he went up there he didn't. He knew that they gave him a great opportunity, so a little thievery was a small thing." In the end, Obama altered his line to "pledging allegiance to the Stars and Stripes."

There was one notable difference between Obama's speech and those he had patterned it after: Obama attacked the establishment in general rather than saving all his fury for the Republican Party. He talked of the "spin masters" and "negative ad peddlers" who want to carve up the country for their own political gain.

ARRIVING AT OBAMA'S MESSY HOTEL ROOM BEFORE THE SPEECH, Axelrod and Gibbs realized that they had not considered an important aesthetic ingredient in that night's success. How should Obama dress? The trim Obama generally looked crisply handsome and presentable in his suits, but how would he look onstage with the predominantly blue backdrop? Obama was wearing one of his dark suits and one of his lightly patterned ties. The suit was fine, but when Axelrod took a close look at the tie, they both felt he needed an upgrade. Obama defended his choice. He especially did not consider Axelrod's opinion of high value, since Axelrod with his penchant for sports attire was anything but a fashion plate. When Obama sought Michelle's opinion, however, she agreed with the two men. So that was that—the search for a new tie was on. Finally Axelrod spotted Gibbs's brand-new baby blue striped tie. This would do well. "But this is my tie," Gibbs protested. "I bought it specifically for tonight." It would now be Obama's tie.

Huddled with his small team behind the stage, Obama, who had been calm throughout the week, suddenly felt a tad nervous. The crowd was certainly juiced and primed for a big performance. In fact, the Democrats yearned for a big performance. Kerry's public uneasiness had not exactly lit a fire under the rank and file. The audience was hungering for someone with charisma, and the buzz around this young African-American lawmaker from the Midwest provided hope that perhaps someone could dazzle them this week. Despite his nerves, Obama maintained an extraordinary mental acuity, especially considering how far he had come in such a short time—from third-place Senate candidate to hyped national keynote speaker. By his side was his volunteer campaign photographer, David Katz, the fresh-from-college young man who had by now snapped thousands of photos of Obama. Knowing that Katz was a talented golfer with a scratch handicap, Obama turned to the young man and said, "I'm gonna go out there and sink this putt." Later, Katz expressed wonderment that Obama could relate to him in such a personal way at such a psychologically intense moment. "That's one of his amazing talents," Katz said. "Here he was, about to deliver the keynote speech, and he had the presence of mind to connect with me at *my* level."

Obama did not fail to delight the crowd. He stumbled a bit in the opening lines, clipping his words on occasion. But after mentioning that his mother was from Kansas, the Kansas delegation erupted in a cheer, and one could see a jolt of energy rush through Obama's body. He had made that special audience connection.

By the crescendo points in the speech, Obama had fallen into just a touch of a black preacher's cadence—and his audience was simply enraptured. "There's not a liberal America and a conservative America—there's the *United States* of America," Obama said in a clear, resolute voice. "There's not a black America and white America and Latino America and Asian America—there's the *United States* of America. . . . We are one people. . . ." Democrats of all races and ages nodded their heads in agreement with his proclamations. Some were crying. Many shrieked and jumped from their seats. Axelrod and Gibbs had ventured onto the arena floor to gauge the speech from that perspective. As Obama wowed the crowd, journalist Jeff Greenfield spotted the two Obama aides a few rows behind him and mouthed to them: "This is a fantastic speech!" Backstage were Green and Katz. As a black man, Green said he felt himself swelling with immense pride. He spotted the usually poker-faced Michelle and noticed that she had tears glistening on her cheeks. A lump formed in Green's throat and chills ran up his spine. "When I looked past the stage," Green recalled later, "and saw how people reacted, when I saw people falling out, people crying, I thought to myself that I had never experienced anything like this, anything this powerful. You know, I'm not sure what this means, but I couldn't help but think, Is he the one? Could he really be the one we have been looking for?"

When Obama finished, Michelle bolted onto the stage in an un-rehearsed moment and patted her husband on the back. They waved to the fawning crowd as she guided him backstage. "Wow!" exclaimed a news anchor on television. MSNBC's Chris Matthews gushed, "I was shivering, it was so good." CNN's Wolf Blitzer observed that Obama "electrified this crowd here." Even former Republican vice presidential nominee Jack Kemp called it a "fabulous speech" on a Fox News program. Backstage, Obama slipped out of his fiery orator's cloak and into the low-key, relaxed uniform he wears from his Hawaiian childhood. He flashed a self-satisfied smile at Green. "I guess it was a pretty good speech, huh?" Obama said.

The next day, as Obama and his entourage ascended on an escalator in the arena, a woman descending on the adjacent down escalator simply beamed upon encountering the Democrats' hottest new star. As the two passed each other, she leaned over and said to Obama, "I just cannot wait until you are president."

21
Back to Illinois

Arriving back in Illinois, Barack Obama was confronted with the reality that he still had an election to win. With Jack Ryan dismissed and the Republican Party foundering, however, it was now nearly August and Obama still had no Republican challenger. On paper, that looked heaven-sent. But this was uncharted terrain. How do you run a campaign when you don't know who your foe is? So Obama kept pushing his staff forward, and his staff kept pushing him. "Jimmy, I don't want to hear any of this 'we don't have an opponent' stuff," Obama told his campaign manager, Jim Cauley. "We have to keep running hard."

To deaden any suspicion that the heady keynote experience might go to his head and shift his priorities away from the voters of Illinois, his staff had planned a statewide campaign tour to begin the weekend after he returned home. Obama knew his Boston week would be crazy, and he preferred a week of mild campaigning. But his staff had other ideas. They had been listening, perhaps a little too closely, to Obama's preaching about not easing up on the throttle. Still running on empty physically after Boston, Obama launched a sixteen-hundred-mile blitz of the state. The Obama caravan visited thirty-nine counties and thirty-nine cities in just five days—a whopping eight campaign stops a day. Obama wanted to reconnect with his family after the hectic convention, so his staff rented a recreational vehicle so he could pile in Michelle and his daughters to accompany him along the trail. But with eight events per day, and a couple of hours' drive between some of these, it was hard to see where his family fit into the frenetic mix. "This was supposed to be a leisurely trip in an RV with my family. Instead, it's turned into the Bataan Death March," Obama lamented as the RV pushed away from Chicago. "I mean, I don't want to dissuade my staff from being aggressive. So it's this delicate balance that I am still trying to figure out. . . . But we've got a campaign to run and we have to get back to reality."

Still, reality had changed unalterably for Obama. His modest fame was expanding into celebrity beyond Illinois and the Washington Beltway. After his roundly hailed keynote speech, he was fast becoming one of the hottest commodities in the Democratic Party nationwide. The publisher of *Dreams from My Father* ran off eighty-five thousand new copies and the book began climbing the best-seller lists. When I talked to Obama as the tour launched, it was obvious that he was ill at ease with this newfound star status. He maintained, quite convincingly, that he just wanted to win the Illinois race and concentrate on being a successful U.S. senator. "I don't intend to be on *Politically Incorrect* anytime soon to talk about whatever issues happen to be in front of the newspapers," he said. "Part of my job is to strike a balance between doing a good job as a legislator, being an effective advocate for voters and still being a decent husband and father. That's a pretty full plate right there."

This thinking might help him focus on the tasks at hand, I thought, but it surely seemed naive. The genie was sprung from the bottle, and there was no way to put it back in. Here was this interesting young star senator headed toward the Washington beast. Expectations would be placed on him from all quarters—the black community, the liberals, the centrists, the party fund-raisers, the media, the campaign contributors, the aides (like Robert Gibbs) who harbored their own lofty ambitions and saw Obama as the conduit to fulfill them. If Julian Green needed five Baracks that day in Boston, his campaign would now need at least that many every day. "In just a few short months, Barack has been shot out of a cannon twice," David Axelrod observed.

The statewide campaign swing was further evidence of the intensity of Obama's overnight celebrity. The Obama fever that swept through Boston hit the same scorching temperature back in Illinois. Everywhere Obama went, in every little quaint town square where he had planned a rally, he was greeted by thick, energized crowds. Publicly, he tried to downplay the breathless attention by talking about his goal at hand—winning the Senate race. Whether standing center stage in a crowded college theater or on the bed of a bronze Chevrolet S10 pickup truck parked in a dirt lot next to a town hall, the ever-disciplined Obama stuck to his message. He told rally after rally that, in his keynote address, he was trying to do nothing more than echo the voices of Illinois residents. More than anything, he said, he sought to bring a more conciliatory approach to America's bitterly partisan political culture. "Apparently, the speech turned out okay Tuesday," he said with an understated grin to a group of about five hundred in a scenic park in Kewanee. "But the reason I was there was because of you, the voters of Illinois. People didn't care that I was from a different town or that I was a different color." With reporters, he endeavored to remain religious to the campaign script of personal humility and public service. "This is all so, well, interesting. But it's all so ephemeral," Obama said between campaign stops in DeKalb and Marengo. "I don't know how this plays out, but there is definitely a novelty aspect to it all. The novelty wears off, and it can't stay white-hot like it is right now."

There was also a part of Obama, whose father abandoned him as a child and whose mother traveled to faraway continents and left him with his grandparents, that clearly fed off the public adulation. Nothing nourished him more than connecting emotionally and intellectually with an audience. In these crowds of God-fearing heart-land Americans, he would typically end his stump speeches with a reference to the Almighty—"God bless you all." But in concluding his first tour speech, as the crowd roared its affirmation, Obama instead blurted out, "I love you all! Love ya!" He then strutted offstage to the strains of booming country music very much befitting the star image that had been thrust upon him. And it wasn't just Obama who was changing. His audience was too. Suddenly their appetite for anything about Obama seemed insatiable. His humorous opening lines about his name being misconstrued as "Yo mama" or "Alabama" were now greeted not with mild chuckles but with howls of laughter. "I vote for people, and not for political reasons, and this man inspires me," gushed David Bramson, a rough-hewn sixty-seven-year-old truck driver from Marengo. Bramson said he would vote again for George W. Bush for president—and the liberal Obama. "He can really get your juices flowing. This is the first Democrat I've seen who doesn't polarize the public," he said. "This guy is real. The others are phony." In Pekin, a woman held aloft a sign: "If you have a dream, vote for Obama."

ACCOMPANYING OBAMA FOR SEVERAL DAYS OF THE TRIP WAS THE senior senator from Illinois, Richard Durbin, who had introduced Obama to the Boston crowd before the keynote address. Durbin was the Energizer Bunny of stump politicians. A liberal Democrat who was one of the few senators to vote against giving President Bush the authority to invade Iraq, Durbin had been simply indefatigable on the campaign trail in the course of his career. "I've learned a lot just watching Dick," Obama said. "He is solid." Durbin's tireless, type A nature is perhaps necessary. Short of stature, with slightly graying hair trimmed in a traditional style, Durbin conveys an almost anti-celebrity persona. He can remind you of the smiling next-door neighbor in khaki pants and button-down checkered shirt, always ready with a friendly wave when you spot him mowing the lawn. His physical appearance was so Everyman-like and his personality so avuncular that he could walk through downtown Chicago in rush hour and hardly be recognized by a single constituent. Before a fundraiser in Springfield during the tour, Obama was busy with phone calls and told Durbin that he would meet him at the event. Durbin replied that he would rather just wait. "I can't go there without you," Durbin explained. "Nobody will have any clue who I am unless I walk in next to you." Yet despite his relative anonymity in Illinois, in Washington, Durbin was considered one of the most serious members of the Senate, and he had earned great respect from his colleagues in the Capitol. When Democrats captured control of the Senate in 2006, Durbin was elected majority whip, making him the second-ranking Democrat on the Hill.

With their contrasting personal styles, Durbin and Obama had worked up something of a two-man comedy routine on this trip, with Obama playing the straight man to Durbin's boisterous standup delivery. Durbin opened each event by telling the less-than-true story that he had prepared this fantastic speech for the Boston audience, but (slight pause) he had given it to Obama instead. "I couldn't be mad," Durbin would say with a wide grin, "because didn't he do a great job with it, folks?!" Privately, Durbin allowed that this joke was older than he was, but it nevertheless was surefire in drawing a laugh. And morning, noon and night, at event after event, Durbin would hit his line perfectly each time. Meanwhile, Obama, with arms folded in his cool, detached posture, would stand next to him and push out a smile, a smile that became more fatigued and more forced on each subsequent telling, as each day passed.

Obama, still a policy wonk at heart, mixed serious politics into his addresses on this trip. He spoke about providing everyday people with a greater voice in Washington and breaking down partisan bickering. He promised to channel the concerns of Illinois residents into sound policy. He assured voters that he was running for their sake, not his own. "Those little small miracles that all of you pull off every day—that is what this campaign is all about," he said. Surprisingly, Obama also took a decidedly anti–Iraq War stance. This could have been tricky in this part of the world, which was red state country, but he pulled it off in his reasonable tone. "When we send our men off to war, we need to make sure we are sending them off to the right war," he told audiences, who reacted enthusiastically.

As the days wore on, miraculously, the crowds seemed to grow even bigger. And Obama's young staff was in no way prepared or equipped to handle them. Outside an aluminum-sided café in Lincoln, Obama stepped from his SUV and was swallowed up by a sea of several hundred admirers, each wanting to shake his hand or give him a hug or take a photo. Mike Daly, Durbin's chief Illinois aide and a veteran campaigner, grew irritated and impatient with this poorly planned scenario. "We've lost control," he said in frustration, as Obama disappeared amid the crowd. "Someone is going to have to go in and fish him out." Sweating and fatigued, Daly stepped out of the pack of humanity and mused about the lasting effect this idol worship could have on Obama. "I think he is grounded internally, but look at this. How can you not let all of this go to your head?" he asked.

After a couple of days of fighting the crowds, Obama was starting to grow more weary of the speeches and the people—especially all the people. As in Boston, everyone wanted a piece of him, sometimes literally. He complained to aides that some women would literally grab his buttocks and physically push up against him. Each event ran well over its time because he would have to sign autographs and shake hands with hundreds of adoring fans, and this extra time had not been built into the schedule. The RV, meanwhile, had been practically discarded as a means of travel because it could not go fast enough on the cornfield-lined country roads to make up the lost time. Instead, aides corralled Sasha and Malia into the RV in the morning and took the children to theme and water parks for the day, reuniting them with their parents at the hotels in the evenings. Obama, however, was hitting fund-raisers and rallies well into the night hours. So much for the family trip. "When are we going to see our children again?" Obama asked Michelle inside the SUV one morning. "I'm not sure," she replied. A few moments later, Obama told his driver that he wanted to stop and pick up a *New York Times*. "Why do you want that?" Michelle asked. "What's in there about you?" Obama seemed rankled by the question. "Nothing that I know of," he said. "You know that I always read the Sunday *Times*." "Oh yeah," she replied. "I guess I've just lost all perspective."

In these private settings, as the exhausting trip progressed, Obama was not always the relaxed, smooth politician from days and weeks earlier. His mercurial moments bubbled up again. He had brought Michelle and his daughters out on the road, but he was seeing little of them. And not only that: With the sheer number of people he had to greet, the worshipping crowds became less ego gratifying and more of a burden. There was no real physical separation between him and the crowds. Finally, on the third day, at a high school in Clinton, a police officer threw one of Obama's young traveling aides a roll of police line tape and told him, "Keep it and use it, son." After that, the aides started cordoning off the crowds to give Obama some breathing room. "We've finally kind of figured out that he is happier when we can keep some of these people the hell away from him," one aide said. Obama, feeling remorseful about this, apologized to audiences that he could not stay longer, sign autographs and meet each one of them.

Obama, at this juncture, had no security presence around him. His driver was a former police officer, but these crowds were too big for one man and a few twenty-somethings to handle. Moreover, Obama's following now came with an emotional element that could possibly draw out the wrong emotion. In Boston, I had noticed that Obama had no security, despite some odd-looking characters who turned out to see him in the days after the speech. So during an interview on the Illinois trip, I asked him and Michelle about the lack of protection. Michelle jumped at the question, saying that she had been talking to the campaign staff about adding security personnel, but they had been resistant because they did not want to make it appear that Obama feared his future constituents. Being asked the question by a reporter legitimized her worry, and she looked over at Gibbs in the SUV. "This is not something that Barack even needs to think about," she said. "I understand you have to achieve a balance between looking out for his safety and not looking like he is afraid of the community he is serving. But we have to find that balance." That night, at a packed outdoor fund-raiser, I told one of Obama's aides about Michelle's reaction and asked if I had gauged things correctly. "They're both totally freaked out by this," he said.

The stress of the trip, and the madness around it, was showing on the ever-affable Michelle. At some moments on the trip, the tension between Michelle and her husband was palpable. No one is more devoted to Obama than his wife, and no one will race to his defense as ardently and as quickly as she will. But her husband's overnight fame was causing her some concern. One morning, as I waited to interview her in the SUV, she dropped a political cartoon into my lap. The artist had sketched a smiling Democratic woman holding up a sign that read: "Dated Dean, Married Kerry, Lust for Obama." Michelle looked at me and said, "This is what I have to contend with." Over time, Obama would smooth over these issues with his wife. This feeling was transitory, but at that moment it was real.

As the caravan made its way back to Chicago on Day Six of the Death March, nearly everyone involved in the tour was ready for its end. It had been a fascinating yet grueling adventure. But no one was more primed for its conclusion than Obama himself. After giving his final speech and posing for a photo with the youthful staff members who assisted him on the excursion, he called over the lead organizer of the trip, a law student named Jeremiah Posedel. The two were standing in the middle of a blocked-off street in a small town just south of the Chicago region. Obama placed his hands on Posedel's shoulders and then fixed a serious gaze directly into the young man's eyes. "You did a great job and I am so appreciative of all the work you've done," Obama told him. "But don't ever fucking do that to me again."

WITH CHICAGO LOCKED DOWN, OBAMA WOULD TAKE MORE OF these extended campaign trips to solidify downstate Democrats and try to convert Republican voters to his cause. But these trips would cause some anguish for Obama as he sought to keep his biggest vice—cigarette smoking—hidden from public view. The whole campaign caravan would have to pull over at a gas station, where Obama would disappear into the restroom, presumably to catch a smoke. "It's embarrassing," Obama said. "We pull over and eight guys jump out of cars just so I can use the restroom." This could place his campaign staff in the awkward position of having to guard their employer's secret nicotine addiction.

Late at night, during one tour of the state, I was riding shotgun in a car trailing Obama's black SUV amid the campaign caravan. At the wheel of our small sedan was press aide Tommy Vietor, a young tousle-haired East Coast native who joined Obama from the presidential campaign of North Carolina senator John Edwards after Edwards folded. Vietor was a smart, eager, computer-geek type who feared one thing: screwing up. He sensed that if he played his cards right, he could find a seat on Obama's rising rocket ship for a long time to come. Specifically, he had his sights set on being Gibbs's chief communications deputy. So Vietor could ill afford to let something outside the script slip to a reporter.

After a long day on the trail, it was nearing midnight and we were headed to a hotel that had been inexplicably booked a long two hours' drive from the final event. As we motored along the dark, flat country roads of Illinois, I spied a small orange-lighted object fly from the passenger window of Obama's SUV and smack into the road ahead of us, briefly bouncing along the pavement until it disappeared beneath our car. Vietor, understanding the magnitude of Obama's well-kept secret and the potential consequences of its revelation to a reporter, immediately turned his head my way to see if I had noticed what was obviously a cigarette butt discarded by Obama. But Vietor and I both said nothing. Only half-awake and tired, I lacked the energy to mention what I had seen and open a conversation about it. Vietor was obviously hoping I hadn't spotted the cigarette and kept his mouth shut.

Several minutes later, however, out flew another orange cigarette butt, which elicited the same reaction from Vietor—a quick worried glance in my direction. After another several minutes, out popped another. Again Vietor turned my way, looking ever more worried as Obama flicked each cigarette from the SUV. "You know, Tommy, I've known for a long time that Barack smokes," I said. "You have?" Vietor asked. "Yes," I responded, "since early in the primary when I got into his SUV and it smelled like a smoky bar on Friday night." "Whew!" he said. "So long as Barack knows *I* didn't fink on him."

Throughout the summer, Obama was locked into a daily schedule that restricted press access, but on these many trips, Obama and I occasionally engaged in candid conversation. During a campaign stop in Springfield, he revealed his occasionally thin skin in one of these chats. He asked why I had been "taking all these jabs" at him in my *Tribune* stories. I was perplexed by what he meant. I thought my coverage had been balanced. But Obama referred to a line from a story that had run a couple of months earlier. Obviously, the line had been eating at him. I had written that while he was a talented orator, his debate skills might be suspect. I said that he had a tendency to be "verbose" in press interviews, occasionally meandering off his message and waxing philosophical about other policy views. He was also not a good "sound bite" politician—not the type who could deliver a quick one-line punch to the gut of an opponent. And the winner of a debate is often the candidate who delivers the most memorable punch. This was mild criticism at best, but Obama obviously was not accustomed to public criticism. A couple of years later, in his book *The Audacity of Hope,* he conceded the verbosity, anti-sound-bite point. He wrote that his elongated musings had probably won him some points with political reporters, however, whom he dubbed a "literary class."

As Obama set Illinois on fire that summer, the Republican Party waged a bitter internal battle about who should replace Ryan on the ballot. Moderates and conservatives had been bickering for years about the direction of the party, and this battle was perhaps the ugliest public incarnation of that disagreement. Various potential candidates' names floated. Even Mike Ditka, the former Chicago Bears head coach, who was a hometown hero, toyed momentarily with the idea of challenging Obama. In early August, the moderates finally succumbed to the party's vocal right wing and allowed the GOP's central committee to bring in conservative firebrand Alan Keyes as its candidate. Keyes, who was then living in Maryland, was a former talk show host who quixotically ran twice for president. A rare African-American conservative, Keyes was best known for his rousing speaking style and his often inflammatory rhetoric, which was steeped in his deeply Christian moral philosophy. His style was scholarly, but extremely controversial, even within the Christian Right. But he was fiery enough and not the least bit reticent about virulently attacking his opponent, whoever he might be. One GOP legislator told Obama that Republicans drafted Keyes to muddy up Obama's image, to "knock the halo" off the Democrat's head. In the end, Keyes would do nothing of the sort.

My first encounter with Keyes came in Chicago's annual Bud Billiken Parade, which runs through the African-American community on the city's South Side. The Billiken parade is touted as the largest African-American parade in the country, running for several miles along Martin Luther King Drive. By now, Obama was a prideful symbol among Chicago's blacks. And on this beautifully sunny afternoon, Obama and Michelle were the king and queen of the parade. Thousands of parade-goers hoisted blue-and-white Obama signs, wore Obama stickers and shrieked in pure joy as his float passed by. They serenaded the Hyde Park Democrat with chants of "O-ba-ma! O-ba-ma! O-ba-ma!" Obama drew such a passionate outpouring from the crowd that even he and his aides were overwhelmed. "At one point, I thought Barack was going to rise up over the people and start saying, 'My children, my children, I have come to free you,'" joked his driver and bodyguard, Mike Signator. "It was just incredible."

Keyes, however, was relegated to the back of the pack, proceeding by foot. And he was anything but welcomed by the crowd. As Keyes shook hands and walked along the parade route, attendees taunted, booed and hissed. One man briefly grabbed his arm and warned him, "Take your ass back to Maryland." As Keyes tried to shake hands between 47th and 48th Streets, a wild-eyed woman ran up to him, lifted an Obama sign above her head and screamed repeatedly into Keyes's face: "Obama for president! Obama for president!" If Keyes, as a black man, had any thoughts of stealing away some black votes from Obama, he could now forget it. He would be lucky to get out of this angry stew without injury.

Keyes's bombastic nature was immediately evident. At the parade, I pulled him aside for an interview. In answering my first question, Keyes, who was wearing a thick gold crucifix around his neck, took dead aim at Obama. He charged that Obama was indirectly supporting the "genocide" of African Americans. Surprised, I asked, "How is he doing that?" Keyes answered emphatically that Obama's endorsement of a woman's legal right to an abortion was killing "thousands of black babies" every year. "We're the first people who have ever been pushed into genocide before our babies are born,"

Keyes said. "So the people who are supporting that position are actually supporting the systematic extermination of black America."

Through the next several months, Keyes's antiabortion rhetoric became no less incendiary. At one point he stated that if Jesus could vote in Illinois, he would cast a ballot against Obama because "Barack Obama has voted to behave in a way that it is inconceivable for Christ to have behaved." Again Keyes referred to Obama's stance in favor of abortion rights.

Keyes was a gifted orator himself, able to deliver a political sermon that could convince his true believers that they were engaged in a battle of good versus evil. He would raise a fist, wag a finger and proselytize like a preacher spreading the tenets of Christianity to the chosen followers.

Nevertheless, the broad populace was unimpressed. Polling showed that Obama was destroying Keyes throughout the state. Still, as with Rickey Hendon in the state legislature, there was something about Keyes that got under Obama's skin. Running across Keyes at a parade on the North Side of the city one weekend, Obama rushed over and tried to talk to him. Obama is someone who loathes conflict, and he thought that he could have a reasonable discussion with this man who had been hurling hateful invective at him. "Barack thinks he can win over anyone," Jim Cauley observed. "He thinks he can go into a roomful of skinheads and come out with all their votes." Before long, Obama and Keyes were engaged in a verbal tussle that was heightened when Obama, trying to calm the situation, put his hand on Keyes's shoulder. The next day's newspapers would feature photos of the altercation, prompting Axelrod to give Obama some worthwhile advice: "You know, Barack, you can't hug a porcupine without getting pricked."

Obama explained in *The Audacity of Hope* that Keyes's attacks on Obama's Christianity and Keyes's readings of Scripture "put me on the defensive." "What could I say? That a literal reading of the Bible was folly?" Obama wrote. "I answered with the usual liberal response in such debates—that we live in a pluralistic society, that I can't impose my religious views on another, that I was running to be the U.S. Senator from Illinois and not the minister of Illinois. But even as I answered, I was mindful of Mr. Keyes's implicit accusation—that I remain steeped in doubt, that my faith was adulterated, that I was not a true Christian."

The rest of the way, Obama kept his head in the game and his hands off the porcupine. That November, in perhaps the most anticlimactic moment of Obama's political ascension, he won the general election by the largest margin of victory in the history of Senate races in Illinois, defeating Keyes by a final tally of 70 percent to 29 percent.

ODDLY ENOUGH, THE FINAL CELEBRATION PARTY FOR OBAMA'S election victory was one of the less compelling moments of the campaign. With the Keyes debacle, Obama's success was all but assured, and most eyes that evening were on the presidential contest between John Kerry and President Bush. When Kerry lost, there was a palpable deflation among Obama's Democratic partygoers. Axelrod, in a moment of idealism, worried about disillusionment among the young people who had volunteered to work on behalf of Kerry and the Democrats. "We can't lose them, but how do we keep them engaged after this?" Axelrod asked. When I looked at Obama and suggested that these young "save-the-world" types, in the description of Jim Cauley, might gravitate toward the Democrats' newest rising star, Axelrod waved a hand. It was too early to think in those terms, he insisted.

Obama's performance on election night was less than stellar, at times revealing his tendency to rebel against the trappings that accompany his career success. As the election results rolled in, his staff had assembled the candidate with Michelle, Sasha and Malia in an upper-level hotel suite for a series of five-minute photo opportunities.

Obama and his family sat on a light-colored couch with a wide-screen TV in the background. Obama flashed his toothy smile throughout the photo sessions and his daughters delighted in the attention for a few minutes before growing bored and asking their parents when the pictures would be over. In between the processions of newspaper and television photographers, Obama squirmed in his seat and appeared more than a little uncomfortable in the contrived atmosphere. Just as when forcing a smile at Dick Durbin's joke again and again, Obama did not suffer well some of the showy, artificial moments of being a politician. "He hates this phony shit," Vietor said as we watched him hug his daughters and smile for the cameras.

Also dampening enthusiasm that evening was Obama's speaking performance. He was so tired from the final frenetic leg of campaigning that his victory speech was rather flat. He had no time to compose anything new, leading him to wander languidly among various familiar anecdotes and talk too long. But his biggest flub was thanking everyone up front rather than after the speech. This led to some of the Chicago television stations shifting away from his speech midstream. But that was of little concern to his advisers. Obama had won the election, and this was all superfluous to that outcome.

Before the election, Axelrod hypothesized that a Kerry victory would be a godsend to Obama because the center of power for Democrats would shift to the White House, lifting the hot glare of the Washington media off Obama, who, after all, would be just an incoming freshman low in seniority. This, in theory, would have allowed Obama to settle into his new job in relative peace. Indeed, Axelrod and Gibbs were already hard at work trying to structure the next phase of Obama's career. In Chicago, political insiders marveled at how artfully Axelrod and Obama's braintrust had managed his campaign amid the truly bizarre circumstances of the race—circumstances that, in the end, worked in Obama's favor. "It was a thing of beauty to watch from the outside, like watching a play that David had written, with all the acts progressing into one another perfectly," a Chicago political consultant said with a sense of deep admiration for Axelrod.

Perhaps the most important aspect of the campaign script was Obama's perpetual clean slate through the race. With his main rivals essentially doing themselves in, Obama was never forced to launch a nasty public attack of any sort. And he never had a major negative attack launched at him. This played perfectly into his mantra of a "new kind of politics" in which combatants can do political battle with civility. But it also raised questions about Obama's toughness: How would he withstand an attack in a future election? I wondered this myself—after all, he seemed fairly sensitive about my mild observation that his intellectual verbosity would be a negative as a debater. He also seemed far too sensitive to the wild attacks of Keyes. "I can't believe how this guy is trash-talking me," he once told Michelle.

Indeed, the talk of the election night party was what the future held for the new senator. Several aides predicted a difficult transition to Washington. "He is the smartest man I have ever met, but I think his first year is going to be really hard," said Amanda Fuchs, his campaign policy director. "He is going to need to learn to delegate more because there will be plenty of smart people around him. He is going to have to learn that he can't do everything himself, especially all his policy. And he is going to learn that Republicans in D.C. are not going to work with him in the same way they did in Springfield, no matter how much he reaches out to them."

The aspect of Obama's life that would be easier was his finances. Obama had leveraged his star power and the blazing sales of his first memoir into a lucrative new book deal, which he signed in December 2004. He netted an advance of nearly two million dollars to write three books, including a children's book with Michelle. Upon hearing of the sum of money he would reap, Michelle had to admit that her earlier judgment had been wrong. Obama's plan for success, which Michelle had likened to so much pie in the sky, had miraculously worked. Her husband had won a Senate seat and now would write a book that would stabilize his family financially for his lifetime. He had climbed the beanstalk and had indeed descended with the golden egg. "I can't believe you pulled this off," Michelle told him. For a man who never sought excessive wealth, Obama now had it. "I am not looking for money," he had told Jim Cauley. "All I'm looking for is a decent house and the ability to send my little girls to whatever school I want." Those goals were now accomplished.

22
The Senator

There are probably folks on the left who want me to be Paul
Wellstone. And I love Paul Wellstone, but I'm not Paul Wellstone.
I don't agree with everything Paul said. And one of the things
you come into office and everybody's projecting—particularly the
way I came in—everybody's projecting their own views onto you.

—BARACK OBAMA

Barack Obama's inauguration week in January 2005 appeared to be a time of joy for him and his family, although the pressures of his celebrity were not to abate. During his first news conference in Washington, a reporter asked earnestly, "What is your place in history?" But most of the week was about enjoying the spectacle of the moment.

After taking the oath of office, Obama, Michelle and the two girls strolled across the Capitol grounds to the Library of Congress, where he would greet a party of well-wishers from both Illinois and Washington. Taking the first steps out of the Capitol, Obama and Michelle clasped hands. "Congratulations, Mr. Senator," Michelle said with a soft kiss. "Congratulations, Madame Senator," Obama responded with a warm smile. With a handful of journalists trailing along to capture the scene, six-year-old Malia looked up at her father and asked, "Daddy, are you going to be president?" It was such an innocent question, but Obama scanned the reporters in his midst and cautiously withheld an answer. Picking up on the cue was Jeff Zeleny, then a Washington bureau reporter for the *Chicago Tribune*.

Zeleny arched an eyebrow. "Well, Senator, aren't you going to answer?" Zeleny asked. Obama again ignored the question. Back in Obama's transitional Senate office, Robert Gibbs received word of Malia's query and made sure to disseminate it to all reporters who might not have heard the potentially clairvoyant anecdote: *Obama's little daughter had asked if he was going to be president!*

As Obama reached the Library of Congress, his party guests were waiting in line to gain entrance. People had come from across Illinois and around Washington to fete Obama. His sister Maya and her husband had traveled from Hawaii. Marty Nesbitt and Valerie Jarrett had come from Chicago, along with an assortment of others. Obama ran up to the Reverend Jesse Jackson, who enveloped Obama's thin frame in a bear hug. Obama had lost nearly ten pounds on the campaign trail that year. "I'm not a toy senator. I'm not a play senator!" Obama said. "I'm a real senator now!" Requests for photographs of the real senator came from all around, and he obliged as many as he could. Jackson, always deft when finding a camera to point his way, hoisted a smiling three-year-old Sasha onto a concrete pillar about four feet tall and posed for photographs with his arm around the bright-eyed girl. After the photo was snapped, Jackson stepped away, leaving Sasha standing atop the concrete pillar by herself. Obama saw his three-year-old daughter stranded on the pillar and rushed over to rescue her. He shot a disapproving glance at Jackson, who was oblivious to Obama's glare and his own lapse in judgment about the little girl's safety.

Inside, amid the party of several hundred fund-raisers and Obama stalwarts, the senator assured his audience that his mission to create a fairer, more just America had only begun. "We are bound at the hip," he said. "We are going to be working hard to make sure that every child gets a decent shot at life and to make sure that every senior citizen is cared for, that the diversity of this country is appreciated and to make sure that we create the kind of nation that these children and your children and your grandchildren deserve.

I promise you that this is not the end of the road." The question was, where did that road now lead?

As a star attraction in Washington, Obama drew applications for staff positions from some of the best talent pools available. Gibbs and Axelrod also had a wealth of connections for finding bright people to run his policy office, legislative affairs, speech writing and other jobs. Fortuitously, one of the most well-regarded chiefs of staff inside the Beltway had just become available because longtime senator Tom Daschle of South Dakota had lost his race. Obama hired Daschle's top aide, Pete Rouse, a Washington veteran for decades, to run his Senate office.

The first order of business for Obama's team was charting a course for his first two years in the Senate. The game plan was to send Obama into the 2007–2008 election cycle in the strongest form possible. No politician had gained the sort of attention that he had won so quickly, and his advisers knew that he had a shot at being a vice-presidential selection or perhaps an outside chance at running for president as early as 2008. Obama, pressed by the Chicago media the day after his election, flatly denied that he would run for the Oval Office in 2008. But history was full of politicians who had reversed those kinds of denials about career advancement.

"The Plan," as his team called it, was formalized on a computer file and was consistently updated as events occurred. It was primarily molded by Axelrod, Gibbs, Rouse and Obama. The Plan was broken into four quarters per year, with his first quarter being dedicated to hiring Senate staff, learning the names and faces of Washington, writing his book, launching his own political action committee to raise money and turning down the volume on his publicity machine ("letting the air out of the balloon," in Obama's words). Lowering his profile was the hardest of these to accomplish. Obama was a highly sought commodity, receiving hundreds of speaking requests each week. Moreover, the year began with the first issue of *Newsweek* magazine splashing Obama's smiling face on its cover with the title "The Color Purple," a reference to Obama's desire to blend America's red-blue politics into a bipartisan hue. Other media also wanted a piece of Obama, and Gibbs spent most of his time turning down interview requests rather than seeking them out. Another problem in tamping down expectations and his media presence was a *Tribune* effort by Zeleny to chronicle Obama's first year in office. The Plan called for Obama's transition to occur with as little media probing as possible, and Gibbs did all he could to thwart Zeleny's access to the senator. "If Mike Tackett wants to know what it's like to be a U.S. senator, he should have run himself," Gibbs said, referring to Zeleny's boss, the *Tribune's* Washington bureau chief. Yet the competitive, indefatigable Zeleny spent the year chasing Obama from event to event and managed to produce a compelling series of stories.

The Plan called for Obama to spend most of the year tending to the home fires of Illinois to make sure his constituents did not feel forsaken. He conducted nearly forty town hall meetings in Illinois in 2005. Most were packed with attendees and gained him positive media coverage in the local community. Obama also took foreign trips in this first year. He had been selected for the Foreign Relations Committee, an excellent spot for a politician with national ambitions to build a foreign policy résumé. He would visit Russia, Eastern Europe and the Middle East, including Israel and Iraq. Then, in 2006, Obama would lift his head from the sand to hit the campaign trail in an effort to raise money and the profiles of Democrats across the country in the party's effort to retake control of Congress. The second half of that year would be capped by a visit to his father's homeland of East Africa and the release of his second book.

Despite this detailed sketch by some of the sharpest political minds around, Obama's first year proved as hard as some of his aides had predicted. The first few months were especially difficult. Understandably, Obama wanted to bring Michelle and the girls to live with him in Washington. He had missed them dearly on the long campaign trail and he wanted to get his "house in order." There was just one problem with this thinking. Nearly everyone counseled him to keep his family in Chicago. Michelle's life—and Obama's refuge from the outside world—remained in Hyde Park. Her mother and her closest friends still lived on the South Side. She still worked at the University of Chicago Hospitals, although she had cut back on her hours during the campaign season. Moving the family to Washington would strip Michelle of this village of support, making her totally dependent on Obama and focusing her life solely around him. Besides, Obama would be traveling back to Illinois every week for town hall meetings, which would have meant leaving Michelle and the girls to fend for themselves in Washington. Axelrod saw this problematic future and turned to his friend Congressman Rahm Emanuel for assistance in heading it off. Axelrod scheduled a dinner between the Emanuels and the Obamas, where using their own experience as an example, Emanuel and his wife, Amy, made the convincing case that it was better to keep the family back in the district rather than bring them to Washington. After some time, Obama realized this was the best course of action and he rented an apartment in Washington and spent Tuesday through Thursday nights there while the Senate was in session. He usually had a full schedule back in Illinois on Saturdays, but Sundays would now be devoted to his family.

There was another ink blotch on this sketch. Writing a book is a full-time job in itself. And writing one while stepping into a new job is a recipe for burnout. Obama had a history of taking on an inordinate number of tasks at once. While he was writing his first book, for example, he was running Project Vote in Chicago. Realizing his full schedule and the time pressures on him, I gave him several months of breathing room before I asked for an interview. When I did see him, on a sunny Saturday afternoon in April, I found the less-than-relaxed, short-tempered version of Obama. We met at a Mexican restaurant in Hyde Park. Obama had a full day of events scheduled, beginning with an early-morning radio program, and was an hour late to our lunch. When he stepped through the door, the face of a middle-aged woman sitting at the first table lit up, and she tried to chat him up about Chicago politics. Obama uncharacteristically showed extraordinarily little patience with the woman, curtly telling her that he was running behind. Later, he told me, "I'm writing a book and I just don't have any time in my life." Throughout the year, Obama was as visibly fatigued as during his long stretches on the campaign trail. Waiting to deliver the commencement address at a Chicago college-preparatory academy on the South Side, Obama nodded off during speeches by the school administrators, and he nearly did the same during his own. Standing at the edge of the stage, holding a microphone, his eyes fell shut momentarily and his knee buckled while he was talking to the crowd.

Obama had finally succumbed to abiding by a tight schedule that was defined largely by others. His friends noticed a different Obama through this difficult period. "Does he look happy to you?" Emanuel asked me that spring. "I think the job looked better on paper to him." Dan Shomon, his former Springfield aide, said his old employer was adapting to the less glamorous aspects of being a celebrity politician— overwork, a dearth of leisure time and a stable of aides making decisions about his life. "Barack is his own man, but he is tightly controlled and in some ways he likes that," Shomon told me in April 2006. "I mean, he has no time for himself and no life. So by being controlled, that does keep people like you away, people who eat up his time. He can't go play poker or get a beer or get enough time with his family— and that's the downside, a big downside. His life is over, and I keep thinking, you know, he really didn't ask for this." Another aide made a similar observation about the pitfalls of political fame: "The Barack you knew in the campaign, the one you could just bullshit with in the car, he's gone now. He no longer exists. He has to watch himself wherever he goes. If he goes somewhere, there is a volunteer there waiting for him and he has to watch what he does and says. He doesn't have time to be himself except when he is with Michelle. He always has to be on."

Another aspect of The Plan frustrated some of his friends, his legislative colleagues and particularly his devoutly liberal followers. In order to keep himself as unscathed politically as possible, he became even more cautious in his political approach, avoiding controversy at all costs. In Springfield, Obama had been an unabashed liberal, even when toiling in the minority party. But if he had larger ambitions, his team believed he could not be fitted too uncomfortably with a liberal straitjacket. This plan closely resembled the largely successful model that Hillary Clinton had followed when she won a Senate seat from New York. Keep your head down, remain noncontroversial and make sure to tend to your constituents back home. One close colleague in Congress became irritated when Obama declined to cosponsor a piece of legislation saying it was "too controversial." "When you find that bill that eighty percent of the people love, you call me," the legislator told Obama. "They're doing the Hillary thing, obviously," the lawmaker said. "And that's just so Washington. Oh sure, yeah, let's just follow the Hillary model."

This aspect of The Plan did present a double-edged sword. In becoming a more conventional Democratic politician, Obama ran the risk of alienating his core supporters on the left. Indeed, he engendered great antipathy in today's most visible forum of political debate—the Internet blogosphere. Early in 2006, angry liberals flogged him online for not objecting to the certification of presidential ballots in Ohio and for confirming Condoleezza Rice as President Bush's secretary of state. Here was this leading critic of the Iraq War arriving in Washington and voting to install in a major cabinet post one of Bush's top advisers through the launch of the war. Obama voted against the nomination of John Roberts to the Supreme Court. But when he voted for cloture for Roberts's nomination and then defended other Democratic senators who voted in favor of Roberts, he was vilified on the liberal blog the *Daily Kos*. Obama replied in a long post to the blog that pacified some on the left simply by his recognition of their online criticism. But he also reiterated that he is no flame-thrower from the left—it is just not in his temperament or in his civil approach to politics. Obama told me in an interview at the end of his freshman year:

> There are probably folks on the left who want me to be Paul Wellstone. And I love Paul Wellstone, but I'm not Paul Wellstone. I don't agree with everything Paul said. And one of the things you come into office and everybody's projecting—particularly the way I came in—everybody's projecting their own views onto you. And so, I think when people are disappointed that I would not certify the objection in Ohio, well, I actually think George Bush won the election. That wasn't a safe move—it was actually my genuine belief that George Bush won the election. When people were disappointed about me voting to confirm Condoleezza Rice—I genuinely felt on the merits that the president, on most executive appointments, deserves some deference in order to put together his team, and that we weren't gonna do better than Condoleezza Rice in this administration when it came to foreign policy. And I actually think that it's proven to be correct, that she is, you know—the bar is low, but she's been a moderating influence on this administration when it comes to foreign policy. So I think that what you've seen are times where I've actually done what I think is best, but you know, there may be some folks who disagree and automatically assume that I've done it for political purposes. And the one thing that is frustrating about my position, and I think our political culture generally—frustrating but not surprising, and I don't blame people for having the

assumption—is that [they assume] whatever I do is political, that I've got my antennae out and I'm making these calculations. And I think we've just gotten so immersed in this cynicism about politics, and as I said, oftentimes for good reasons, that people think whatever you do must be motivated by politics.

Despite this, Obama voted with his party 95 percent of the time in 2005, according to *Congressional Quarterly*; just eight senators voted more consistently Democratic than he did. Of his hundreds of votes during his first year, Obama agonized over a dozen or so. On those difficult matters, he called his top staff into his office and led freewheeling debates about how to proceed. "We have enough strong personalities here that we definitely get two people on opposite sides of the issue," said his legislative director, Chris Lu. "And he just loves to have us argue it out in front of him." Lu was one of the moderate voices in this atmosphere of smart young staffers. Robert Gibbs, the pragmatic southerner who considers political calculations foremost, was the other strong voice pulling to the center in these debates. Lu maintained that Obama was his own man when it came to making tough votes. The staff might leave the room thinking he would vote one way, but he could just as easily go the other. Overriding the arguments of Lu and Gibbs, for instance, Obama was one of thirty-four senators opposing the Military Commissions Act, which gave the military special interrogation capabilities when questioning "high-value detainees." This was consistent with Obama's history in Springfield of protecting civil liberties, a belief that is rooted in his in-depth study of the Constitution. "If you want to position yourself in the middle, you'd vote for that bill," Lu said. "But he fundamentally believed that bill was troubling, that it would basically cut out habeas corpus rights for detainees. As a law professor and somebody who understands the role of due process, who understands sort of the historical origins of habeas corpus, it offended him. . . . And so no matter what Robert and I said about the scary world we lived in or how this vote would play out, you know, in some future thirty-second attack ad, he didn't care."

THE ENGLISH PLAYWRIGHT SIR TOM STOPPARD ONCE DESCRIBED HIS dance with the press to the *New York Times Magazine* in this way: "My reticence is a form of conceit, not of modesty. It has to do with not making myself available." Indeed, Obama's plan to not make himself available through much of 2005 had the opposite effect, as one might expect. By turning down interview requests from television networks and other major media, as well as ducking the *Tribune*'s Zeleny, Obama only magnified the media's pent-up desire. While derided by some of his closest political friends, the decision to maintain a cautious path in his first year turned out to be masterful in keeping his stock high. Obama made sure that his public schedules reflected every speaking appearance in Illinois, but privately, he was also busy raising money for the party and his newly formed Hopefund political action committee, which had nearly two million dollars by the end of 2005. A *Tribune* poll late in the year found him with an amazing 72 percent favorable rating among his Illinois constituents.

Legislatively, Obama had few major accomplishments in 2005, although he did advance some measures that would gain him a measure of publicity back in Illinois. His lack of seniority in the Senate—he reminded all audiences that he was ranked ninety-ninth out of one hundred—shielded him from huge expectations on this front. After a series in the *Chicago Sun-Times* described the slow process of government pay for Illinois veterans, Obama jumped on the issue and pushed the administration to equalize funding. On his town hall tours through Illinois, he noted his work to hike the production of ethanol by increasing the number of gas stations that sell the

E85 ethanol blend. He also became a leading early voice on the threat of the avian flu pandemic.

Obama's greatest legislative success was teaming with Republican senator Richard Lugar of Indiana on a bill that expanded U.S. cooperation to reduce stockpiles of conventional weapons and expanded the State Department's ability to interdict weapons and materials of mass destruction. In the spring of 2005, Obama had traveled to Russia with Lugar to inspect nuclear weapons stockpiles. Robert Gibbs said that his failure to more vigorously publicize Obama's work on the bill, which was signed into law by President Bush in January 2007, was his greatest public relations failure during Obama's early Senate tenure. "It was an important law and it was my fault that Barack didn't get more credit for it," Gibbs said.

Early on, Obama alienated some colleagues with his desire not to speak out more forcefully against Republican policies, but he wanted to stay true to his mantra: He was a healer, not a divider. In this vein, besides Lugar, Obama reached out to other Republicans by proposing immigration reform with Senator Mel Martinez of Florida and a more rigorous review of government contracts related to Hurricane Katrina with Senator Tom Coburn of Oklahoma. Again Obama alienated some colleagues by declining to make phone solicitations during the so-called Power Hour of fund-raising for the Democratic Senatorial Campaign Committee, but he made amends by campaigning arduously for Democrats across the country·in the 2006 election cycle.

Obama's two major moments in 2005 came when he did lift his head from the sand.

The first of these was his June 2005 commencement speech at Knox College, a small liberal arts school in western Illinois. A thunderstorm threatened, but the dark sky only sprinkled on the gathering as an admittedly nervous Obama delivered the first major address of his Senate tenure, a speech that would reaffirm his liberal tendencies and encapsulate the heart of his governing philosophy.

Citing various moments of American history, with an emphasis on the emancipation of slaves and the civil rights movement, Obama offered his theory as to why America has endured and prospered for more than two centuries. It is not simply the workings of the free market, or the tireless work ethic of the country's laborers, or immense individual ambition—America's success is built on "our sense of mutual regard for each other," Obama proclaimed, and "collective salvation" is the surest way to ensure the country's continued prosperity. And, Obama added, the best means for looking out for each other is through government—strengthening public schools, providing health care to all citizens, devoting time to community service rather than "focusing your life solely on making a buck." The global economy is threatening the country's economic health, Obama said, and the only way for America to continue to compete is to invest in itself:

> . . . there are those who believe that there isn't much we can do about this as a nation. That the best idea is to give everyone one big refund on their government—divvy it up into individual portions, hand it out, and encourage everyone to use their share to go buy their own health care, their own retirement plan, their own child care, education, and so forth.
>
> In Washington, they call this the Ownership Society. But in our past there has been another term for it—Social Darwinism— every man or woman for him or herself. It's a tempting idea, because it doesn't require much thought or ingenuity. It allows us to say that those whose health care or tuition may rise faster than they can afford—tough luck. It allows us to say to the Maytag workers who have lost their job—life isn't fair. It lets us say to the child who was born into poverty—pull yourself up by your bootstraps. . . .
>
> But there is a problem. It won't work. It ignores our history. It ignores the fact that it's been government research and invest-

ment that made the railways possible and the internet possible. It has been the creation of a massive middle class, through decent wages and benefits and public schools—that has allowed all of us to prosper. Our economic dominance has depended on individual initiative and belief in the free market; but it has also depended on our sense of mutual regard for each other, the idea that everybody has a stake in the country, that we're all in it together and everybody's got a shot at opportunity—that has produced our unrivaled political stability.

Obama had studied this theory of social Darwinism several times in his life. His influential Occidental professor, Roger Boesche, would see echoes of his political theory class in the speech. The reference to social Darwinism is also found in Mario Cuomo's keynote address in 1984, when Cuomo attacked Ronald Reagan's trickle-down principles as cruel to the working class. Obama's Knox speech was written in conjunction with his staff speechwriter, a young man named Jon Favreau whom Gibbs had recruited on the basis of their time together on the Kerry campaign. In crafting a speech, Favreau grabs his laptop and sits with Obama for about twenty minutes, listening to his boss throw out chunks of ideas. Favreau then assembles these thoughts into political prose. Favreau describes Obama as a master storyteller and a student of history, so Favreau is constantly looking through history books to draw parallels between the past, the present and a vision for the future. Even after a major speech is composed, however, Obama writes comments in the margins and tweaks passages right up until the time of delivery. The night before, he often stays up until the wee hours working on the final draft. As a professional long-form writer and a perfectionist, he can wrestle for hours whittling his elaborate thoughts into fifteen- and twenty-minute speeches. "It's hard for him to give up words," Favreau said.

Obama's Knox address drew almost no immediate press attention, largely because it was delivered at a small school in Illinois on a weekend. Only local press attended the event. But thanks to the power of the Internet, the speech sped through cyberspace, drawing cheers from liberals, who were beginning to worry that their newest prophet was not the one they had originally envisioned. Conservatives, however, looked at this speech and realized that even though Obama patiently gave heed to their arguments, and even though he was an eloquent and inoffensive Democratic voice, he was not one of them.

OBAMA'S SECOND MOST SIGNIFICANT PUBLIC MOMENT OF 2005 occurred in his self-imposed role of racial bridge-builder, a role that can cause deep worry among his cautious aides. Building bridges to Republicans, after all, can win moderate votes; talking too directly about race can alienate both blacks and whites. Yet as an African American, Obama realized that he could not escape the issue; overnight, he had become the country's highest-ranking black politician.

In September, against the wishes of some of his advisers, most of whom are white, Obama agreed to appear on ABC's Sunday-morning political affairs show, *This Week with George Stephanopoulos,* to speak on the tragedy of Hurricane Katrina. It was one of only two interviews he granted to major networks in his freshman year. The storm had killed more than a thousand people and left tens of thousands homeless amid the floodwaters of New Orleans and the coast. Traveling with Lugar in Europe at the time, Obama was personally affected by the enduring television images of thousands of mostly black, mostly poor people who had lost their homes and their livelihoods in the storm. As President Bush and local officials dithered and panicked, and as some Democrats and black leaders cried racism, Obama stood out as a voice of subtlety and reason.

Stephanopoulos prefaced his first question with the famous remark from hip-hop artist Kanye West that "George Bush doesn't care about black people" and asked Obama if racism was at the root of the federal government's seemingly lackadaisical response to the disaster. Obama responded in a typically measured fashion that alienated neither blacks nor whites and yet struck at the heart of the matter. He acknowledged that race was a component in the aftermath of the hurricane, but he did this without using the shrill and angry language that gives whites such great discomfort:

> I think that Kanye West expressed a great deal of anger, anguish that exists in the African-American community. I think that the entire country felt shame about what had happened. Now, my general attitude has been that the incompetence by . . . the Department of Homeland Security and by this administration was color-blind, but what I do think is that whoever was in charge of planning was so detached from the realities of inner city life in a place like New Orleans that they couldn't conceive of the notion that somebody couldn't load up their SUV, put one hundred dollars' worth of gas in there, put [in] some sparkling water and drive off to a hotel and check in with a credit card. There seemed to be a sense that this other America somehow was not on people's radar screen, and that, I think, does have to do with a historic indifference on the part of government towards the plight of those who are disproportionately African-American.
>
> . . . I think that in the African-American community there's a sense that the passive indifference that's shown towards the folks in the Ninth Ward of New Orleans or on the West Side of Chicago or in Harlem, that that passive indifference is as bad as active malice, and I think that the broader American community and white America in particular would make those distinctions and those fine lines. I think that the important thing for us now is to recognize that we have situations in America in which race continues to play a part, that class continues to play a part, that people are not availing themselves of the same

opportunities, of the same schools, of the same jobs, and because they're not, when disaster strikes, it tears the curtain away from the festering problems that we have beneath them, and black and white, all of us should be concerned to make sure that that's not the kind of America that's reflected on our television screens.

Obama's careful language was in stark contrast to the anger and frustration vented by black leaders of older generations. The Superdome became a temporary home to tens of thousands of storm evacuees who were stranded without supplies for days. Princeton University scholar Cornel West (ironically a hero of Obama's) said caustically, "From slave ships to the Superdome was not that big a journey." The Reverend Jesse Jackson likened the Superdome in New Orleans to the "hull of a slave ship." Jackson was a close friend and supporter of Obama's, so I asked him what he thought about their different approaches toward race. The question clearly irritated him. He said he worried that the careful language of younger black leaders would cede ground that had been acquired in the long struggle to equalize the playing field for blacks and whites. (In short, he meant that many of his own crusades for racial justice might be undone.) "Every speaker has the right to use his own style and try to assess his own angle," Jackson said. "But I was on I-90 in New Orleans when we saw people by the thousands and they were dying in people's arms. They were putting people in buses by blocks, women here and children there. And I said it looked like the hull of a slave ship. And it did."

Jackson was particularly perturbed by an editorial in the editorially conservative *Chicago Tribune* that compared the words of Obama and Jackson. The newspaper harshly criticized Jackson for viewing the tragedy in racial terms while Obama saw the issue not as entirely racial but as a matter of social justice among all economic classes. Jackson's "racism charge is simplistic and ridiculous," the newspaper said. "But it also could prove dangerous if it fosters the impression that government emergency plans aren't what's really in need of fixing. . . . What played out in New Orleans was more about economic class than race. The Senate's only African-American understands the distinction—and the need for the nation to address it with more than inflammatory rhetoric."

Thinking back on the editorial agitated Jackson, who shuffled his feet and wriggled in his chair. Obama "said he didn't see race, he saw class. I saw race because I was there. It was impossible not to see race if you were there," Jackson said, his tone growing almost defensive. "The *Tribune* editorial board took the position that an enlightened young guy saw class, but an old guy saw race. But the whole world saw what it was. It was poor class *and* black race in the hull of a Louisiana ship."

Jackson then worried that young blacks who had not gone through a tumultuous racial past might be more immune from seeing racism when it was before them. He also said that the conservative bent of the country since the Reagan era gave black politicians less ability to talk boldly about racial injustice. "I was jailed [for] trying to use the public library. I remember blacks being drafted for World War II and you couldn't vote," Jackson said. "I think Barack chooses to walk a very delicate balance, but sometimes it is not your walking that is the issue, it is what is beneath your feet. The right wing radically shifted the earth. It is . . . contract compliance and affirmative action that has made [Obama] possible. When the laws of the last fifty-one, fifty-two years that made his advance possible . . . they are taken away, you have to fight."

Jackson's racial anger and Obama's conciliatory tone represented the debate within the black community about how to approach modern race relations. As Jerry Kellman pointed out, there is no cause that Obama felt more deeply about in his heart than advancing the situation of African Americans. So far, Obama's political acumen has placed him in the perfect niche of white and black appeal that has eluded almost every black politician before him. Historically, black officials hewing too closely to so-called black issues, such as safeguarding government programs for the poor and challenging the Republican Party commitment to a fairer society, often found themselves losing support among whites. Conversely, black politicians reticent about venting racial anger and who challenge African Americans to study and correct their own deficiencies often lose support in their own community. Obama has followed both of these paths and thrived nonetheless. Thus, in political terms, Obama has struck gold when it comes to race. Instead of being torn asunder trying to please each racial camp, he has strung a tightrope between the two and walked it with precision. Obama shrugged his shoulders when I offered this theory to him at the end of 2005:

> I think there is a generational shift taking place in how core values that are important to the African-American community are expressed in a way that builds bridges with other communities. I think there's a majority in the African-American community who recognize that we have a multiplicity of voices, and not everybody's going to serve the same role—that Reverend Jackson or Reverend [Al] Sharpton is going to have a different role to play than someone like myself, who's representing all sorts of people. I just think that I am the most prominent of a new generation of African-American voices . . . [and] I actually have felt very comfortable speaking on issues that are of particular importance to the African-American community, without losing focus on my primary task, which is to represent all the people of Illinois. And I haven't felt contradictions in that process. I think that on every issue, whether it's a racially tinged issue or a foreign policy issue or a social issue, if I'm speaking honestly, if I'm speaking what I think, then usually things turn out all right.

CHAPTER

23

South Africa

*It is from numberless diverse acts of courage and belief
that human history is shaped. Each time a man stands up
for an ideal, or acts to improve the lot of others, or strikes
out against injustice, he sends forth a tiny ripple of hope,
and crossing each other from a million different centers of
energy and daring, those ripples build a current which can sweep
down the mightiest walls of oppression and resistance.*

—ROBERT F. KENNEDY, "DAY OF AFFIRMATION" SPEECH,
CAPE TOWN, JUNE 1966

*I realize that I offer these words of hope at a time when hope seems
to have gone from many parts of the world. As we speak, there is
slaughter in Darfur. There is war in Iraq. . . . And I have to admit,
it makes me wonder sometimes whether men are in fact capable of
learning from history, whether we progress from one stage to the
next in an upward course, or whether we just ride the cycles of
boom and bust, war and peace, ascent and decline. . . . And then I
thought that if a black man of African descent would return to his
ancestors' homeland as a United States Senator, and would speak
to a crowd of black and white South Africans who shared the same
freedoms and the same rights . . . then I thought: things do change,
and history does move forward.*

—BARACK OBAMA, "A COMMON HUMANITY
THROUGH COMMON SECURITY" SPEECH,
CAPE TOWN, AUGUST 2006

Barack Obama's journey to Africa had been planned since early 2005, shortly after he took the oath of office for the U.S. Senate. Scheduled for August 2006, it was one of the final pieces of The Plan, the two-year outline to keep Obama's star rising and his political power at its highest ebb. As with The Plan, the trip was devised by his top political minds—Chief of Staff Pete Rouse, media consultant David Axelrod, Communications Director Robert Gibbs and Obama himself. The "congressional delegation" trip, or CODEL in the official parlance of Washington, was designed to be various things: a fact-finding mission for the new senator, a family visit to his paternal relatives in rural Kenya and, perhaps most important, a public relations splash. The hope among Obama's team: to raise the senator's profile nationally and internationally; to solidify his support among a key constituency, African Americans; and to bulk up his foreign policy credentials.

Obama's trip, in many ways, would echo the excursions of two other iconic Democrats, both of whom took high-profile trips to Africa and reaped political benefits in the African-American community back home. President Bill Clinton, still beloved among blacks in the United States, was greeted deliriously over his twelve-day trip to Africa in 1998. And, in particular, Senator Robert Kennedy's 1966 journey to South Africa, where he forcefully denounced apartheid, sent a clear message to blacks in the United States. Kennedy's trip is the venture that Obama's most resembled—two young, charismatic, idealistic senators with presidential aspirations reaching out to desperately poor blacks on the globe's most often ignored continent. Images of Kennedy being mobbed by African blacks were beamed back to America through newspaper and television. "I believe there will be progress," Kennedy told residents of Soweto. "Hate and bigotry will end in South Africa one day. I believe your children will have a better opportunity than you did." And Kennedy's "ripple of hope" speech (actually titled "Day of Affirmation") is considered by some RFK biographers to be his best.

Because of the trappings that accompanied Obama's incredible star power, the African enterprise was much more successful as a major media hit than as a mission to imbue a first-term senator with greater knowledge about Africa. The trip took on a special fascination among the press because of the astounding market success of *Dreams from My Father,* of which a large portion was devoted to Obama traveling to Kenya in his early thirties to study his paternal African heritage and connect with his Kenyan relatives. In August 2006, the national media and various segments of the American public were enthralled with Obama's life story, and this was another way for them to explore his history and, consequently, another way for Obama and his aides to advance the rapidly growing legend around that unique ancestry. Thus, the trip became the focus of enormous media attention. Needless to say, with a swarm of Kenyan, American and international reporters documenting his every public move, it proved difficult for Obama to have anything close to a "normal" CODEL.

This is not to say that Obama's goal in traveling to Africa was not rooted in a certain idealism. Even before he was elected, he had visions of visiting that continent as a senator. In addition, conversations I had with Obama along the 2004 campaign trail made it abundantly clear that the atrocities of Darfur's civil war were a deep source of concern for him. In those conversations, Obama was hesitant to prescribe a specific solution for the civil war, but he believed that the African conflict deserved greater attention in U.S. foreign policy. As such, he also told a roundtable of journalists at the Democratic National Convention in July 2004 that the two places he would most certainly visit after his election were the Middle East and Africa. Also, as a senator, Obama was successful in passing an amendment to a 2006 Iraqi spending bill that increased aid to the Republic of Congo, one of his few legislative accomplishments as a new member of the minority party.

So in charting Obama's first two years in office, Obama and his advisers carved the Africa trip in stone. The idea, Gibbs told me in March 2005, just a few months into Obama's term, was to send the senator into the 2007–2008 national election cycle with his public image as strong as "humanly possible." Gibbs was not specific about whether that meant readying Obama for a presidential run or as a viable vice-presidential selection for whomever the 2008 Democratic Party nominee turned out to be. Gibbs was not specific because, in early 2005, Obama's long-term political fortunes as a senator remained a mystery, and at that point it would have been viewed as arrogant to have 2008 presidential aspirations, even if that was the case for Obama, Gibbs or other advisers. Furthermore, if a presidential run was the ultimate hope, there was no way to gauge if Obama's celebrity would remain strong enough to make a 2008 bid for the White House politically viable. But his advisers certainly were charting a bold course to strengthen and expand Obama's national reputation quickly, and those larger career decisions would come as events unfolded, all dependent on the execution and outcome of The Plan.

"Kenya will just be crazy—the media, the people, everything will be insane," Gibbs told me over a breakfast plate of eggs Benedict in a Chicago restaurant back in March 2005, a year and a half before the trip. As usual, his instinct was dead on the mark. The fifteen-day trip was organized to include visits to five countries, but the bulk of the journey was to be spent in South Africa and then Kenya. After Kenya, Obama had planned brief visits to the Congo, Djibouti and the Darfur region of Sudan, site of the bloody conflict that was killing thousands of Sudanese a month and displacing millions more. But Kenya, the homeland of his father, was the physical and emotional centerpiece of the CODEL. Since Obama's election to the U.S. Senate, Kenyans had adopted him as one of their own, and his rapid ascent to political power in the United States had made him a living folk hero in the East African nation, especially among his father's native tribe, the Luo. A beer named for Obama had gone on the Kenyan market after his 2004 convention speech (Senator beer); a school in rural Kenya was named in his honor; and a play based on his *Dreams* memoir had been staged earlier in 2006 at the Kenyan National Theater. Thus, Obama's brain trust expected large, enthusiastic crowds once he reached Kenya. And they were not to be disappointed.

On my way to Africa, I encountered Obama in a bookstore in the Amsterdam airport on the layover between my flight from Chicago, his from Washington and our connecting flight to South Africa. He wore his typical uniform designed for anonymity—a light charcoal gray synthetic jacket and a Chicago White Sox baseball cap fixed low over his eyes. We exchanged greetings and I did not try to engage him in a long conversation, realizing that we would be seeing each other every day for the next two weeks. Instead, I went into my campaign posture of giving him space, largely because of what I had seen awaiting him at the gate for the plane: nearly a dozen journalists, a handful of them toting video equipment. The media insanity was about to ensue. For the next couple of weeks, it would seem, Obama's every utterance and mannerism would be captured on video or audio.

OBAMA'S AFRICAN ADVENTURE BEGAN IN CAPE TOWN, THE PICTUResque city at the far southern tip of the continent. His first morning opened rather inauspiciously. At our hotel, the Table Bay—a modern, upscale facility that anchored a sprawling mall complex on the Cape Town harbor—an embassy official greeted him by asking if he had ventured out the night before with some of the media and other members of his CODEL. "I can't hang with these guys in their twenties and thirties," a tired-looking and raspy-voiced Obama answered somewhat tersely. By then, Obama had completed his second book, *The Audacity of Hope*. But a year filled with late nights of writing, a day job as a senator and weekend duties as a husband and father had taken its toll on him physically. Now, after another night in a faraway hotel at the end of a seemingly endless plane ride (actually, it was twenty-two hours), Obama tried to suppress his routine morning grumpiness. In any case, Obama no longer drank and was never one to grab a beer in the hotel lobby while on the road. At the end of the day, he would disappear into his hotel room and watch ESPN or touch up his speech for the next day or, more likely, both.

That morning, the reporters and videographers in the entourage got their first taste of the lack of organization behind the media end of the trip. The American-based reporting gaggle was already about a dozen deep, including several magazine writers, two documentary film crews and newspaper reporters from the *Tribune* and *Sun-Times* in Chicago and the *Post-Dispatch* in St. Louis. Yet despite all these inquiring minds, there was no physically produced schedule for the day's events. A couple of reporters openly groused about this state of affairs, and it was obvious that Gibbs had no clear idea how to control or appease all of us. He explained the night before that Obama had been permitted by Senate ethics officials to bring along only two Senate staff members on the CODEL—himself and Mark Lippert, Obama's foreign policy adviser. The lack of advance planning would soon wear on all involved, including Gibbs. But this was another example of Obama's lack of real power in Washington. Democrats were in the minority, and he had no leverage to convince the Republican administration that his trip was different from that of the rest and that he would need additional staffing, particularly to handle the media. Axelrod suggested that he hire a professional public relations firm from campaign funds to help organize the trip. "But the lawyers wouldn't let us do it," complained Axelrod, who was not on the trip. The result was that Gibbs told the reporting entourage in scattershot fashion what would be happening next.

The first event that day was the most significant: a cruise to Robben Island, where Nelson Mandela spent eighteen of his twenty-seven years in prison. This would be a day of symbolism—a black American politician visiting the solemn site where Mandela was incarcerated for leading, and ultimately winning, the fight against an unjust, virulently racist society in South Africa. If Obama were lucky, this story line would play across the globe on major networks and in major journalistic publications. And to this point in his Senate career, Obama had not been short on luck.

As the ferry pushed off from the Cape Town harbor, Obama settled into a seat next to his guide for the day, Ahmed Kathrada, an apartheid-era African National Congress leader who was jailed for eighteen years on Robben Island, much of that time alongside his friend Mandela. Kathrada's current appearance belied his youth as a rebel. He was slight of stature, bespectacled and wore white Nike running shoes and a maroon fleece jacket, which gave him the look of an innocuous tourist rather than a retired antiapartheid activist. As the low morning sunshine illuminated Kathrada and Obama in a yellowish glow, still photographers snapped pictures and documentary film crews scurried about. Furry boom microphones hovered overhead as Kathrada provided Obama with a historical overview of the prison site. Obama initially shot a wary glance at the big microphones but soon went about his business as just another celebrity tourist to the island, a place that had been visited by such luminaries as Bill Clinton and Oprah Winfrey. Over the course of the day, Kathrada would tell Obama that guards kept the roughly fifteen hundred prisoners in nearly complete societal isolation, refusing, for example, to tell them that Americans had landed on the moon. They were permitted only to send out one five-hundred-word letter every six months. Obama also learned that a caste system based on the shade of one's skin had been in place in the prison. Lighter-skinned prisoners of Asian heritage like Kathrada, who were called "Asiatics," were treated slightly better than darker-skinned African blacks.

Once the boat docked, Obama and Kathrada led the march of media and other interested parties up to the uninhabited prison about fifty yards away. The spotlessly clean facility was constructed of gray stone, quarried on the island by the former prisoners, and had been slightly renovated into a museumlike showpiece. The two men stepped down the narrow hallways and quickly reached Mandela's cramped prison cell. Photographers and reporters pushed together outside the door to document the moment inside, hoping to hear anything that Obama might utter and grab a clear photo of him inside the cell. Just then, Pete Souza, the veteran *Chicago Tribune* photographer, with a keen eye for the dramatic, scrambled away from the pack and into the prison yard, where he hopped up on a gray wooden bench just outside the small, barred window to Mandela's cell. Souza later explained that he had remembered a famous photo of Clinton visiting the cell that had been shot from that external vantage point, and he sensed that the same image of Obama would be perfect. Several others followed Souza's lead and ran after him in hot pursuit. A photographer inside the cell mentioned the Clinton shot to Obama, prompting the senator to respond with "Oh, really?" Talking with Kathrada, Obama had already taken clear notice of the history behind him—and now he suddenly took notice of the historic media opportunity before him. With Souza outside shooting through the window, Obama straightened his shoulders, pushed his jaw forward and squinted his eyes into a serious gaze. Souza's photo in the *Tribune* the next day, which ran across the globe on wire services, offered a pensive-looking Obama peering out the window from behind the steel bars. Several other photographers filed a similar captivating image. Though there were several more hours of public appearances, with that serious pose, Obama's work for this day had been done.

His second day in Cape Town again revealed his deft political touch, although it was more cerebral and less theatrical in nature. He visited a community health center that mostly treated AIDS patients, consulted with an outspoken AIDS activist and shared a private moment with a beloved global figure, Nobel Peace Prize winner Desmond Tutu.

In 2006, South Africa was suffering through one of the most

severe AIDS epidemics in the world, with one in five people in the nation—nearly five million—infected with the virus, according to the United Nations. South Africa's leaders had come under heavy criticism for promoting its spread through unsound public statements that flew in the face of scientific evidence. A former South Africa vice president, for example, had recently conceded that he had unprotected sex with a woman suffering from AIDS. And not only that, the politician claimed that a shower afterward would reduce his risk of infection.

The health center was located in Khayelitsha, a poor township amid miles and miles of tin-roofed shantytown shacks in stark contrast with the modernity of Cape Town. Outside the clinic, Obama was pressed by reporters to speak about South Africa's AIDS crisis and what should be done to quell it. Obama had been seeking a meeting with the country's president, Thabo Mbeki, who was one of the politicians who seemed least concerned about the deadly impact of AIDS on his constituents. Mbeki had publicly questioned whether the HIV infection led to AIDS, a scientific fact known the world over. Here, Obama was caught in something of a dilemma. How broadly should he criticize the current government and risk scuttling his potential meeting with Mbeki?

Obama chose to come out swinging. He charged that the government was in "denial" about the crisis, and he advocated a "sense of urgency and an almost clinical truth-telling" about the spread of the disease. "It's not an issue of Western science versus African science," he said. "It's just science, and it's not right." He then dropped that day's major headline: He would take an AIDS test when he reached Kenya in hopes of erasing the stigma behind the disease among Africans. AIDS is spread primarily by heterosexual sex in Africa, yet most Africans choose to die rather than be tested. With these controversial proclamations, it now looked unlikely that Obama would meet with Mbeki to lobby him to address the AIDS crisis. Yet however ephemeral his statements were that morning, Obama gave voice to a crisis that was killing hundreds of South Africans per day. Few world leaders had spoken out so vigorously on the handling of the crisis by the South African government. "It sends this message of political leadership, of being prepared to be open about HIV," said Zackie Achmat, one of South Africa's most notable AIDS activists. "We wish more politicians were that honest."

The afternoon meeting with Desmond Tutu was a low-key affair. It was held in Tutu's office inside a rather prosaic stretch of two-story, yellow-brick commercial buildings that looked as if they would fit comfortably into a nondescript office park in suburban middle America. Tutu wore a gray cardigan sweater and gray pants. In a brief appearance before reporters, he lavished praise on his celebrity visitor. He told Obama, "You're going to be a very credible presidential candidate." To this, Obama replied with his "Aw, shucks" demeanor, although he didn't seem at all rattled by such a prominent figure envisioning great things for him. Tutu joked, "I hope that I would be equally nice to a young white senator." After a chuckle from Obama, Tutu added: "But I am glad you are black."

Back in Cape Town that evening, Obama delivered a fairly non-controversial forty-three-minute address before an attentive audience culled by a progressive think tank. Gibbs handed reporters a copy of the speech, but as he often did, Obama deviated from the prepared remarks almost immediately. "Well, he stayed with it through the first ten words," Gibbs said to me with a roll of his eyes.

In this speech, titled "A Common Humanity through Common Security," Obama stressed his familiar theme of an interconnected humanity. But here in South Africa, the common bond was not just among good-hearted Americans but among well-meaning people stretching across borders and across continents. He cited Mahatma Gandhi and Martin Luther King's influence on the antiapartheid movement in South Africa, and how that movement, in turn, spurred activism back in the United States in the 1970s and 1980s. He said modern threats such as AIDS, nuclear proliferation, terrorism and environmental degradation should bind people together across the globe, not divide them. He offered few specifics as to how that should occur, but asserted that there should be an "overarching strategy" to coordinate cooperation among nations, with the United States playing a leading role. He called on America and South Africa to partner to help weaker nations "build a vibrant civil society." His penultimate moment came when he observed that his very presence in Africa provided living proof that humanity was moving forward. He closed with that favorite quote from Reverend King about the arc of the moral universe slowly bending toward justice.

Obama was still fatigued, and consequently he walked fairly dryly and slowly through the well-written text. A magazine writer asked me later if I had ever seen Obama look this tired, and thinking back to that commencement address when he was so sleep-deprived that his knee buckled onstage, as well as other such occasions, I replied that I had. But as is often the case, his energy level spiked when he finished with the prepared speech and took audience questions. At the conclusion, audience members, many of whom had never heard of Obama, showered him with hearty applause. Several attendees whom I interviewed said they were in full agreement with his hopeful message, but one noted that it fell well within a conventional political framework. "It was very interesting and he is very level-headed. I certainly think he is a very good ambassador, a very able politician," said David Wheeler, a retired university instructor. "He put across the position of the United States as being beneficial to the rest of the world." It was also worth noting that Obama's underlying message was that a black politician from the United States who had African roots might just be beneficial to the rest of the world as well.

BY THE THIRD DAY, FATIGUE WAS SETTLING OVER THE MEDIA EN-
tourage, and frayed nerves were evident from even the most patient

individuals in the group. Immediately following Obama's speech the night before, we had left Cape Town and driven a couple of hours to Pretoria. The next morning, after a short night's rest, reporters gathered in the Pretoria hotel lobby and readied themselves to be hauled off to that day's events, which at this point were unknown to them. This fact had already irritated several reporters, who had wanted to know how to prepare for the day. As the media grumbled, an embassy official appeared. He informed us that Obama had no public events scheduled for that day. This did not shock me or anger me, since Gibbs had told me before we departed that there would be "down days." But it did come without fair warning. The news particularly unsettled several newspaper reporters, whose editors most likely were expecting a story to be filed every day. What were they to file today?

Patience with Gibbs and the lack of advance work was now extremely thin. It did not go unnoticed that he did not deliver this unwelcome news himself, but sent an embassy official to face the media. "He's a total obfuscator," a documentary filmmaker said of Gibbs. (One documentary film crew had been hired by Axelrod. The second was on contract with Hollywood actor Edward Norton's production company.) Not only had precise scheduling been absent, but there had been virtually no access to Obama in private moments. Unstaged moments often make the most tantalizing scenes of a documentary. But as I had learned two years before in the Senate campaign, Obama intensely guards his personal time, what precious little of it there is. And in his weary state, he certainly would not agree to cameras invading his hotel room or his traveling vehicle. Nor would he countenance a writer sitting next to him and gauging his private moods. My long-standing relationship with Obama perhaps gave me the best opportunity for direct access to the senator, but ever since Gibbs appeared on the scene in the general election campaign and began his long, tightly controlled reign over media relations, I had learned to live with greatly restricted access compared with the early Senate campaign days. And by now, I had also reconciled myself to the fact that no amount of pushing Gibbs would change this reality. Indeed, even the documentary crew most sympathetic to Obama—the group hired by his own media consultant to produce flattering footage that would be used as campaign material—was irritated about the lack of private access to the senator.

Fortunately for the newspaper reporters, the day soon provided some real news—none of it good for Obama. He learned that, indeed, President Mbeki would not meet with him. The official reason: an Iranian delegation was in Johannesburg for a summit to discuss their country's decision to move forward with a uranium en-richment program. "It would look inappropriate if the president were to meet with Obama with the Iranians here," an official with the U.S. Embassy said. Obama later speculated that his harsh words about the government on the AIDS crisis did not help his cause. The second bit of bad news was that Obama was forced to cancel his visit to the Congo because of violence surrounding a presidential runoff election. These events occurring in tandem emphasized that, despite the media glorification of this trip, Obama had no actual power to affect global policy. Indeed, even if he were a senator idolized by America's progressives and canonized in the media, he really was not a major player on the world stage. At least not yet.

To satiate the unfed American reporters, who hadn't seen the senator all day, Gibbs made Obama available for a news conference in our Pretoria hotel in the early evening. Obama arrived in black suit jacket and white shirt, but quickly noticed that the assembled re-porters were dressed down. We had been tourists most of the day, after all. To fit in with the reporters, Obama slipped off his suit coat and dropped into a soft chair in his stock white dress shirt. He slowly rolled up his sleeves to look even more casual. And for the first time on the trip, he looked fresh and physically rejuvenated. He certainly had gotten some badly needed rest that day, as well as a badly desired visit to the gym. I learned later that he had worked out and followed it with a long nap, two things that Gibbs most likely did not want reported back in the American newspapers. One could envision that headline: "Obama Lands in Pretoria, Takes Nap, Hits Gym."

Obama was his typically collected and well-spoken self during the news conference. Nevertheless, he tightened considerably when questioned by the two reporters who had covered him most aggressively back in Washington—Lynn Sweet of the *Sun-Times* and Jeff Zeleny, then of the *Tribune*. Since arriving on the continent, these two fiercely competitive journalists had been in a fitful contest to send home the most scintillating tidbits about Obama's adventure, and both had been working and cajoling Gibbs mercilessly. Sweet was constantly in his face, while Zeleny plied him with drinks at the bar late into the evening. Sweet was pulling multimedia duty and was a perpetual ball of chaos. She not only filed daily stories but authored a blog for the *Sun-Times* website *and* sent back both video and still photography. Not being trained in television media, she produced video dispatches that had the feel of narrated vacation footage. Moreover, her constant battles with the wobbly tripod that held her video camera provided amusement to all around. The thirtyish Zeleny penned daily stories and, along with Souza, compiled several handsome audio-video packages for the *Tribune* website. Sweet, a veteran Washington reporter whose demanding manner could border on abrasive, had long tested the nerves of Obama. He had once hung up on her in a phone interview. And Zeleny, in addition to pushing Gibbs for information, was not shy about stepping up to Obama whenever a pertinent question struck him. The often imperious senator seemed to maintain a level of respect for Zeleny's professional dedication, but at the same time it was apparent that he preferred to own his personal space at all times, and Zeleny did not mind invading it. For all of this tension, Obama's Africa visit received mostly positive and nearly play-by-play coverage on the websites of both newspapers, leading his critics to charge in web postings, quite incorrectly, that Zeleny and Sweet were, in fact, media toadies for the senator.

In the press briefing, Obama told reporters that he had been careful not to criticize the United States too harshly while he traveled abroad, but said he could feel in South Africa "some negative impressions outside our borders that we're going to have to deal with." He said America's decision to invade Iraq was responsible for that. "I think the perception is that not only did we act unilaterally, but that we have essentially determined that our interests and concerns and viewpoints are the only ones that are relevant," he said. "You hear a lot of discussion that the United States dictates its foreign policy as opposed to cooperating with other nations. So I think there is a lot of work that we're going to have to do in the coming years to recover the levels of legitimacy that I think we had." He also addressed questions about how he felt bringing a media circus with him to visit his Kenyan relatives. He had last visited Kenya fourteen years before while researching his *Dreams* memoir, and he had come alone. He was far from alone now. "I'm going there as a United States Senator, but this gives me an opportunity to reconnect and find out what's going on and find out what folks need," he said, sidestepping the question. "My anticipation is that I will be able to help in the future in terms of projects and ideas that they want to pursue. But no matter what happens, there is always going to be some level of discomfort just because there is this huge gulf between life in the United States and life in Kenya." Obama also said that he worried that his visit would be "hijacked" for political gain by some Kenyan politicians, particularly the Luo tribe, to which his father belonged. This, as it would turn out, was a legitimate fear.

Day Four in Africa jumped headfirst into activity. We drove to Soweto, a Johannesburg suburb that gained international attention in June 1976 with the Soweto Uprising, mass riots spurred by the white government's decision to force black students to be educated in the

Afrikaans language rather than in English. Soweto is now a middle-class suburb of blacks that houses a museum dedicated to the uprising and its most famous victim, Hector Pieterson, a thirteen-year-old killed when police opened fire on protesting students. With Hector's sister Antoinette as his guide, Obama toured the Pieterson museum, which is largely ignored by the locals but draws a good number of tourists. A few American tourists who patronized the museum recognized Obama, shook his hand and asked for autographs. The museum workers, meanwhile, asked reporters who he was.

With media crews buzzing around them, Antoinette solemnly walked Obama along the museum's exhibits. They gazed at photographs of Mandela and other images from the antiapartheid movement. When they reached the most dramatic moment of the tour, Obama knew exactly what to do. The two stopped in front of a wall-sized print of the iconic photo of the lifeless body of Antoinette's younger brother as he was carried from the protest scene in the arms of another young man. The riveting image, taken by a news photographer, was publicized around the world and helped to galvanize the international community against apartheid. Though the focus of the photo is on the limp dead teen, the viewer's eyes also wander to seventeen-year-old Antoinette running alongside the young man holding her dead brother. Her mouth is agape and her right hand is raised helplessly into the air. In a "feel-your-pain" moment reminiscent of Bill Clinton, Obama slid his long slender arm across Antoinette's shoulders and pulled her against his thin torso. She reached around his waist and pulled him tighter. The two lingered in front of the huge photo as flashbulbs feverishly flickered behind them. "That was the shot there, man," the *Tribune*'s Souza observed. "Just a great shot, and Obama knew it."

Outside, through a light rain, Obama offered a short speech as he stood with Antoinette before a memorial to her slain brother. Obama often pays tribute to the leaders of the civil rights movement in the United States by saying that their efforts paved the way for his success. Here in Johannesburg, he did much the same, noting that his first political activism came in college when he protested apartheid and advocated divestment of American funds from South Africa. "If it wasn't for some of the activities here I might not have been involved in politics," he said.

The next quick stop was a museum in Soweto dedicated to Rosa Parks, the black seamstress who helped launch the civil rights movement in the United States by refusing to give up her seat to a white person on a bus in Montgomery, Alabama. Obama glad-handed the curious onlookers throughout the museum, which truly resembled a small library, and then ran across something that prefaced events to come later in the trip—a framed black-and-white photograph of Robert F. Kennedy during his seminal trip to South Africa in June 1966. Kennedy, standing atop the roof of a car amid a sea of black South Africans, was leaning forward and extending a hand to the enthused crowd. In the coming days, there would be scenes similar to this one for Obama—only they would play out in his homeland of Kenya. Here in South Africa, he was barely recognized. Seeing the photograph, Obama could not help himself. He glanced down at the image and a half smile grew from a corner of his mouth. "You know," he said to a person in the entourage, "my desk in the Senate is the same desk that Robert Kennedy had." Whether Obama had meant to draw a parallel or not, the image was drawn.

CHAPTER
24
Nairobi

This is where he belongs. He just goes there to work [in America],
but he should and will come back home to be one of our own.
—A KENYAN WOMAN

Obama's arrival at the Jomo Kenyatta International Airport in Nairobi the next day bespoke the utter madness that was to mark
Obama's six-day Kenyan adventure.

The difference between the cultures of South Africa and Kenya
was immediately evident. Nairobi is a city of more than three million people, but the first thing one notices after arriving at the airport from South Africa is the lack of white people. And the whites
who were there, like me, were immediately approached aggressively
by any number of smiling Kenyans and offered assistance, by carrying a bag or giving directions or supplying a taxi. This assistance
was for a fee, of course.

Another noticeable difference in Nairobi was the ubiquitous
presence of uniformed police officers, many of them toting assault-
style rifles. The atmosphere was far less Western-oriented, more
fragile and clearly more dangerous than in Cape Town. Nairobi
might be Kenya's capital and the center of culture, business and politics in all of East Africa, but it had been pushed into becoming a
modern urban mecca far too fast. In the early 1960s, the city's infra-
structure had exploded into place after Kenyans won independence
from colonialist Britain. This had resulted in some neighborhoods
appearing completely modern, and even middle-class or better, by

Western standards. But not far away were sprawling slums without potable water, indoor plumbing or electricity. Roads curved for no apparent reason, and traffic lights seemed to barely contain the autos speeding along the streets. Paved sidewalks were nonexistent, with pedestrians walking along uneven red-dirt paths in close proximity to moving traffic.

Kenya was a functioning democracy, but it still operated heavily on a tribal caste system. Political parties were divided along the various ethnic tribal lines. Bitter rivalries existed among these tribes, and the resultant political warfare had severely hindered economic and civic growth. Obama's father hailed from the Luo tribe, which made up about 13 percent of the country's population and had a strong farming background, mostly in the western region.

Only a few minutes after my arrival at the airport, I noticed that certain crimes were either overlooked or, perhaps, encouraged. Once I claimed my luggage, I went to confirm my flight reservations to ensure that arrangements for my departure six days later were in order. As I waited in line, two white European men in their fifties talked with a tall, lithe Kenyan woman who appeared to be in her early twenties. The men squeezed the attractive young woman's behind, ran their hands up and down her thighs and then offered her a small wad of money, which she stuffed into a pocket in her tight white jeans. The men looked as if they were measuring cattle, but she didn't seem the least bit offended and, ultimately, walked away arm in arm with each of them. To be sure, such a transaction might occur in airports in any number of cities around the globe. But what was most revealing: This one went down just several feet from a cluster of uniformed police officers in berets, dark uniforms and with assault weapons in their hands.

The senator had traveled from South Africa on a U.S. government jet, and by good fortune or bad luck, some of us in the media were at the airport when he landed. Obama's staff had hoped to keep his entrance a secret, but of course Kenyan politicians had tipped favored reporters to his evening arrival time. Most of the American reporting gaggle would have missed it too if it had not been for an unsettling occurrence. Axelrod's camera crew was held up at the baggage entrance trying to get its video equipment through security. Kenya's reputation for rampant corruption was well known, and back in the States, the leader of the documentary crew, Bob Hercules, had sent money ahead to a Kenyan "fixer" to ensure that their equipment would make it through the customs agents. Nevertheless, Hercules soon found that the payment—a bribe, if you will—had not secured safe passage. (Another media crew from Chicago also paid a bribe to get its equipment through. Together, the bribes equaled about eighteen hundred dollars in U.S. currency.) As Hercules and his crew haggled with airport officials to get their equipment released, the environment outside the terminal suddenly changed. An eerie silence washed over the evening dusk and about a hundred people started gathering in small groups along the roadway and in the medians. They were strangely quiet, expressed little emotion and were all looking toward a building in the front of the airport where a couple of dozen security personnel had gathered. In a hushed voice, a man explained, "Obama is here." Ah yes, the young prince was returning to his father's homeland.

A moment later, Gibbs popped out of the building, quickly surveyed the scene and disappeared back inside. So I headed toward the building, where a clutch of media and police had amassed, and soon grew aware that being white might actually be a plus—it might get me through the thicket of bodies that had been cordoned off by police. What else could a white man with a notepad and camera be other than a Western-based news reporter? Sure enough, police allowed me and a magazine reporter past the first media barricade.

With Gibbs's help, we wended our way through the mob of people inside the building before Gibbs put a hand on my upper back and pushed me into a small back room where Obama was enduring a quick photo "spray" with select Kenyan media. The media hit had been thrown together on the spot, and it showed. There was no focus to it. Obama had only agreed to sit down for the shot when the foreign minister told him that he wanted to "take care of my guys in the press," Gibbs later explained. "Barack didn't want to say no, so we sat down and did it. It wasn't our idea."

The photo spray lasted all of two or three minutes. Some people carried cameras and some did not. It was difficult to distinguish the journalists from the onlookers, the plainclothes authorities from the civilians. People without the proper clout were physically escorted out of the room. Those with journalistic clout, including myself and a couple of other American journalists in the Obama entourage, were permitted to stay. Dressed in one of his navy blue suits and a light blue shirt, Obama was sitting on a chair with the Kenyan deputy foreign minister. Cameras whirled all around. Obama smiled and tried to look relaxed, but I could see by his rigid jaw that his Hawaiian calm was eluding him. When Gibbs abruptly announced that the spray was ending, Obama tried to ease the tense and chaotic atmosphere by telling the assorted gathering, "You'll be tired of me by the end of the week."

Outside, another hundred or so people had gathered along the streets leading up to the various terminal buildings. They stood under palm trees and along curbs and one man hoisted a little girl onto his shoulders. I stepped across the small street from the building and watched as Obama made his first public appearance as a U.S. senator in his father's homeland. To my surprise, when he came through the doors, the crowd reacted with near silence. They simply stood and watched in quiet reverence. Obama, with a government official at his side, stepped quickly toward an awaiting white Ford Explorer parked at the curb just a few steps from the building. Gibbs had instructed him not to stop and take questions, or even acknowledge the cameras and gathering crowd. But Obama walked up to the vehicle and could not help but look out to the people. Discarding Gibbs's advice, he seemed to realize that it might appear impolite to altogether ignore the crowd around him. Besides, as a skillful politician, it is deeply ingrained in Obama's psyche to acknowledge an audience amassed for his benefit. Finally, one photographer yelled at him, "Wave!"

So Obama raised a crooked arm and waved stiffly, like a wiper across a car's windshield, or like the infamous Richard Nixon's bon voyage wave as he stepped onto the plane after resigning the presidency. Obama then flashed a forced smile and ducked into the SUV. The vehicle burned rubber as it sped away, with a twelve-car convoy piloted by embassy officials and police in tow. The scene more befitted a visiting head of state than a junior member of a foreign country's legislature. I breathed a heavy sigh and felt the adrenaline rush begin to subside. This was clearly not going to be the same laid-back atmosphere as in South Africa, where our subject could roam the streets in relative anonymity and events seemed more orchestrated than organic.

Obama moved swiftly to the hotel in the speeding caravan, but the rest of rush-hour traffic was stymied, thanks to Kenya's widely acclaimed guest. The bus carrying my grouping of the media gaggle took an hour to reach the Nairobi Serena Hotel, even though it was just a few miles away. Roads were closed all through downtown to allow Obama's motorcade easy access, and this severely jammed up traffic. At the hotel that evening, the first order of business for Obama: interviews for the Chicago TV media. Each of Chicago's major network-affiliated stations had sent a reporting and camera crew to cover the Kenya visit. International press, including writers for *Time* and *Newsweek* magazines, had also arrived. David Axelrod's old media chum Mike Flannery from Chicago's Channel 2, a CBS affiliate, headed Obama's interview list. The relationship between Flannery and Obama extended back to at least the Senate campaign, when Flannery's coverage of Blair Hull's marriage files and drug use contributed to the burial of Hull's candidacy. "How's it going so far, Robert?" Flannery asked Gibbs upon spotting him in the hotel lobby. "Oh," replied a harried Gibbs, "I'm like a one-legged man in an ass-kicking contest."

Gibbs, endeavoring to bring a sense of order to the chaos, held a 9:15 P.M. briefing for reporters in a casual meeting room on the hotel's ground floor. The vibrant aroma of after-dinner coffee, one of Kenya's primary crops, emanated from the restaurant area of the hotel. Reporters filled a long table, some couches and a handful of chairs. Gibbs finally had preprinted daily schedules for us. There were a number of new faces in the press corps, including international wire services and, most notably, the Chicago crews. Gibbs warned that a frenetic atmosphere would be the norm. "I'm not sure if you folks were at the airport," he said. "But we're going to find that even when things are not advertised, some Kenyans will gather." *Some* Kenyans—that would prove the understatement of the week. "What we learned today—expect the unexpected," Gibbs said in concluding the briefing. "Now the fun begins."

OBAMA'S FIRST OFFICIAL KENYAN FUNCTION OPENED THE NEXT morning, and it highlighted the deified nature of his presence to many Kenyans. Michelle and their two daughters had arrived the evening before, and the family appeared at the Nairobi State House for a morning ceremony welcoming the senator. The event was held outside the State House under a tent. Dozens of embassy employees, both black and white, wore orange-and-yellow T-shirts with OBAMA IN THE HOUSE emblazoned on the front. Songs had been composed for Obama's visit, and a group of clapping and finger-snapping Kenyans harmonized over these lyrics: "When you see Obama has come to Kenya, this day is blessed." As Obama opened his speech, he was interrupted by a friendly, but misplaced voice. Eight-year-old Malia shouted to her father, "Daddy, Daddy, look at me!"

No one could have been more pleased to see Obama, yet felt less blessed, than Christopher Wills, an Associated Press reporter based in

Illinois. Wills had covered Obama when he was still in the state legislature and the burgeoning Obama phenomenon was still relatively confined to progressives and blacks in the United States. Thus, the AP honchos in New York and London made no objection to Wills being the lead reporter on the trip to Africa. Wills promised his editors in Illinois that he would write a couple of newsy feature stories from Kenya. By the time Wills arrived in Nairobi, however, the dynamics had shifted greatly within the AP. The wire service's London bureau had finally recognized the significant media buzz that Obama's journey was drawing worldwide. As a consequence, the AP's Nairobi bureau chief was nagging Wills by cell phone to supply half-hour updates on Obama's every move, giving Wills a severe case of the jitters. This unexpected turn of events came after Wills had undergone an agonizing experience with the Kenyan embassy in the United States to attain the proper travel credentials. Obama is always mindful to cultivate friendly relationships with the reporters who cover him, and he is happy when there are a good number of them around. He is happiest, though, when they are kept at a safe distance. So when Obama spotted Wills amid the media gaggle, he made sure to acknowledge him. Or perhaps, less cynically speaking, Obama simply spotted a familiar face and it comforted him. With Obama, as with many of the best politicians, it is never perfectly clear whether he is being politic or merely human. In either case, in contrast to the regal nature of the proceedings around him, Obama yelled out from his crowd of Kenyan government dignitaries, "Chris Wills! You made it! You got your visa!" A slightly bewildered Wills didn't seem to know how to react to the unexpected shout-out. He responded: "Uh, yes, Senator. Thank you."

Obama met that morning at the State House with senior government officials, including Kenyan president Mwai Kibaki. After triumphing in the December 2002 elections, Kibaki's National Rainbow Coalition (nicknamed Narc) took control of the government in 2003, ending nearly four decades of rule by Kanu, the

Kenyan African National Union. Kanu was widely viewed as corrupt and had been accused of land-grabbing and raiding public coffers for private enrichment. Kibaki won the office on a pledge to rid the country's institutions of corruption and revitalize its economy. But more than three years later, corruption persisted, the economy was largely stagnant, and Kenyans who had been optimistic at Narc's electoral success were again pessimistic about the direction of their country and its institutional leadership. "People have just kind of given up on the government," a veteran Kenyan journalist, Dennis Onyango, told me. "They feel we'll never get what we want." Obama, in his meeting with Kibaki, discussed with the president the importance of clean government. The American senator maintained that investment from overseas would never arrive if Kenya's government and business communities remain soaked in graft. He cited the airport bribes paid by Chicago media crews as evidence that corruption remained pervasive, that it was a corrosive element in everyday society and that it negatively impacted Kenya's international image. Kibaki replied that he was working to stamp out corruption and promised to look into the airport bribes.

Obama's next stop was a meeting with government officials and business leaders at a restaurant secured in a plaza behind huge wrought-iron gates. Judging by the secluded design of the restaurant, which made it easy to protect with a couple of guards at the closed gates, I assumed that it was the site of many high-level lunches among top Kenyan politicians.

It was here that we would first witness the intensity of emotion that Kenyans felt toward Obama, the emotion we had read about beforehand in various media accounts. Those news stories had not been overblown. Outside the restaurant, workers of many pursuits, all hungry for a glimpse of their American hero, had left their jobs to crowd atop balconies, huddle in doorways and press against the iron fences. Obama met privately for lunch with local officials and, as we waited, reporters fanned out to interview the Kenyans amassing outside the gates. The interviews bore out the state of idolatry surrounding Obama among the Kenyans.

To some, he was a native son who had risen to great power in the world's most influential nation, and because of this he gave them hope that they or their children could persevere and succeed in their own daily lives. To others, he was an all-powerful political figure who could put Kenya on the worldwide radar. To still others, he was nearly a deity, an ethereal figure who would bring riches and all good things to Kenya from the promised land of America. This last group believed that Obama truly belonged in Kenya, not America.

A forty-year-old woman named Catherine Oganda maintained that Obama ultimately would choose to leave America and live in Kenya: "This is where he belongs. He just goes there to work [in America], but he should and will come back home to be one of our own." I asked why she believed that, and she continued: "Because the father is a Kenyan. You know, your father is your bloodline; it's not your mother—it is your father. So you belong where your father comes from, in your fatherland. Kenya is in his blood." A fifty-year-old man named John Nyambalo had a slightly different take, but one that was no less divorced from reality. He saw Obama as a living representation that the United States had overcome racial intolerance. "If the Americans can select a senator like Obama," he said, "that means that Americans embrace the whole world and they are true democrats. There is no racism there."

After lunch, our caravan headed to the memorial that had been erected at the former site of the U.S. Embassy, which had been car-bombed in 1998, killing nearly two hundred and fifty souls. The deadly bombing, later linked to the Islamic fundamentalist terror movement that struck the United States on September 11, 2001, had helped to create a bond between the United States and Kenya. Both countries suffered from the attack. Dozens of people stood at the entrance of the memorial site waiting for Obama, with Michelle and their two young daughters among them. Obama shook the hands of a long row of current embassy staff on his way up toward the site, with Michelle nearly last in line. The last few introduced themselves to Obama, and then Michelle smiled and held out her hand and offered the same, as if she were just another member of the greeting party. "I'm your wife, welcome," she said with a warm smile. "Hello, Wife," Obama said with a playful grin.

Gleaming in a brilliant sun, the memorial itself stood at the far end of a plaza just beyond a small fountain. It was rather unassuming, giving the appearance of an elongated headstone on a burial plot—a concrete block in the shape of a half-moon rising from the plaza's bricked surface. Its facade was a sheet of brown marble with the names of the deceased etched in it, as well as the following epitaph: "May the innocent victims of this tragic event rest in the knowledge that it has strengthened our resolve to work for a world in which man is able to live alongside his brother in peace." A couple dozen photographers and TV reporters were assembled at the far end of the plaza, readied for the shot of Obama at the memorial. Through the trees that guarded the memorial site, located in the heart of Nairobi, I could see a large crowd assembling in the streets—more Kenyans with hopes of catching just a passing glimpse of Obama. Also witnessing the scene were workers in a seven-story office building that overlooked the park. They leaned out of big steel-framed windows and peered down on the proceedings with rapt attention. With his right arm wrapped around Malia's waist and Sasha standing at his left elbow, Obama sat down at a white-clothed table and signed an official guest book before heading over to the memorial with his family in tow. He carried a wreath and laid it gently at the foot of the tombstone. Then he turned to a small group of officials huddled to his right as photo and video crews knelt and stood not far to his left, their cameras clicking away.

Scanning the epitaph, Obama bowed his head and offered his own words of consolation and remembrance: "The tragedy that happened here is a reminder that, ultimately, all of us suffer from conflict and, ultimately, all of us suffer from terrorism. But we have to redouble our resolve, as the memorial says, to find ways to live in peace and to find ways to resolve our conflicts in a way that does not result in the kind of tragedy that occurred here. We will not forget what's happened here. We want to make sure that all of us are vigilant in terms of preventing it from ever happening again."

After the brief ceremony, Obama was taken inside a nearby building to chat with embassy and other government officials. Malia and Sasha, dressed in bright pink tops and white skirts, were set loose to play in the memorial area that doubled as a small corner park. I hadn't bothered a tired-looking Michelle when she first arrived at the hotel the day before, so this seemed like a good opportunity to reconnect. She was strolling around, shaking hands and eyeing her daughters as they ran about happily. But our chat was cut short. Immediately after we exchanged greetings, a roar erupted from beyond the trees. Its sheer volume startled Michelle. She leaned her upper torso far backward and a stunned look crossed her face. "Oh my goodness! What was that?!" she exclaimed. "That," I said, "is for your husband. He must have come out." The wondrous look slightly receding from her face, she replied innocently, "Oh, my! For Barack?" Clearly, Michelle was in no way prepared for this overheated response to her celebrity husband.

We both headed for the narrow exit to the memorial and Michelle was gobbled up into a pack of security personnel. The crowd in the streets, consisting mostly of men, had reached a state of euphoria. They were cheering in full throat, standing atop cars, dancing, whistling and screaming and waving their arms wildly. Police had established a perimeter at the edge of the street. Yet even though the crowd seemed wild and uncontrollable, no one had stepped a single foot past the Kenyan officers, as if an invisible wall held them in check. The people were chanting in unison: "Obama, come to us! Obama, come to us!" I looked for the senator and spotted him to the right along the perimeter with several security officers packed around him for protection. He was feverishly shaking hands with members of the fawning crowd in a surreal press-the-flesh moment. With each step he took toward the street, closer to the frenzied mass of people, the chanting rose a notch in volume. "Obama, come to us! Obama, come to us!"

I watched the senator from a safe zone inside the perimeter about ten paces behind him. Incredibly, the scene was growing ever more chaotic as Obama worked his way closer to the belly of the throng. A horse carrying a police officer, spooked by the noise and instability of the crowd, bucked his front legs into the air and nearly kicked me in the head before the officer reined him in. "Be careful! Don't get yourself trampled!" warned a perpetually tense Jennifer Barnes, the embassy's media liaison. As I wandered closer to the edge of the perimeter, within a few feet of the first row of people, a woman from the crowd suddenly lunged toward me and grabbed my left bicep. Before I could pull away, an officer swung his black billy club and cracked the woman square on her forearm. Her arm fell limply to her side and the officer pushed her back into the sea of people with his club, swiping his club casually, like a chef pushing a pile of crushed onions across a cutting board. I decided that I better keep a safer distance from the crowd. The scene was so full of heightened emotion that even the most innocent acts became hyperreality. Bill Lambrecht, a reporter for the *St. Louis Post-Dispatch,* handed a woman a piece of paper from the embassy that contained background information on Obama's visit. Lambrecht figured she could have it as a souvenir. Five men immediately jumped the woman and successfully ripped the paper from her hands. Lambrecht shook his head in discouragement upon seeing what his nice gesture had wrought.

Just then, as I turned backward and took a few steps, I saw Obama heading in my direction. It became apparent that I was about to be sandwiched between two surging walls of humanity—the crowd before me and the crowd behind Obama. Unfortunately, I had no security detail guarding me from harm. To escape the fate of trampling, I galloped sideways to an opening. Standing in this safe spot was Bob Hercules, the documentary director sent by Axelrod. "This is intense," Hercules said. "It's like the messiah has returned."

I spied Gibbs, who was surveying the proceedings from an opening a few feet behind Obama. His arms were folded and he wore a satisfied half smile on his face. Gibbs should have been smiling, I thought. This is exactly the kind of messianic outpouring of idol worship, exactly the kind of crazy mob scene that he had envisioned a year and a half earlier when he first mentioned the Africa trip to me over eggs Benedict in downtown Chicago. The legend of Obama was growing by leaps and bounds. Indeed, The Plan was going exactly according to script.

AFTER A COUPLE OF MINUTES OF THIS CHAOS, GIBBS GRABBED Obama's sleeve and advised him to head back to an awaiting SUV. This move delighted security personnel guarding Obama, who were struggling to shield him from the crowd and had reservations about the press-the-flesh session from the outset. As they escorted Obama back toward the parked caravan of vehicles, Gibbs noticed a young man holding aloft a black-framed, eye-catching portrait of Obama. It was painted in brown, black, gold and white and it featured a profile of Obama with an ethereal glow engulfing him. Beneath him were the words *Waruaki Dal,* or "Welcome Home." Obama had already made his way back into the safety of his SUV. But seeing this young man, Gibbs realized there was more to milk from the madness of this media event. Gibbs grabbed the arm of the young man and tried to lead him back to Obama's SUV. Instead, Gibbs ran into a team of security personnel reluctant to let the man pass. An intense discussion between Obama's media maestro and the security officers ensued. Security eventually relented to Gibbs's persistent demands— but only after patting down the young man aggressively. Gibbs extracted Obama from the SUV and introduced him to Gregory

Ochieng, a man in his twenties who hailed from a rural village near the farm of Obama's paternal relatives. About a half-dozen TV cameras rolled as Obama graciously accepted Ochieng's painting and thanked him profusely for the gesture. The meeting resembled a young fan meeting an athlete or a musician whom he idolizes. Within hours, that encounter, along with the surreal scene that preceded it, was beamed across the globe and appeared in newscasts worldwide.

Ochieng, a member of the Luo tribe, told several of us in the media that he felt a deep connection to Obama because Obama's father had been a Luo. "He is my tribesman," Ochieng explained. "I feel happy that a Kenyan is representing us in the U.S. as a senator. So when I heard he was coming here, I thought of doing something that was unique." Asked by reporters if the encounter lived up to his expectations, he said with a broad smile, "It is better." Gibbs's instincts were note perfect. Again he had struck PR gold.

If the overwhelming outpouring from thousands of Kenyans was not enough to signify that Obama's trip was something wholly unique, the next event would solidify it. Media members literally filled a ballroom at the downtown Nairobi Grand Regency Hotel for Obama's news conference. Every journalist in Africa seemed to be in attendance. Wearing a gold-lined lapel pin with the flags of Kenya and the United States molded together at the center and then spread like wings on a bird, Obama stood behind a brown wooden lectern with more than one hundred representatives of the media before him. He emanated a slightly regal air that, in this setting, felt far more presidential than senatorial.

Obama opened the news conference by recognizing that this visit was remarkably different from his previous trips to Kenya as a private citizen. This time, he sought to be a "bridge between the two nations." "Part of my role," he said, "is to communicate how much the American people appreciate the Kenyan people and how much they value the partnership that the United States has with

Kenya. Part of it, I think, is also to listen and find out what is on the minds of the Kenyan people. . . . Part of my goal is also to maybe highlight some of the values and ideals of the United States that I think might be helpful to the Kenyan people as they pursue their development."

Toward that last point, Obama delivered a message of morality, as he is wont to do. He suggested that the very cultures of some African nations needed to change. He said he perceived a movement backward, toward a political system riven by tribalism and a reluctance to acknowledge internal problems to the outside world:

> I think that there's a tendency—which is understandable, given the history of colonialism—to not want to speak out against fellow Africans and to be protective of even some of the mismanagement, the corruption that takes place. And I think we've moved beyond that; I think the time is now, where we have to understand that nobody in Africa wants to be bullied, nobody in Africa wants to see the products of their labor expropriated by a government that is not representing them properly. Nobody wants to be tortured to death because of speaking their mind. Nobody wants to have to pay a bribe in order to get a business or get a job or just go about their daily business. And it's incumbent upon us, when we see those things happening, to speak up. . . . I think ultimately, that kind of honesty will improve governments everywhere.

Arriving back at the Serena Hotel, Gibbs hastily called a news conference in a small courtyard area for the traveling press, mainly so that TV reporters would have informal interview footage of Obama discussing the day's events. As Gibbs scoured the hotel to make sure everyone was aware of the Q&A, Obama made small talk with the reporters, who were anxiously waiting to query him about the wild scenes of the day. After a bit of talk about baseball and music, Obama began to grow impatient, as Gibbs still had not appeared. "Where is Gibbs?" Obama said to no one in particular. "Where is the animal trainer? Where is the whip and chair?" This remark did not go over well with several reporters, who frowned at the suggestion that they were, at the very least, uncouth and, at worst, so hopelessly uncivilized that they needed to be tamed. A magazine writer sensed this elitism in Obama and included it in her piece about the Africa trip in *Elle* magazine a few months later. The writer, Laurie Abraham, conceded that she had been charmed by Obama's writings and media appearances. But mostly, she had been drawn to the brash, idealistic, soul-searching Obama whom she encountered in his *Dreams* memoir. After following him closely for several days in Africa, she had shelved this sense of Obama in favor of a more pragmatic vision of an elected official entering middle age and making compromises for political benefit. She found a more calculating man—a politician, surprise!—and an occasionally aloof one at that. "When he is not working a crowd, he can seem so sublimely cool and confident that his manner veers toward haughtiness," Abraham wrote.

Others in the traveling entourage experienced this less appealing element of Obama's persona as well. One videographer told me that he grew weary of being treated like something of a nuisance on the excursion: "I feel demeaned when Obama says stuff like that. We're over here working our asses off. We're not leeches. We were invited here, even pitched on this trip. I think we've been very respectful of his space." Indeed, the American press had been far less aggressive in physically pursuing Obama than Kenyan reporters, who had no qualms about throwing an elbow to push another photographer out of the way or pushing up to the front of the line. Obama's privately haughty manner and wary posture toward reporters seems rooted in two things: an internal conceit that formed in his character after being treated as a special human being as far back as childhood, and an understandable antagonism toward those individuals who, since he gained celebrity status, have impinged on what little privacy still exists for him.

Throughout the interview with the American press, Obama tried to downplay the effect he would have on Kenyan society. "Kenya is not my country. It's the country of my father," Obama said. "I feel a connection, but ultimately, it's not going to be me, it's going to be them who are climbing a path to improving their new lives."

Siaya: A Father's Home

It's not just God we praise, but Obama too.
—ELDERLY KENYAN WOMEN CHANTING TO OBAMA

This was Obama's third trip to the small rural compound of his father's family. He visited just after college in 1983 and then again after Harvard Law in 1991 in order to research his *Dreams* memoir. The family farm was located near a town called Kolego in the Siaya District of Nyanza Province. This western province sat on the edge of Lake Victoria, the world's second-largest freshwater lake, which is about the size of Ireland. Siaya was home to various modest farming and trading villages and its residents were largely poor, with more than half living on less than a dollar a day. The district's largest city was Kisumu, which, at roughly a quarter of a million residents, was Kenya's third most populous town. The region was inhabited almost exclusively by Luo, the tribe that Obama was astonished to learn as a child lived in mud huts and subsisted by farming their own land. Kisumu was about a hundred and seventy-five miles east of Nairobi, which could be a drive of seven or eight hours through the wickedly difficult terrain separating the remote villages and towns in western Kenya. The traveling press and Obama took separate flights on commercial airlines to Kisumu. We then joined a caravan of vehicles that somehow would navigate the crater-marred, red-dirt roads that extend into the far rural reaches of Siaya.

The atmosphere surrounding Obama's previous visit to the family compound had been vastly different from that of this CODEL. Back then, Obama had boarded a train in Nairobi and ridden it through the night and the morning across the mountainous Rift Valley. When he arrived in Kisumu, he had walked half a mile to a bus station by himself. With his older half sister, Auma, as his primary guide, he had been greeted by a handful of relatives at the station. "Obviously there's been a big shift in terms of my travel accommodations," Obama told reporters in Kenya. "The last time I arrived in my grandmother's village, there was a goat in my lap and some chickens." He dropped into Kisumu this time by way of a forty-minute morning plane ride aboard an East African Airways jet.

The Kenya Airways flight carrying the press gaggle landed at the Kisumu airport before Obama's plane. The airport was actually a one-runway airstrip, with flights landing and departing hourly. Separated from the runway by only a group of thin trees and a sidewalk, a squat, dirty yellow building held the cramped, cluttered offices of three commercial airlines. I was surprised to find only a few interested onlookers, given that Kisumu was one of Kenya's largest urban areas and given the crowds we had seen in Nairobi. But when Obama's flight landed, a mass of people seemed to materialize from thin air. Obama was dressed as he would have been for any number of campaign swings through Illinois—dark blue blazer, white shirt, beige pleated plants, casual leather shoes. As he and Michelle stepped down from the plane, the media descended on them, with half a dozen boom microphones magically appearing over their heads. The enthusiastic crowd of a couple of hundred turned out to consist mostly of Peace Corps workers from the United States. Obama strode up to the airport building as media members pushed against one another to gain the best possible footing. "How's everybody doin'?" Obama asked in his elegantly cool manner. But when the media kept pace with him as he entered a holding room labeled Government and VIP Lounge, Gibbs was forced to intervene and give Obama some breathing room to chat with local officials. First Gibbs shouted for everyone to leave the room. Then he began yanking them out one by one. "This is a prelude to a big mess," said one of the documentary filmmakers. "Lord, help me," Gibbs mumbled under his breath as he escorted one camera crew after another out of the room.

Obama's public day was not yet ten minutes old, and Gibbs was already fighting it.

Obama was loaded into an SUV and the traveling press was herded into two small rickety buses, affectionately named Samson and Delilah by their owner, a local man, and his son, who eked out a living by ferrying tourists around Kisumu and Lake Victoria. With these buses, appearances did not lie. A few minutes into the trip through Kisumu, our bus sputtered and lost power. As it glided into a gas station, its passengers groaned, mostly out of fear of losing the caravan and missing out on Obama's main event at the family farm. Had we traveled all this way only to miss the final act? Jennifer Barnes, the U.S. Embassy official riding with us, was livid. "Why would you show up for a daylong trip with no petrol!" she shouted at the driver. (This, in fact, was the second time a vehicle carrying the press contingent had run out of gas. A taxi in Pretoria met the same fate after dinner one night, albeit that was in a much better neighborhood than this.) The driver seemed to take forever filling up the tank, prompting an exasperated Gibbs to exclaim, "Fill this fucking thing up!" Once it moved again and we caught up to the motorcade, Gibbs commanded, "Jesus, now don't lose him!"

We wended through town after town toward the New Nyanza Provincial General Hospital, where Obama and Michelle were to take their AIDS tests. It was Saturday, and along the sides of these roads were makeshift markets selling everything from fruit to American T-shirts to Air Jordan basketball shoes. Every Kisumu street was lined with waving, hollering and overjoyed people, hoping to catch a glimpse of the Obama motorcade. Again—there were thousands of them. I noticed that some of these Kenyans, desperate for any item of American culture, were wearing our hand-me-downs. One boy walking near a railroad track wore a T-shirt that displayed the 2006 Super Bowl champion Seattle Seahawks. But Seattle lost that game to the Pittsburgh Steelers. The shirt obviously had little value in the United States and was shipped to Africa.

Arriving at the hospital, we encountered managed bedlam. Thousands upon thousands had turned out to see Obama. And to our dismay, the senator's car had disappeared into the sea of people ahead of us. He was to be dropped off close to the hospital while our decrepit Samson and Delilah parked a couple of football fields away. Reporters scurried out and began marching toward the hospital in high gear, again worried about missing a key moment of the day.

The closer we got to the hospital and its surrounding park area, the thicker the crowd became. Soon, people seemed to occupy every square inch of land. They stood on rooftops of the hospital buildings, sat along balconies and incredibly had climbed into trees and dangled from the limbs. Most reporters were separated trying to make their way through the dense throngs. A couple of journalists were pick-pocketed of recording devices and other gear. The Kenyans were simply rapturous, chanting rhythmic verses and screaming for their hero, Obama. Some wore T-shirts bearing Obama's name or image. Others held aloft pictures of him or waved flags adorned with his name or face. Nearly everyone in this remote part of Kenya spoke or sang in Luo, and we American reporters had to ask Kenyan media members to translate for us. One chant was taken from a 2002 election when former Kenyan president Daniel arap Moi lost power in the reform movement. The verse went: "Everything is possible without Moi." But on this occasion, it had been altered to "Everything is possible with Obama." A row of four older women, their eyes rolling back into their heads, looked almost possessed as they writhed and shimmied to an indigenous African rhythm.

As on the day before in the streets of Nairobi, an invisible barrier had been formed by authorities carrying billy clubs and assault rifles.

This opened a wide safe zone in the park area for an AIDS trailer and the dozens of media members. Obama and Michelle made their entrance in a cluster of people and headed for a mobile clinic operated by the U.S. Centers for Disease Control and Prevention to take their AIDS tests. Though on a far smaller scale than in South Africa, this part of Kenya nevertheless had the highest incidence of AIDS in the country, with about one in seven people infected. Gibbs and Obama said they were told that his public AIDS test could spur hundreds of thousands of Kenyans to take a test themselves. Obama and Michelle disappeared inside the mobile unit to thunderous applause. After the husband and wife were pricked in their fingertips and blood samples drawn, Obama stepped outside and stood atop the trailer steps to address the enthused audience.

He took a microphone in his left hand and held aloft his right arm to speak and subdue the crowd. Instead, the frenzied masses drowned him out and began to push ahead. Police, apparently worried that people would be trampled or smothered, allowed the surge to move forward a bit. Obama urged calm. "Stop pushing, no pushing," he implored through the microphone and loudspeaker. His words went unheeded and Obama pulled the microphone away from his mouth. "This is intensity," he said calmly to himself. "I am going to take a seat." Obama sat down on the top step and his eyes caught mine. I smiled and said, "Wow. This is way beyond LeBron, baby." Obama returned a slight smile. "This is . . . ," he said, pausing to find just the right word—an innocuous word, a cautious word. He finally settled on "interesting." "This is interesting," he said again.

AFTER THE BLOODLETTING FOR THE SAKE OF THE AIDS CRISIS, THE Obama caravan rumbled off to another of Obama's causes—a multimillion-dollar CARE Kenya project in Central Ugenya that benefits orphans. Obama partially financed the project, which was geared toward empowering older women who care for orphans.

The project provided them with financing to buy such things as sewing machines. To reach the event, the convoy navigated the uneven and unpaved roads into what seemed like one of the most remote parts of the world. We chugged by collection after collection of tiny green farms marked by thatch huts with mud or tin roofs. Dust floated everywhere, partially swept up by the long brigade of vehicles and partially lifted up naturally from the occasional barren field in the verdant and hilly terrain. Driving through this obstacle course of far-flung red-clay roadways took a special talent. It was also not for the faint of heart. The roads were barely wide enough for two oncoming vehicles to pass each other. The final half-dozen feet on either side of the road fell into a steep grade that would make Samson and Delilah lean so far to one side that we seemed to be almost parallel to the ground. Drivers hurtled their machinery toward oncoming traffic at unbelievably fast speeds considering the treacherous road conditions. It was a classic game of chicken as to which vehicle would give ground, and it was assured that this maneuver would take place at the last instant before a possible collision. Seemingly in defiance of the laws of gravity, our buses managed not to tip over on the journey. After dozens of miles of this rough, white-knuckle travel, it was easy to see how Obama's father, a poor driver who was a heavy drinker, died in an auto accident while navigating this dangerous terrain.

When our buses finally arrived at the CARE Kenya project, it was clear that we were in the far reaches of Africa, a place largely bereft of clean running water and electricity. The event was set amid a deep green thicket of trees and dust. About a hundred and fifty rural Kenyans, many in traditional African garb, gathered to partake in what can best be described as a spiritual ceremony in Obama's honor. Nearly all of those in attendance were women and children, just as most of the people back in downtown Kisumu were men. The women generally remain in the rural regions, tending to family farms and child rearing, while the men travel to town for urban jobs and send money back home. The ceremony resembled a religious experience. A handful of the older women danced and pranced in song, at one point showering Obama with these words in Luo: "It's not just God we praise, but Obama too." Obama danced with the native women and addressed those assembled through a jerry-rigged loudspeaker system powered by a car battery hidden behind a thick tree. In a short speech, he called this "a wonderful homecoming." He looked as relaxed in this atmosphere as he does anywhere.

As Obama played politician, I couldn't help but notice Michelle, who was sitting behind him with a bored-looking Malia and a fidgety Sasha leaning up against either side of her. Michelle's facial expression registered somewhere between a scowl and a frown. She was unmistakably trying to process how exactly she had found herself and her family being feted as some kind of deity in the middle of remote Kenya. When the spotlight landed on her, however, she brightened and, as usual, played the good wife of a senator. She even stood up and danced in a circle with the native Luo women. But Michelle was still trying to adjust to this new world of Obama glorification. When the event concluded, the *Tribune*'s Zeleny and I asked her how she was coping with this strange new environment. "I haven't digested it all yet. It's all a bit overwhelming," she confessed. "It's hard to interpret what all of this means and what it means to us as a family." When a few more reporters gathered around, Michelle realized she was beginning to sound impolitic, and she changed gears. "This doesn't really make me think of Barack Obama and his fame and fortune," she said, seemingly searching for some larger meaning behind this odd trip. "It makes you think of what you can do to help here. . . . The spectacle is interesting, but in the end, this has to be all about more than Barack Obama."

BACK IN THE CARAVAN, WE WERE OFF TO THE MAIN EVENT OF THE day—and perhaps the entire trip: Obama's visit to his father's farming compound, the specific physical roots of his paternal African heritage. Unbeknownst to Obama's trip planners or staff, presidential candidate Raila Odinga had assembled thousands of Luo for a political festival in Obama's honor. For Odinga, Obama's visit was a political gift. For the entire trip, Odinga endeavored to gain as much publicity for himself as he could by cozying up to this beloved figure from the United States who happened to have a Luo heritage. Odinga even had T-shirts printed with the image of him and Obama and the humble declaration AFRICA'S GREATEST SONS.

In an effort to remain above Kenyan tribal politics, Obama's staff had been coy about revealing the senator's schedule to Odinga. Still, the time of his visit to the family compound was widely known to locals. Kisumu, in fact, had been in a carnival atmosphere for several days, with nightclubs celebrating Obama's trip in raucous fashion. So there was no way to get around the orchestrated political assembly, which had all the trappings of the same kind of event in the United States. Local officials and their minions, who most likely had some connection to political patronage of the Luo, gathered at Odinga's request to see Obama. A group of uniformed schoolchildren had also amassed outside the schoolhouse named in Obama's honor. They stood in a long row, swaying their hips and waving their arms rhythmically to a song composed specifically to honor their American visitor. The lead voice was a boy about twelve years old who crooned in a beautifully melodious manner as the other children harmonized behind him. To the melody of the American song "This Land Is Your Land," the children had supplanted American geographic points of interest with Kenyan spots: "This land is your land, this land is my land, from Lake Victoria to the Coastal Province, from Nairobi to the Rift Valley, this land is my land alone." They stood amid buildings of the primitive-looking Senator Obama Kogelo Secondary School. From the financial donations of Obama, the school had purchased chalkboards and wooden desks and various types of science equipment. The problem is, the science experiments would undoubtedly require running water, and the schoolhouse appeared to have none. The school featured just three small buildings, with one classroom each, on the edge of a big hill overlooking a small green-brown valley. The classrooms were less than inviting, just concrete floors and worn wooden tables and chairs. Most unfurnished basements in the United States look more modern and better equipped for schooling. "Hopefully, I can provide some assistance in the future to this school and all that it can be," Obama said at a quick dedication ceremony, careful to lend support but not to overpromise.

The magnitude of the upcoming political ceremony caught Obama and his small staff off guard. The large turnout would have outdone most such gatherings on the South Side of Chicago in the heat of election season. The vibe, however, was strangely similar. America might be an ocean away from Africa, but in a democracy, politics is still politics. A reporter asked Gibbs facetiously, "Does this meet or does it wildly exceed your expectations?" Gibbs only chuckled in response. Wearing brightly colored robes over their Western-style suits and ties, the Luo Council of Elders sat in white lawn chairs underneath a long row of canvas tents set up to shield them from the hot sun and ubiquitous dust. Aside from their colorful garb, they looked like any group of American politicians waiting their turn to speak. Standing with his slender brown arms folded was Michael Adara, a thirty-five-year-old information technology worker in Nairobi. He said he had made the excursion from the big city to this rural outpost to see the senator in the flesh. "The thing that attracts me and other people to Obama is that he traced his roots to right here," Adara told me. "It lets us feel that we can all trace our roots and find our real home." Adara wore a black T-shirt with an image of Obama's face and the words THE SENATOR, THE DIPLOMAT, THE POLITICIAN, THE LUO.

Speaking in Luo, Odinga, the presidential candidate, opened the festivity with a wordy talk about the meaning of Obama's visit and what one surmises was a stump speech. As a member of the challenging Luo tribe, he called for an end to government corruption, a sentiment that drew sustained applause. It was then Obama's turn. He stepped onto a wooden table, grabbed the microphone and, sensing the political rally atmosphere, dropped into an extemporaneous variation of his own campaign speeches from back in his Senate race. The theme, once again, was the interconnectedness of all people and the unifying nature of the human condition. But in adjusting to the crowd here in Kogelo, Kenya, instead of talking about his own life, he substituted his father's life story. Speaking plainly in his midwestern drawl, Obama said:

> As I was driving up here, I thought about my father. Some of you may be aware, I didn't know him that well. He actually did come back here to Kenya. I was the one who stayed back, stayed back home, stayed back in the United States. We corresponded. We spoke. But I did not grow up with him. It wasn't until as an adult that I came to visit this area. I remember the first time that I came, I thought to myself that even though I grew up on the other side of the world and even though I had not had a day-to-day connection, when I came here I felt the spirit among the people that told me I belonged. Everybody was so warm and so gracious and so friendly and hospitable. One of the things that you realize about this area is that even though a lot of times people don't have a lot, they are willing to give you what they have.
>
> There is a generosity of spirit in this community, which is extraordinary. As I traveled through here, one of the things I realized is how remarkable it was the journey my father had traveled. He grew up around here. He was taking care of goats for my grandfather. And maybe sometimes he would go to a school not so different from Senator Barack Obama's school, except maybe it was smaller and they had even less in terms of

equipment and books. And teachers were paid even less and sometimes there wasn't enough money to go to school full time. Yet despite all that, because of the health of the community, the community lifted him up and gave him the opportunity to go to secondary school and then go to a university in America and then get a Ph.D. from Harvard and then come back here and work with many of the individuals who are here today. It's a story of what's possible when a community comes together and supports its children.

After Obama concluded, the moment of the day was at hand. He was off to visit his father's compound and his grandmother just a few hundred yards from the school site. The only problem: how exactly to get there. Obama tried to shake some hands at the political function, but as usual, the staff hustled him into an SUV. Reporters, meanwhile, broke into two camps. One group headed for Samson and Delilah. As for me, I kept my eye on Gibbs, who advised that we should just walk rather than board a bus that would fight crowds to travel less than a mile. Gibbs made a small attempt at organizing the reporting crew. But suffering from the same fatigue that we all felt as the day wore on, he quickly gave up. "It's like herding cats," he said in frustration. "They'll find their way, I hope."

After about a fifteen-minute walk, we arrived at the small farming compound. There were several small clay buildings with tin roofs amid a spread of grass, weeds and dirt. Chickens and goats wandered here and there, as if they were the tenants of the community taking a stroll around the grounds. Small farming fields of rice were on either side of the compound. Most people living in this part of the world subsisted daily on the food they produced at home—rice, eggs, cabbage. A clutch of Kenyan reporters and other international press were already assembled awaiting Obama's arrival. A few dozen relatives of the Obama clan were there as well. One thin elderly man who was dressed all in white—cotton shirt, pants and matching hat—displayed a blue-and-white "Obama, Democrat for Senate" button on his chest. He leaned on a wooden cane and looked around in bemusement at the bizarre scene: media crews carrying modern boom microphones and digital cameras to a place where televisions and computers did not exist. Most of Obama's relatives had donned their best dresses and suits. They were, after all, gathered for a weekend family function, although one covered by scores of reporting crews, many from another continent. The "spectacle" that Obama had mentioned was in full display.

Obama's caravan arrived at the compound, but Samson and Delilah and most of the American reporting gaggle had not. When Obama exited the vehicle to greet his grandmother, media and relatives swarmed around the two. (His grandmother is not actually a blood relative. She is the woman who raised his father. The circumstances surrounding his blood grandmother's disappearance from the family are murky.) Obama embraced "Granny," as she is known, while reporters and photographers and security and other official staff pushed and shoved each other, either to protect Obama or to document the moment, depending on the job they held. Kenyan photographers were particularly aggressive. They threw elbows to open a clear shot at Obama and his grandmother.

Amid the frenzy, Obama and his grandmother strode slowly up a slight grade toward the main house, which had a new tin roof and fresh coats of blue and white paint, thanks to Obama's recent financial help from afar. About halfway up the small hill, Obama stopped in midstride and recognized that he was missing his own companions. "Where are my wife and children?" he asked plaintively. They had fallen behind the mass of marchers surrounding him. Sasha then appeared before him and he scooped her into his arms. With chaos all around, she looked frightened and grabbed her father tightly around the shoulders and neck. "I have you, Sasha," her father said soothingly. Obama started walking again and I noticed that Pete Souza, the *Chicago Tribune* photographer, had finally made it to the compound. Souza had missed Obama's arrival because the media buses were fighting the crowds, just as Gibbs predicted, and he was exasperated. "This is out of control! Just absolutely out of control!" he observed. "I mean, I barely got one shot." As for myself, I was busily taking notes and endeavoring, once again, not to get trampled by the pack.

Reaching the main house, Obama and his grandmother disappeared inside for their first visit in nearly fourteen years. The family reunion was initially scheduled to last nearly two and a half hours, after which Obama was to get some time alone at his father's and grandfather's gravesite, located on the compound. But because of the unforeseen events and chaotic atmosphere, after about forty minutes Obama emerged from the house and stood, arm in arm, between his eighty-three-year-old grandmother, Sarah Hussein Obama, and Auma. They all fielded questions from the press. He said the family had eaten porridge and chicken. Asked if his grandmother had given him any words of wisdom, Obama answered without hesitation: "Don't trust reporters." When the questions were over, an American television correspondent based in Africa asked Obama to autograph his copy of *Dreams*. This drew disapproving stares from print journalists, who considered such a request to reveal a severe lack of objectivity.

Obama was then supposed to visit his father's grave, the site of his emotional climax in his *Dreams* memoir. But such a private moment was unattainable in this atmosphere. Instead, aides pulled him away from the press and guided him into the SUV for the trip back to the Kisumu airport and then Nairobi. A U.S. military liaison, who was part of Obama's traveling entourage, surveyed the madness and shook his head. "It's a fucking circus," he said. "I feel bad for Obama."

CHAOS WOULD FOLLOW OBAMA INTO THE NEXT DAY, WHEN THE CODEL would take us to one of the bleakest places on the planet.

Kibera (pronounced Kee-bear-a) is recognized as the largest single slum in all of Africa, and thus in all the world. Between seven hundred thousand and a million impoverished souls are packed into a tract of urban land that is just two-and-a-half square kilometers. Situated in the southwest quadrant of Nairobi, Kibera was first settled extensively in the 1920s when British colonizers allowed a group of soldiers from what is now Sudan to establish homes on a wooded hillside on the outer reaches of Nairobi. The ethnic group, called Nubians, had fought for the Allies in World War I as part of the King's African Rifles. Even though the British allowed the Nubians to live on the land, the English never gave the group official title to the territory. The Nubians established a community that they called Kibra, meaning "jungle" or "bushes." Nevertheless, with no legal claim to the territory, they were essentially squatters.

This lack of recognition of Kibera followed through the entire twentieth century. Even after independence from Britain in the 1960s, the Kenyan government never officially recognized the community. No title deeds were issued. No sewage or water lines were constructed. No real power was bestowed upon the poor, who despite their privation swelled the population of Kibera. Most residents moved there from rural villages to seek better schooling in Nairobi or find jobs in the large city. Even as the population surged, however, the community remained in the shadows of Nairobi. Much like poor neighborhoods in the United States, Kibera was rarely visited by Kenyan politicians save for election time when they are seeking votes from all constituencies. Interestingly enough, Illinois's senior U.S. senator, Dick Durbin, had toured Kibera several months before. Durbin's visit, however, attracted none of the fanfare that Obama would, even though Durbin at the time held much greater power in the Senate as assistant minority leader. When I mentioned to a Kenyan journalist that Durbin had recently visited Kibera, he said that Durbin's visit had garnered virtually no media attention.

Kibera was every bit as distressed as it had been billed. Many residents lacked basic services, such as clean running water and plumbing. Sewage and garbage were dumped into the open; dwellings were made of canvas and tin with corrugated roofing; and some children appeared less than fully nourished. The inhabitants, however, were positively gleeful at Obama's visit. And like elsewhere, they turned out in droves. The motorcade could only move at a snail's pace through the densely populated community because people had filled the streets and swarmed around the vehicles. A colorful mural painted on a wall on the village outskirts paid homage to Obama. It featured a man sitting on a barstool at an "Obama joint."

Accompanied by Michelle, Obama attended two organized events in the slum: one that discussed a program involving microfinancing of small businesses in Kibera and a second that outlined a program to educate young people about HIV/AIDS prevention. But it was outside the events where Obama made the biggest media impact. Crowds had flooded around the small tin-roofed building housing the first event, and authorities had to push open a walking path for Obama and his wife to reach their motorcade afterward. As he stepped outside onto a dirt path leading to his SUV, Obama grabbed a bullhorn and raised it to his mouth, but the overzealous crowd drowned him out. He looked down, smiled and began again. "Hello!" he screamed in Luo. "Everybody in Kibera needs the same opportunities to go to school, to start businesses, to have enough to eat, to have decent clothes," he told the residents, who madly cheered his words. "I love all of you, my brothers, all of you, my sisters. I want to make sure everybody in America knows Kibera. . . . Everyone here is my brother! Everyone here is my sister! I love Kibera!"

Obama's next morning opened with a tree-planting ceremony for the sake of the environment. Shovels in hand, Obama, his wife and his daughters dropped an African olive tree and surrounding dirt into the middle of Freedom Park, a wide-open downtown Nairobi green space. Throughout the short event, a handful of photographers edged in front of their peers, jockeying for position, drawing the ire of

Obama's two-man security crew from the U.S. government because the photographers initially resisted moving back. As the disagreement subsided, one security man complained to the other, "Fuckin' asshole journalists, man." His partner nodded in agreement. The park had been cordoned off to outsiders, and another crowd of about five hundred Kenyans had amassed outside the fencing to see Obama. He grabbed another bullhorn and gave them a quick hello before ducking back into his dark SUV and being whisked back to the Serena Hotel for lunch. A fatigued-looking Gibbs shrugged. He turned to a few reporters and said, "Let's walk back to the hotel. You know, you never see the real streets of a country when you go on these kinds of things. I'd like to feel the real streets of Kenya for ten seconds." The moral here: Even ruthless political operatives have a softer side and want to experience the real world on occasion.

That afternoon, after all the spectacle of the trip—the staged events, the press-the-flesh moments, the trite banter with reporters—Obama stepped into perhaps his most comfortable environment, a situation that almost always inspires him. He gave a speech at a college. Obama's address before a crowd of about a thousand students and academics at the University of Nairobi was carried live on the country's largest television network and rebroadcast twice. The setting was a large, rather prosaically designed campus auditorium that Souza, the *Tribune*'s photographer, immediately saw as having limited artistic potential. Instead, Souza left the room and captured images of students who had stopped whatever they were doing to listen in rapt attention to Obama. Loudspeakers carried Obama's words into courtyards and cafeterias and study rooms, where students appeared transfixed by the American senator. As for Obama, his timing was perfect, his voice was as rich as ever and thus his performance was at peak level. It was hard not to witness this kind of speech by Obama and not envision a presidential run. His presence, confidence and moral clarity filled the room.

His address was a call for Kenyans—in particular, young

Kenyans—to work toward ending the country's culture of corruption and ethnic politics. He asserted that positive change is almost always brought about by idealistic youths rather than older adults who have internally reconciled society's injustices. In calling for an end to corruption, Obama spoke in generalities about who was responsible for this ill, and he sidestepped assigning specific blame to a political party or leader. "Here in Kenya, it is a crisis, a crisis that is robbing an honest people of the opportunities they have fought for, the opportunity they deserve," he said. "Corruption has a way of magnifying the very worst twists of fate. It makes it impossible to respond effectively to crisis, whether it's the HIV/AIDS pandemic or malaria or crippling drought."

The senator received warm and occasionally enthusiastic applause during the speech. But this audience of Nairobi's most well-educated youth was clearly more discerning about Obama than the roaring crowds in the streets. Some students with whom I spoke after the address said they could sense a definite charisma about Obama, but they were disappointed by his lack of specific remedies for Kenya's endemic problems. They wanted him to take a harder line on the current political leadership. "There are people here with so much hatred toward the government that they wanted a direct attack," said Dennis Onyango, a senior writer for the *East African Standard*. "They wanted him to name names." But Obama was diplomatic about the source of the country's corruption problems, which was on message in his mission for the trip. He strove to voice a strong anticorruption stance, and yet he did not want to point fingers and be dragged into local tribal politics. This is ground he has trod before in the United States—providing an overarching voice of moral authority without stepping into the fray and choosing sides. "He sounded very much like a politician," one student told me. "He was eloquent, but it was a politician's talk."

This, effectively, was the end of my African excursion with Obama. For the next two days, Obama and his family went on safari in the Masai Mara region of Kenya, again with the American media entourage in tow. Obama's staff had initially tried to keep the safari a private family outing, either to protect the image of Obama as continually working on the trip, or to give him some downtime with family. But as Obama was discovering more each day, the press was not going to let go of him anytime soon, so he relented and, despite occasional discomfort with the prying media, he even showed a bit of mercy for the weary American journalists. "Barack figured the reporters had traveled all this way and they were entitled to some relaxation and fun, too," Auma said.

After the safari, Obama was off to Chad to speak with refugees of the horrific civil war in bordering Sudan. He had tried to visit Sudan itself, but he has always been outspoken about the atrocities occurring there, and the Sudanese government would not grant him a visa. At that point, more than two hundred thousand people had been killed in the Darfur region and two million more had been dislocated in the fighting between rebels and the government.

Beyond the sheer human misery before him, Obama would find the visit to the refugee camp wholly disheartening. One traveling journalist said Obama was "furious" with his aides because he had been allotted only ninety minutes to speak to the refugees. The refugees' stories were translated from Arabic to French to English, consuming a great deal of time and making it difficult for reporters to hear. Obama, who had studied the Darfur situation intensely, learned only so much from the interviews. While the event most likely produced good television and newspaper copy, it did little to advance Obama's knowledge about the conflict—and this was what apparently infuriated him. In his days as a community organizer, he was accustomed to listening to poor South Side residents pour out their troubles for hours on end. But here, he didn't even receive the Cliffs Notes version. In a post-Africa interview, Obama told me that he found the entire African journey both "wonderful" and, in characteristically diplomatic terms, "a little bit frustrating":

It was a little bit frustrating that, you know, now that rather than taking trips, I have to take CODELs, which means a lot of official business, a lot of pomp and circumstance, a lot of press. Which, you know, means that I can't sort of wander off and explore these countries in the way that sometimes are the best ways to learn. But, you know, obviously, the Kenya portion of the trip, in particular, evoked a response that I hadn't expected, to that magnitude. But on the other hand, it spoke maybe to, you know, the influence that I can exercise. That's gratifying, in the sense that, like, when Michelle and I took that AIDS test, you know, the CDC said maybe half a million people might now take an AIDS test as a consequence of you taking it. When I gave the speech on corruption, you know, it was broadcast nationally, I think two or three times. I think I changed the debate inside Kenya for weeks after my visit. And so, you know, it was, I think, gratifying to feel as if I had used my bully pulpit effectively while I was there.

In the end, Obama's trip to Africa was a learning experience for him—and for his family. It was yet another example of just how pervasive and intrusive the media will be in the life of a politician who maintains a profile of this stature. It was also yet another lesson in the enormous expectations of devoted followers not only in the United States but in Kenya. It brought to light that America remained the focal point of the world for many nations like Kenya. But even though Obama had risen to great heights in American politics as a senator, his power to "leave the world a better place" was still limited.

Upon her return to Chicago, Michelle admitted that she was "overwhelmed" by the "magnitude of everything." She and Obama tried to leaven the enormous outpouring on their behalf with humor. "Barack and I joked the whole way that we have an armed escort now, and when we went in before, we just walked around from shop to shop. To have it elevated like this was kind of surreal. . . . There is part of you that is embarrassed by the scene of it. Part of you just wants to say, 'Can we tame this down a little bit? Does it have to be all this? This is out of hand.' That is my instinct and I know that is his instinct too—do we really need all this?"

For Auma, now a social services worker outside London, the madness of the trip and the intense idol worship from the Kenyans raised great concern. She still fears that her brother might be headed down a path littered with all the same land mines that contributed to her father's premature death. To her, the restless life journeys of father and son have eerie parallels. After receiving an elite American education, her father returned to Kenya with incredible expectations thrust on him by his family and the Luo tribe. Ultimately, Barack Sr. could not satisfy all these intense desires and was overwhelmed by them. Obama certainly has studied his father's life in depth and he appears to have learned from that story, Auma told me. But she is still concerned. "My father tried to live up to all of those expectations and I think Barack needs to learn from my father's mistakes," Auma said. "I think he is learning, but he just needs to set realistic goals for himself and set out to achieve them. Barack is like my father in that he is driven to perfection with regards to his work and he just needs to give himself a little slack. I am proud of Barack and I love him. But I worry about him."

Lebron Revisited

*People have always had a tendency to give Obama a pass. It's
like no other politician I've seen. They feel like he is on this
important mission. And maybe he is.*

—FORMER AIDE DAN SHOMON

Within weeks of his return from Africa, Barack Obama began
thinking seriously about running for president in 2008. The reception he had received and the publicity he had generated back home
was exhilarating, if not intoxicating. "How bad can you feel when
everybody is telling you that you should be president?" observed
Nate Tamarin, a former aide. A White House bid this early was not
the timetable that Obama would have chosen—he still had two
young daughters at home whom he adored and missed—but there
was such a strong political wind at his back that he and his ambitious advisers simply had to give a presidential run serious consid-
eration. "We definitely looked at it as something that was now
plausible," Robert Gibbs said. "It would not be a whimsical thing."

Obama had been in the public eye through most of 2006, receiving almost universally glowing treatment in the national media. Earlier, in March, he had appeared at the annual Gridiron dinner, where
Washington journalists and politicians gather to make light of each
other. In typical fashion, he impressed reporters with wit, intelligence and poise. Even President Bush took note of the Washington
love affair with Obama. "Senator Obama, I want to do a joke on
you," the president told the Gridiron audience. "But doing a joke

on you is like doing a joke on the pope. Give me something to work with. Mispronounce something." After seeing him at the Gridiron, *New York Times* columnist Maureen Dowd, one of the country's touchstone liberal columnists, suggested that Obama should think about the White House. She opined that the Democrats should not cast him aside because of lack of national experience. "The Democrats should not dismiss a politically less experienced but personally more charismatic prospect as 'an empty vessel,'" she wrote. "Maybe an empty vessel can fill the room." The idolatry and flattery from the mainstream media in 2006 was constant, unlike anything that politicians generally experience for a day, much less months or years on end. *Men's Vogue* put Obama on the cover, and, to the magazine writer, the senator let slip his thoughts of the Oval Office and his overabundance of confidence. "My attitude is that you don't want to just be the president," Obama said. "You want to change the country. You want to be a great president." Profiles of him appeared everywhere, from *New York* magazine to the national newsweeklies to a spread of him and Michelle in *Ebony*. The media could not get enough of him, and neither could his growing legions of followers. A Washington political consultant called Obama the "Black Jesus." Even his drug use as a teenager became a laugh line and won him praise for candor. Asked about smoking pot by the *Tonight Show*'s Jay Leno, Obama said casually, "I inhaled; that was the point." His supporters seemed to forgive him for this and much more—for his unbridled ambition, for raising millions in campaign dollars from established interests, for tacking to the center, for speaking mostly in the same broad, general themes. "People have always had a tendency to give Obama a pass," former aide Dan Shomon said. "It's like no other politician I've seen. They feel like he is on this important mission. And maybe he is."

Up to now, The Plan had been working to near perfection.

So after Africa, Obama began talking to people earnestly and deliberately about an Oval Office bid. His first round of discussions came with his immediate inner circle of Michelle, David Axelrod and Gibbs. Then he branched out to others, such as Jeremiah Wright and Jesse Jackson, Newton Minow and Abner Mikva, Penny Pritzker and Valerie Jarrett, Cassandra Butts and Marty Nesbitt. In all these discussions, Obama heard little to discourage him from pushing forward. "There are an awful lot of people urging him to go," Axelrod told me over breakfast. "There aren't too many people waving yellow caution flags." In September, when Obama was the main speaker at Senator Tom Harkin's annual steak fry fund-raiser in Iowa, the first presidential caucus state, the future became clear to Minow. By happenstance, the former counselor to Adlai Stevenson and John F. Kennedy had watched Obama's speech to more than three thousand Iowans on the C-Span public affairs network. He soon called Obama. "I saw John Kennedy and now I have seen you—and I haven't seen anything quite like it in between," Minow told Obama. "You ought to go for it now."

This was perhaps all the more interesting because, in truth, Obama had not performed well in Iowa. When he arrived at the Indianola fairgrounds, it was typical Obama-mania, as adoring fans engulfed him everywhere. Gibbs made sure to alert reporters that a "Draft Obama for President in 2008" petition was circulating in the crowd. But when Obama spoke, he left some of these veteran Democratic activists a bit bored. After years of red-meat speeches from hard-charging Democrats such as John Edwards and Howard Dean and home-state hero Harkin, Obama's lecturing manner and professorial prose more evoked the cerebral Adlai Stevenson. Obama also made the same mistake as he had on his election night—failing to prepare a speech ahead of time and instead letting it all come to him on the stage. When his crescendo lines failed to draw the appropriate responses, he reached back for another crescendo line. This made his speech wander from anecdote to anecdote without tying all the themes into a coherent whole. It also made the speech, at nearly forty minutes, far too long. "When it comes to leaders of the Democratic Party, I think he is up there," said sixty-two-year-old Irene Wesley of Ames. "But he just needs to be maturing a little bit, learn to relate to a crowd a little bit more and come up with some positive accomplishments in the Senate. He just needs a little maturity and he will be there." Others, however, gushed over Obama. The influential *Des Moines Register* columnist David Yepsen opened his column the next day, "Oh Oh Oh Obama." Yepsen later wrote, "This guy looks like a winner." And overall, most of the Iowans seemed to adore him. So if Obama could get this kind of positive reaction in Iowa after a sub-par performance, his advisers mused, what would happen if he brought his A-game?

In October came the much-anticipated release of Obama's second book, *The Audacity of Hope: Thoughts on Reclaiming the American Dream*. A major publicity campaign was undertaken, with Obama launching a national two-week book tour and appearing on seemingly every major network talk show, capped by an appearance with Michelle on *Oprah*. Book sales boomed, and *Audacity* eventually dislodged John Grisham's latest work from the top of the *New York Times* bestseller list. This satisfied Obama's stridently competitive nature. He had asked his advisers, "What can we do to make it number one? I want to be number one."

The book's content was not nearly as raw as that of *Dreams from My Father*. After all, *Audacity* was a work from a man in his mid-forties who, by this time in life, had made concessions and reconciliations to an imperfect world, both for his own survival and his own advancement. In fact, much of the book wrestles with how a politician can hold on to his ideals amid a scrutinizing press, a media culture that feeds on conflict and a political system that makes it necessary to raise big money from special interests and wealthy donors. In this way, for a book from a politician with presidential aspirations, *Audacity* was rather candid, and it again put Obama's uniquely personal writing voice on vivid display. Obama acknowledged internal struggles, including an insecurity about his role in his own family; he even took the blame for most domestic problems. He fretted about losing his own voice because a politician can fall victim to "a committee of scribes and editors and censors" who "take residence in your head." He said that he sometimes worried that his idealism was overcome by personal vanity. And he conceded that, as a consequence of modern political necessity, he now spent far more time in a rarefied world of moneyed and intellectual elites and much less time with ordinary people.

This was very different from the wandering, occasionally angry young man in *Dreams*. "There was more of an edge to me back then, I suppose," Obama said of the angst-ridden prose found in *Dreams*. "You grow a little older, you know, and become more forgiving of yourself and others." In *Audacity*, Obama provided a somewhat sentimental vision for the future of America, a vision that, of course, called for more civility and unity in our culture. Much like his political rhetoric, however, the book lacked detailed, real-world specifics about how to accomplish that. Overall, it received positive reviews, something that his staff had worried about before its release. They did not want him to come across to the East Coast media elites judging his political future as an unserious, self-serving politician.

But it was the monster sales of the book that provided the linchpin for Obama's next move. As Obama's book hit the top spot in sales, books written by the major Democrats considering the presidency, Hillary Clinton and John Edwards, languished far down on the bestseller lists. Also, Obama's book tour was pure madness. Thousands gobbled up tickets for the events in all corners of the country. At a book signing in Portsmouth, New Hampshire, after Obama delivered a centrist-oriented speech, fifty-two-year-old Penny Reynolds wasn't sure if Obama was ready to lead the nation, but she said his message of unity and bipartisanship made her an instant fan. "He's not really saying anything different than anybody else except that we should all try to work together, enough of this partisan stuff," she said. "And it's sad to say, but that's really refreshing right now." Obama's advisers took this kind of talk as confirmation that, beyond his physical and emotional appeal, Obama's message of political consensus was striking a lasting chord with people. With a widely unpopular Republican president in the White House who had governed in a rigidly ideological manner, Obama the good-natured consensus-builder seemed to be the antidote for these faithful Democrats. "The combination of the book, the reaction he got campaigning and the election results—it all sort of validated his message," Gibbs said. "It made [running for president] something that we could no longer avoid thinking about."

THE DAY AFTER THE NOVEMBER ELECTION, IN WHICH DEMOCRATS took control of Congress, Obama and his advisers began meeting in Axelrod's office in Chicago's West Loop to discuss seeking the Democratic nomination for the presidency. Vital to these discussions was Michelle. In a *Chicago Tribune* interview at the end of 2005, Michelle had complained that because of her husband's onerous work schedule, she often felt like a single mother and she worried about an erosion of Obama's paternal connection to his daughters. Axelrod and Gibbs knew that these kinds of press quotes would be highly damaging to a presidential candidate. They also knew that Obama would need full support from home if he were to run. So Michelle was brought into the discussions as a full partner. Her voice carried as much resonance as that of anyone in the room. By December, the Michelle hurdle had been cleared. She asked only that, as a prerequisite for her approval, Obama quit smoking, which he agreed to do. "If Barack really wants this, Michelle will support him and do what's necessary," said Cassandra Butts, Obama's friend from Harvard. "That's always been their relationship."

In an interview in December 2006, Michelle sounded very much as if she had weighed the pros and cons of the decision, as well as the consequences of that decision on her family. She had risen out of the fog of the overwhelming African adventure, yet still marveled at the American public's yearning for more of her husband. "You keep waiting for people to be like, okay, you are tired of him now, right? You've had enough." When I said the public seemed to want even more, she said only, "I know."

By now, an image of Michelle was evolving in the media, and it was not entirely positive. Obama had consistently portrayed her as both a solidifying force ("my rock at home," he would say) and the scolding wife who kept his ego in check. She also gained a spate of publicity and raised some eyebrows for being promoted to a lucrative vice presidential position at the University of Chicago Hospitals, where she was now earning more than a quarter of a million dollars a year. In addition, she sat on the board of a food supplier to Wal-Mart Stores Inc., a position that earned her another fifty thousand dollars a year, although her husband had been a critic of the labor practices of the megaretailer, giving the appearance of hypocrisy. So if she were to be considered as First Lady material, a more three-dimensional, positive image of Michelle would be necessary. And she would also need to hit the campaign trail with and without her husband. "I know that my caricature out there is sort of the bad-ass wife who is sort of keeping it real, which is fine," Michelle said, adding that her husband's career pressures on the family remained an issue. "There is still the part of me that, if we do something this big, our kids are still really little, and what I am not ready to sacrifice is their livelihood. But . . . I am going to be the person who is providing them with the stability. So that means my role with the kids becomes even more important. What I am not willing to do is hand my kids over to my mom and say, 'We'll see you in two years.' That's not going to happen. . . . There has to be a balance and there will be a balance. You just have to make your mind open to it, in some way, shape and form." (As the campaign kicked off, Michelle would quit the food supplier's board and cut back on her hours at the university.)

Michelle indicated that she had thought deeply about the prospect of losing her husband, presumably to an assassin. I mentioned to her that a newspaper editor in Nairobi had asked me, "Don't you worry that as a black man in America the skinheads will kill him?" She confided that he was now in a fragile spot as a major black politician. "I don't worry about it every day, but it's there. And it's a nonstarter," she said. "So if we take this next step, there would have to be a comprehensive security plan in place. . . . It only takes one person and it only takes one incident. I mean, I know history too. So it's still an issue." She said that her own career ascension has been tied to this possibility. Michelle worried, for example, that if something happened to her husband she would lose her prime means of financial and emotional support. "I do think about the fact that my husband is in a high-risk sort of position right now. And I need to be able to take care of myself and my kids. I have to be in a position that if anything unexpected or unfortunate happens, where are all those people who are being critical of my credentials or my ability to serve on boards, where are they going to be if I have to take care of my kids? There would be great sympathy and outpouring if something were to happen, but I have to maintain some level of professional credibility not only because I enjoy it, but I don't want to be in a position one day where I am vulnerable with my children. I need to be in a position for my kids where, if they lose their father, they don't lose everything."

A COUPLE OF WEEKS BEFORE INTERVIEWING MICHELLE, WHEN Obama was still being coy about his final presidential decision, I asked Obama how he viewed the 2008 election cycle. In his typical understated way, he said that the trend of the country—its severe antipathy toward the Bush administration—had given the Democrats a "great opportunity" to win the White House. "I think the Democratic nomination in '08 is worth something. Yeah, I do. But

I think that over the next two years, the Democrats have to show the country that they are listening and that they are interested in crafting a set of commonsense practical solutions."

Obama gave little indication that he was anywhere but on the path toward running for president. If there were any doubt left, his visit to the early-primary state of New Hampshire in December erased it. He was the main speaker at a fund-raiser that drew more than fifteen hundred guests. That visit effectively pushed at least one other potential candidate out of the race, Senator Evan Bayh of Indiana, who had drawn just a couple of dozen people to an event the preceding weekend. This was similar to what followed the Harkin steak-fry in Iowa, where former Virginia governor Mark Warner witnessed firsthand the Obama phenomenon. As Warner spoke from the stage, a good number of the attendees had their heads turned backward to watch Obama wend his way through the fawning crowd. Soon thereafter, Warner announced that he would not run.

But as the wheels were in motion to launch a campaign, Obama— coincidentally or not—was hit with the most damaging news story of his career. With his newfound wealth, in 2005, he and Michelle had purchased a mansion in Hyde Park for more than one million six hundred thousand dollars. Michelle, in particular, had wanted a good-sized home because, she said, the couple wanted a roomy refuge from the public trappings of fame. But purchasing a piece of property right next door, on the very same day, had been the wife of Antoin "Tony" Rezko, an old friend and financial contributor of Obama's who had been indicted just months before on federal fraud charges. Reporters for the *Tribune,* where the story broke, also found that Obama and Rezko's wife engaged in a series of financial transactions to redivide the properties and improve their parcels. The relationship between Rezko and Obama went back twenty years. Obama had met him when Obama was in law school and Rezko's development partners had tried to hire him. At his north suburban home, Rezko had hosted a fund-raiser for Obama in 2003 that helped fund early parts of his Senate candidacy. Since then, Rezko had been a regular contributor to Obama's campaigns. The two had also been social friends, with the Rezkos dining out a few times a year with the Obamas.

As soon as the story was aired, Obama expressed contrition and openly conceded bad judgment, calling the transaction "bone-headed." His radar with Rezko had broken down, he said, and if he had to make the deal again, he would not have. "Look, I came up through politics in Chicago and Cook County and Illinois," Obama said. "And this is the first time that I've ever, in ten years, having risen from knowing nobody to being a U.S. senator, where people suggested anything that I'd done was inappropriate. And so, I'd have to say that, you know, that would indicate that I must have a pretty good radar, because that's a pretty good track record."

Back at the *Tribune,* reporters saw the hand of Obama's shrewd political advisers in the deal's sudden uncovering. The paper had gotten hold of the story through an anonymous tip. As the Blair Hull experience showed, it's always better to get any negative press for your candidate out of the way as early as possible so it does not break at the height of the election season. Obama had been bloodied for a few days in the Chicago media, and surely future political opponents would point to the Rezko deal as untoward, but the story was out there for all to see—and Obama seemed little the worse for it.

WITH MICHELLE ON BOARD, WITH AXELROD AND GIBBS ASSEM-bling a national political operation, with Democrats across the country in thrall over Obama, the decision was made. In January, Obama announced on his Internet website that he was forming a presidential exploratory committee and would announce his final plans in early February. This methodology gave the candidate two media hits surrounding his announcement, two for the price of one.

Top aides were not without concerns about Obama's preparedness. The speed of his ascendancy has been unprecedented and it was uncertain how that would affect him in the long run. He neared burnout in his first months in the Senate. And his life had not slowed from warp speed since he first decided to run for the Senate in 2003. Could this breakneck pace be sustained through a bruising presidential contest? "He is in fantastic shape, but I wonder about his physical stamina," said one of his consultants, Pete Giangreco. "It takes just an incredible amount of physical stamina out there."

Indeed, running for president is like no other experience a human can endure. It is so bizarre and so surreal that journalists Mark Halperin and John F. Harris have dubbed modern presidential politics "the Freak Show." In their book *The Way to Win*, the two posited a theory that advanced modern technology and the breakdown of old media like newspapers and network television have made presidential contests into a ruthless blood sport in which referees no longer exist. This, in turn, exacerbates partisan fighting and is largely responsible for the polarization of the voting public. "The supreme challenge for any presidential candidate is keeping control of his or her public image in the face of the Freak Show's destructive power," the authors wrote.

I asked Axelrod in December whether he thought Obama was ready for this Freak Show. His message, after all, was conciliation. What happens if things devolve into a muddy free-for-all? Axelrod's candor was surprising. "I don't know," he said. "What do you mean?" I responded. "Do you think he can handle it?" "I don't know," Axelrod said again. "One thing about running for president is that—and he knows this—it's like putting an X-ray machine on yourself twenty-four hours a day, because . . . at the end of the day, the American people know who you are. But with Barack, he's kind of a normal guy in a lot of ways. He likes to watch football on Sundays. He treasures his time with his kids and Michelle. I think he has an inner toughness, and that is reflected in the road he traveled to get where he is, because you know, he didn't exactly start off in an optimal place. And he has, I think, struggled through a lot of challenges to make himself what he is. I think there's this impression that here's this Harvard-educated, stem-winding intellectual, but he is a guy who was raised by a single mother who wasn't there to help all the time because she couldn't be. And you know, he fought his way through a lot."

I wasn't sure whether Axelrod was trying to sell me or himself with this speech. And I am not sure it even mattered. There was no turning back. Axelrod's young political talent had cast himself in the Freak Show.

LESS THAN TWO MONTHS LATER, ON A FRIGID FEBRUARY DAY OUT-side the Old State Capitol in Springfield, Illinois, I stood on a grandstand riser filled with newspaper, wire service and television photographers waiting for Barack Obama to tell the world he was running for president of the United States.

Two more risers were equally jammed with media. And big-name national journalists were scattered throughout the crowd of fifteen thousand true believers assembled down below. As I put pen-cil to paper, my thoughts rambled back over the past three years that I had spent following this likable, idealistic yet utterly mercurial pol-itician, and I couldn't help but recall the small moment when Obama showed me off to a stranger who honked at him as we walked down Martin Luther King Drive in Chicago. Just three short years ago, Obama seemed proud to have a single reporter interested in him. Now, on this day in Springfield, he had thousands of journalists worldwide ready to listen, and examine, his every word. And he had thousands more enthused spectators overjoyed at viewing this small slice of American history.

Perhaps the only moment that came close was that keynote ad-dress in Boston, when I had been sitting in a vast arena wondering how Obama would perform. On this day, I had to admit, I had no wonder. Having watched him deliver impassioned speeches to blacks on Chicago's South Side, to Latinos on its Near West Side, to rural whites in downstate Illinois, to people in cities across the United States and to poor villagers in remote Africa, I knew exactly what to expect. A call for mutual citizenship, a call for a new generation to lead, a call for an end to the Iraq War. Obama had stayed up into the wee hours the night before, crafting and rehearsing these words. It was twelve degrees outside, but this was the Hawaii native's true element—preaching to the masses the gospel of Barack Obama; the gospel of a common humanity, the gospel that, if everyone would just join together behind him, he could be the one to make the world a better place.

As his friend Marty Nesbitt told me that day in Boston, Obama is like that basketball player on his high school team back in Ohio, always able to elevate his game when the situation demanded it. With Michelle on his arm, a supremely confident-looking Obama strode onto a long black catwalk that led to a wooden podium, situated front and center. The handsome couple, each in black winter overcoats, held hands as they walked forward, waving to the cheering crowd and stoking the electrified atmosphere. Nearing the podium, Obama let go of Michelle, who stepped down to allow her husband to take the stage by himself. The crowd heaved into a chant, "O-ba-ma! O-ba-ma!" The senator walked up to the podium, his most natural setting, and I couldn't help but say to myself, "Here comes LeBron, indeed."

NOTES

CHAPTER 1: THE ASCENT

4 reference to article on red and blue America: Thomas Edsall, *Washington Post,* April 2003.

6 "He definitely has this . . . ,": Robert Gibbs discussion with author, December 2006.

7 "He's always wanted . . . ,": Valerie Jarrett interview with author, May 2005.

9 "Hours before he gave . . . ,": Clarence Page interview, *The Chris Matthews Show,* NBC News, October 2006.

10 "It's like nothing . . . ,": Julian Green interview with author, January 2007.

10 "We originally scheduled . . . ,": John Lynch at Manchester, New Hampshire, fund-raiser, December 2006.

10 "I can't, for example, walk . . . ,": Barack Obama interview with author, December 2006.

10 "For us . . . he hasn't . . . ,": Auma Obama interview with author, August 2006.

11 "It's like you are carrying . . . ,": David Axelrod interview with author, December 2006.

11 "David always worries . . . ,": Axelrod confidante in discussion with author, January 2007.

11 "Let us transform . . . ,": Obama at a presidential announcement in Springfield, Illinois, January 2007.

13 "People don't come . . . ,": Bruce Reed quoted in "Destiny's Child" article by Ben Wallace-Wells, *Rolling Stone,* February 2007.

CHAPTER 2: DREAMS FROM HIS MOTHER

17 "After all, you don't . . . ,": Barack Obama, *Dreams from My Father: A Story of Race and Inheritance* (New York: Random House, 1995).

18 "I wrote four hundred . . . ,": Obama discussion with author, January 2004.

19 "When my tears . . . ,": Obama, *Dreams from My Father.*

20 "There's still a great deal . . . ,": Michelle Obama discussion with author, February 2004.

23 "New-fangled gadgets . . . ,": Madelyn Dunham interview with author, October 2004.

23 "International law . . . ,": Dunham interview with author, October 2004.

24 "His mother was . . . ,": Ibid.
24 "It's always hard to talk . . . ,": Obama interview with author, August 2005.
24 "Everything that is good . . . ,": Obama during Mom's Rising event in Washington, D.C., September 2006.
24 "I think sometimes . . . ,": Obama, *Dreams from My Father*.
25 "It was this desire . . . ,": Ibid.
25 "She was extremely brilliant . . . ,": Dunham interview with author, October 2004.
25 "I remember one time . . . ,": Maya Soetoro-Ng interview with author, October 2004.
26 "Her feet never . . . ,": Dunham interview with author, October 2004.
26 "vaguely liberal,": Obama, *Dreams from My Father*.
26 "It wasn't at all . . . ,": Soetoro-Ng interview with author, October 2004.
27 "With her friends . . . ,": Obama interview with author, August 2005.
27 "She was not exactly . . . ,": Dunham interview with author, October 2004.
28 "I assume . . . ,": Soetoro-Ng discussion with author, October 2004.
29 "I am a little dubious . . . ,": Dunham interview with author, October 2004.
29 "the first large wave . . . ,": Obama, *Dreams from My Father*.
30 "What my father became . . . ,": Auma Obama interview with author, August 2006.
30 Obama speech reference: University of Nairobi in Nairobi, Kenya, August 2006.
31 "A lot of grandiose . . . ,": Dunham interview with author, October 2004.

CHAPTER 3: JUST CALL ME BARRY

32 "Hawaii was heaven . . . ,": Obama interview with author, August 2005.
33 "I learned how to eat . . . ,": Obama, *Dreams from My Father*.
33 "I think [Indonesia] . . . ,": Obama interview on *Fresh Air*, National Public Radio, August 2004.
33 "Men take advantage . . . ,": Obama, *Dreams from My Father*.
33 "The world was violent . . . ,": Ibid.
34 "To be black . . . ,": Ibid.
36 "We never suffered . . . ,": Dunham interview with author, October 2004.
37 "Hawaii is such . . . ,": Soetoro-Ng discussion with author, October 2004.
39 "My father's absence . . . ,": Obama interview with author, December 2003.
40 "Every man is trying . . . ,": Obama interview with author, December 2003.
41 "He wasn't this . . . ,": Keith Kakugawa, *Nightline*, ABC News, April 2007.
42 "He was just . . . ,": Bobby Titcomb interview with author, August 2004.

42 "You know, in Hawaii . . . ,": Ibid.

43 "I suppose I provided . . . ,": Dunham interview with author, August 2004.

43 "I was trying . . . ,": Obama, *Dreams from My Father.*

44 "Some of the problems . . . ,": Obama interview with author, October 2004.

45 "Hey, don't touch the 'fro!": Soetoro-Ng interview with author, August 2004.

45 "On the basketball . . . ,": Obama, *Dreams from My Father.*

45 "I used to know . . . ,": Dunham interview with author, August 2004.

45 "dabbled in drugs . . . ,": Obama discussion with author, February 2004.

46 "This was really . . . ,": Dunham interview with author, August 2004.

46 "Just call me Barry . . . ,": Eric Kusunoki interview with author, August 2004.

47 "I recall his sincere . . . ,": Chris McLachlin interview with author, August 2004.

48 "I got into a fight . . . ,": Obama interview with author, October 2004.

48 "There's still something . . . ,": Julian Green discussion with author, October 2004.

48 "He would have been . . . ,": Arne Duncan interview with author, January 2006.

49 "I recall that . . . ,": Suzanne Maurer interview with author, October 2004.

49 "We agreed . . . ,": Ibid.

CHAPTER 4: THE MAINLAND

50 "I'll admit . . . ,": Dunham interview with author, October 2004.

50 "childhood dream": Obama, *Dreams from My Father.*

50 "It could seem . . . ,": Soetoro-Ng interview with author, October 2004.

51 "We're disenfranchised . . . ,": Dunham interview with author, October 2004.

51 "Just an all-around . . . ,": Obama discussion with author, October 2004.

51 "Maya's just . . . ,": Nora Moreno-Cargie discussion with author, October 2004.

52 "I just love . . . ,": Ibid.

54 "Hawaii has ideas . . . ,": Soetoro-Ng interview with author, October 2004.

54 "She just didn't . . . ,": Ibid.

55 "At best, these . . . ,": Obama, *Dreams from My Father.*

56 "They weren't defined . . . ,": Ibid.

56 "California blacks . . . ,": Don Terry discussion with author, November 2003.

57 "Obama is a . . . ,": Scott Fornek discussion with author, February 2004.

57 "When we ground out . . . ,": Obama, *Dreams from My Father.*

57 "I noticed that people . . . ,": Ibid.

58 "The schools . . . ,": Obama discussion with author, February 2004.

58 "He seemed to have gotten . . . ,": Dunham interview with author, October 2004.

58 "Moving to the mainland . . . ,": Soetoro-Ng interview with author, October 2004.

58 "Do you mind . . . ,": Obama, *Dreams from My Father.*

59 "put down stakes . . . ,"; Obama interview with author, October 2004.

59 "I had two . . . ,": Obama interview with author, August 2005.

60 "Barack put himself . . . ,": Green interview with author, November 2004.

60 "My philosophy . . . ,": Roger Boesche interview with author, June 2005.

61 "Nietzsche calls . . . ,": Ibid.

61–62 "I knew that . . .", "There were people who . . .", and "Those two years . . . ,": Obama interview with author, August 2005.

62 "I would imagine myself as . . . ,": Obama, *Dreams from My Father.*

63 "Could this be . . . ,": Jerry Kellman interview with author, March 2006.

CHAPTER 5: THE ORGANIZER

65, 66 "One of the things . . . ," and "All I had . . . ,": Kellman interview with author, March 2006.

66 Roseland material adapted from "Rich 90s Failed to Lift All," *Chicago Tribune,* David Mendell/Darnell Little, August 2002.

67 *Rules for Radicals,* Saul Alinsky (New York: Random House, 1971).

68 "Once I found an issue . . . ,": Obama, *Dreams from My Father.*

68 "It wasn't until . . . ,": Obama interview with author, October 2004.

69 "The first day . . . ,": Jeremiah Wright interview with author, November 2006.

69 "disaster . . . ,": Obama interview with author, October 2004.

70 "I'm here to do . . . ," and "Baby Face,": Kellman interview with author, March 2006.

71 "He was our . . . ,": Callie Smith interview with author, October 2004.

71 "Jerry Kellman is . . . ,": Obama interview with author, August 2006.

72–74 "Barack wanted to serve . . . ," and Kellman material: Kellman interview with author, March 2006.

74 "So I figured . . . ,": Obama interview with author, August 2006.

75 "I came to realize . . . ,": Barack Obama, *The Audacity of Hope: Thoughts on Reclaiming the American Dream* (New York: Crown, 2006).

76 "Trying to hold . . . ," and Wright material: Wright interview with author, November 2006.

77 "It's a great book . . . ,": Obama interview with author, August 2006.

77 "Faith to him . . . ,": Wright interview with author, November 2006.

78 "slumbering giant . . ," and continuing quotation: Barack Obama, "After Alinsky: Community Organizing in Illinois," *Illinois Issues,* University of Illinois at Springfield, 1990.

CHAPTER 6: HARVARD

80 "Ain't nothing gonna . . . ,": Obama, *Dreams from My Father.*

82 "Everywhere black people . . . ,": Ibid.

82 "I just can't . . . ,": Titcomb interview with author, October 2004.

84 "He was mature . . . ,": Michael Froman interview with author, January 2007.

84 Some Harvard material from "In Law School, Obama Found Political Voice," Jodi Kantor, *New York Times,* January 2007.

85 "command centre . . . ,": "Pinkos and Pistols," *The Economist,* April 2002.

85–86 Laurence Tribe material: Kantor, *New York Times,* January 2007.

86 "But you know, once . . . ,": Cassandra Butts interview with author, December 2006.

87 "Diversity is enriching . . . ,": "The Rudeness of Race," *Chicago Tribune Magazine,* February 11, 1996.

87 "You know, whether we're . . . ,": Butts interview with author, December 2006.

88 "Conservatives and libertarians . . . ,": The Federalist Society mission statement, 1982.

89 "was populated by . . . would-be . . . ,": Brad Berenson quoted in "In Law School," Kantor, *New York Times,* January 2007.

90, 91 "Barack always floated. . . ," and "He did show great . . . ,": Berenson interview with author, January 2007.

91 "On the *Law Review* . . . ,": Obama interview with author, October 2004.

92 "He wanted to be mayor . . . ,": Butts interview with author, December 2006.

CHAPTER 7: SWEET HOME CHICAGO

93, 94 "He sounded too good . . . ," and "I thought . . . ,": Michelle Obama interview with author, January 2004.

95 "this kind of stability": Kellman interview with author, March 2006.

95, 96 "We always felt . . ." and "Dad said . . . ,": Craig Robinson interview with author, November 2005.

98, 99 "I call him. . . ," and "I told him . . . ,": Michelle Obama interview with author, January 2004.

99, 100, 101 "My mom and I . . . ," "He was tall . . . ," "Barack's game . . . ," and "Barack was like . . . ,": Robinson interview with author, November 2005.

101 "Barack treated . . . ,": Michelle Obama interview with author, January 2004.

102 "If you are biracial . . . ,": Kellman interview with author, March 2006.

103 "There are times . . . ,": Obama interview with author, October 2004.

104 "What I notice . . . ,": Michelle Obama quoted in "Her Plan Went Awry, But Michelle Obama Doesn't Mind," Cassandra West, *Chicago Tribune,* September 2004.

104 "You always think . . . ,": Obama, *Dreams from My Father.*

105 "Leave your name . . . ,": Judson Miner quoted in "Obama Got Start in Civil Rights Practice," Mike Robinson, Associated Press, February 2007.

106 "The courts are . . . ,": Obama interview with author, October 2004.

CHAPTER 8: POLITICS

107 "Upon my return . . . ,": Obama, *Dreams from My Father.*

108–110 Alice Palmer material: Sunya Walls, "Alice Palmer Withdraws from Race for Re-election," *Chicago Weekend,* Chicago Citizen Newspaper Group, January 1996.

109 "Since she endorsed . . . ,": Obama quoted in the "Inc." column, *Chicago Tribune,* Judy Hevrdejs and Mike Conklin, December 1995.

110 "I've since discovered . . . ,": Alice Palmer quoted in "Candidate Not What He Seems, Foes Insist," Salim Muwakkil, *Chicago Sun-Times,* February 1996.

112–113 "The political debate. . . ," and "Any solution . . . ,": Obama quoted by Hank De Zutter in the *Chicago Reader,* December 1995.

114 "vacuous-to-repressive . . . ,": Muwakkil, February 1996.

115 "at the swearing-in . . . ,": Illinois politico discussion with author, January 2004.

115, 117, 118 "I am thinking . . . ," "I said . . . ," "We are driving . . . ," "He's not . . . ," and "It was just a great . . . ,": Dan Shomon interview with author, January 2005.

119 "Buy him a black . . . ,": Michelle Obama interview with author, January 2004.

119 "So I asked . . . ,": Shomon interview with author, January 2005.

120 "My grandmother . . . ,": Obama discussion with author, August 2004.

CHAPTER 9: THE LEGISLATOR

121 "The fact that . . . ,": Kirk Dillard interview with author, September 2004.

122 "It wasn't like . . . ,": Shomon interview with author, January 2005.

122 "The first few years . . . ,": Rich Miller quoted in "How Obama Learned to Be a Natural," Edward McClelland, *Salon.com,* February 2007.

123 "He didn't think . . . ,": Obama associate interview with author, 2004.

124 "This sets the standard . . . ,": Obama quoted in Associated Press, May 1998.

125 "just give Barack hell . . . ,": Kimberly Lightford interview with author, March 2007.

126 "I have been advised . . . ,": Rickey Hendon interview with author, September 2006.

127 "Obama is interested . . . ,": Don Wiener opposition research report, 2004.

127 "He is idealistic . . . ,": John Bouman interview with author, September 2004.

127 "I always found . . . ,": Joe Birkett discussion with author, October 2006.

128 "The most important . . . ,": Obama interview with author, September 2004.

128 "Members of both . . . ,": Dillard interview with author, September 2004.

131 "Barack is viewed . . . ,": Donne Trotter quoted in "Is Bobby Rush in Trouble?" Ted Kleine, *Chicago Reader,* March 2000.

131 "There are whispers . . . ,": Kleine, March 2000.

132 "rising star . . . ,": editorial, *Chicago Tribune,* March 2000.

132 "What was interesting . . . ,": Kellman interview with author, March 2006.

132 "What's fascinating . . . ,": Obama interview with author, October 2004.

132 "What I've found . . . ,": Obama discussion with author, January 2004.

134 "I'll never forget . . . ,": Obama discussion with author, February 2004.

134 "There was just . . . ,": Shomon interview with author, January 2005.

136 "I'm angered, frankly . . . ,": George Ryan quoted in "Ryan Comes Up Short," Rick Pearson/Ray Long, *Chicago Tribune,* December 1999.

136 "Sen. Barack Obama . . . ,": editorial, *Chicago Tribune,* March 2000.

136 "This vote was . . . ,": Bobby Rush, quoted in *Chicago Tribune,* December 1999.

137 "I cannot sacrifice . . . ,": Obama in press conference, January 2000.

137 "Proving the political . . . ,": David Mendell, "Obama Defends Decision to Miss Anti-Crime Vote," *Chicago Tribune,* January 2000.

138 "Less than halfway . . . ,": Obama, *The Audacity of Hope.*

138–139 Congressional debate material: Rush-Obama-Trotter debate video courtesy of WTTW-TV, Chicago.

139 "Just what's . . . ,": Ibid.

CHAPTER 10: THE NEW ROCHELLE TRAIN

141 "Bobby just ain't . . . ," and "It made me realize . . . ,": Obama interview with author, October 2004.

141 "Barack, you didn't . . . ,": Wright interview with author, November 2006.

142 "In his race . . . ,": Abner Mikva interview with author, October 2004.

144 "I was broke . . . ,": Obama press interviews in Boston, July 2004.

144 "He is motivated . . . ,": Shomon interview with author, January 2005.

145 "I'm sure . . . ,": Dunham interview with author, October 2004.

146 "There were a range . . . ,": Obama interview with author, October 2004.

147 "Emil is driven . . . ,": Obama discussion with author, February 2004.

147 "I am blessed . . . ,": Emil Jones Jr. discussion with author, February 2004.

148 "He always talked . . . ,": Kellman interview with author, March 2006.

149 "I am thinking . . . ,": Shomon interview with author, January 2005.

CHAPTER 11: THE CANDIDATE

150 "I put Michelle . . . ," and "Suddenly Adelstein's . . . ,": Obama interview with author, December 2004.

151 "The big issue . . . ," and "Ultimately . . . ,": Michelle Obama interview with author, December 2006.

152 "So I told Dan . . . ,": Obama interview with author, December 2004.

153–154 "Barack says he wants . . . ," "I guarantee . . . ," and "So we all said . . . ,": Marty Nesbitt interview with author, October 2005.

155 "We had known Barack . . . ,": Penny Pritzker interview with author, February 2006.

157 "We were pulling . . . ,": David Axelrod discussion with author, December 2004.

157 "If I were you . . . ,": David Axelrod from author interview with Obama, December 2004.

158 "There was no way . . . ,": Obama interview with author, December 2004.

158 "But the problem was . . . ,": Obama confidante discussion with author, early 2005.

158 "She felt that [Obama] . . . ,": Axelrod interview with author, December 2004.

159 "My name should . . . ,": Obama from author interview with Wright, November 2006.

160 "He had gone . . . ,": Wright interview with Manya Brachear of the *Chicago Tribune,* October 2006.

161 "I didn't grow up . . . ,": Obama interview with author, December 2004.

CHAPTER 12: THE CONSULTANT

165 "Guys I Never Want . . . ,": Ed Rollins, *Bare Knuckles and Back Rooms* (New York: Broadway, 1996).

165 "I just wanted to go . . . ,": Axelrod interview with the author, June 2005.

167 "Dad never shared . . . ,": David Axelrod, "The Truth about My Father's Death," *Chicago Tribune,* June 2006.

167 "It was just all about . . . ,": Axelrod discussion with author, January 2004.

169 "David was in his glory . . . ,": lobbyist discussion with author, December 2004.

170 "He was a meal . . . ,": Axelrod discussion with author, December 2004.

170, 171 "I would have made . . . ," and "You know, I hear . . . ,": Axelrod interview with author, December 2004.

173 "He came in . . . ," and "When he speaks . . . ,": Bettylu Saltzman interview with author, April 2005.

174 "there will be . . . ," and "I knew that this . . . ,": Obama interview with author, December 2004.

176–177 "That's the speech . . . ,": Obama interview with author, August 2006.

178 "Bought himself . . . ,": Obama quoted in "Legislator in Race to Unseat Fitzgerald," Pearson/Chase, *Chicago Tribune,* January 2003.

178–179 "My involvement was . . . ," and "In a classic way . . . ,": Axelrod interview with author, December 2004.

CHAPTER 13: THE RACE FACTOR

180 Jones–Obama conversation: Jones interview with author, January 2007.

181 "Being in the majority . . . ,": Obama interview with author, December 2004.

182 "This Senate thing . . . ,": Shomon in interview with author, January 2005.

183 "I'm sure . . . ,": Axelrod discussion with author, November 2004.

183 "We were dead . . . ,": Nate Tamarin interview with author, December 2004.

184–185 Jim Cauley quotes: Cauley interview with author, December 2004.

186 "As a politician . . . ,": *Chicago Sun-Times,* Laura Washington column, September 2003.

187 "There is going to be . . . ,": Obama discussion with author, August 2005.

187 "It's been mentioned . . . ,": David Katz discussion with author, June 2005.

188 "I am not running . . . ,": Obama at Mars Hill Baptist Church, Chicago, November 2003.

188–189 "We just assume . . . ,": Obama at South Suburban Chicago campaign office, January 2004.

189 "It was the first . . . ,": Cynthia Miller interview with author, April 2005.

190–191 "The anger and frustration . . . ,": Michelle Obama interview with author, August 2005.

CHAPTER 14: THE REAL DEAL

193 "Don't blame Pam . . . ,": Obama discussion with author, November 2003.

195 "Hey, I'm doing . . . ,": Obama to passerby in Chicago, November 2003.

199, 200, 202 "Well go ahead . . . ," "I think politics . . . ," and "When I see . . . ,": Obama interview with author, December 2003.

204 "It worked on two . . . ,": Axelrod interview with author, December 2003.

205 "My moment . . . ,": Cauley quoted in "Race Against History" by Noam Scheiber in *The New Republic,* May 2004.

CHAPTER 15: HULL ON WHEELS

207 "Barack has taken . . . ,": Tom Balanoff interview with author, January 2004.

208 "The legislature . . . ,": Obama at Senate debate, January 2004.

210 "Don't you think . . . ,": from Illinois political consultant discussion with author, June 2005.

211 "Noah's Ark . . . ,": Mark Blumenthal interview with author, January 2007.

211 Hull material adapted from "Hull Proves Money No Object in Bid for Senate," Mendell, *Chicago Tribune,* February 2004.

214 "You know . . . ,": Axelrod discussion with author, February 2004.

215 Hull divorce files: public records from Cook County Circuit Court.

216 "Each day we . . . ,": Chase discussion with author, March 2004.

217 "Barack was concerned . . . ,": Axelrod interview with author, August 2005.

217 "I sure hope . . . ,": Pete Giangreco discussion with author, March 2004.

217 "Some politicians . . . ,": "The 'IT' Factor," *Chicago Sun-Times,* January 2004.

218 "For god's sakes . . . ,": Obama aide discussion with author, January 2004.

CHAPTER 16: THE SMALL SCREEN

220 "There's no doubt . . . ,": David Mendell, "Obama Banks on Credentials, Charisma," *Chicago Tribune,* January 2004.

220 "You have Christina . . . ,": Matt Hynes discussion with author, January 2004.

221–222 *Chicago Reader* reference and Bob Secter quote: Michael Miner, "The Kiss and the Cover-up," *Chicago Reader,* March 2004.

222 "Sounds like . . . ,": Obama discussion with author, February 2004.

223 "the favorite son . . . ,": Obama at South Suburban Chicago campaign office, January 2004.

223 "I knew Dan Hynes . . . ,": Illinois politico discussion with author, August 2005.

224 "They're running . . . ,": Anita Dunn discussion with author, February 2004.

224 "You know, Chico . . . ,": Cauley discussion with author, February 2004.

225 "My father was . . . ,": Obama in various campaign speeches, 2003–2004.

226 "It's getting harder and harder . . . ,": Obama discussion with author, February 2004.

226 "My general attitude . . . ,": Obama interview with author, December 2004.

226 "musicians riffing . . . ,": Axelrod interview with author, December 2006.

227 "I am my . . . ," and "The arc of . . . ,": Obama in various campaign speeches, 2003–2004.

227 "I'm going to be . . . ,": Paul Simon discussion with author, December 2003.

228–231 Campaign ads courtesy of David Axelrod.

230 "Barack is extremely intelligent . . . ,": Axelrod interview with author, March 2004.

231 "man for this time . . . ,": editorial, *Chicago Sun-Times,* February 2004.

231 "proven record of . . . ,": editorial, *Chicago Tribune,* February 2004.

231 "Obama is on fire": Jason Erkes discussion with author, March 2004.

231 "We can't get beat . . . ,": Secter discussion with author, February 2004.

233–234 "Mr. Obaaaaama . . . ," and "Can you believe . . . ,": unidentified women, St. Patrick's Day Parade, downtown Chicago, March 2004.

234 "You just wait . . . ,": Obama discussion with author, March 2004.

CHAPTER 17: A VICTORY LAP

235 Campaign ads courtesy of Hynes campaign.

236 "There was nowhere . . . ,": Blumenthal interview with author, January 2007.

236 "Let's face it . . . ,": Erkes discussion with author, March 2004.

237 "He stayed . . . ,": Dan Hynes, Senate debate, February 2004.

238 *Tribune* story reference: Mendell/Chase, "Candidate Without Faults Is a Rarity," *Chicago Tribune,* March 2004.

239 "Hey, this story . . . ,": Obama discussion with author, March 2004.

240 "He did . . . ,": Axelrod discussion with author, March 2004.

240 "We have all . . . ,": Cauley discussion with author, March 2004.

241 "I don't want this . . . ,": Obama discussion in campaign SUV, March 2004.

241 "It was a straight arrow . . . ,": Blumenthal interview with author, January 2007.

242 "Ambition has . . . ,": Shomon discussion with author, June 2005.

243 "Well, you are . . . ,": Jarrett conversation recounted to author, May 2005.

243 "He's really pretty excited . . . ,": Michelle Obama quoted in Eric Zorn column, *Chicago Tribune,* March 2004.

243 "The most surprising . . . ,": Axelrod interview with author, June 2005.

245 "Truthfully, it feels . . . ,": Leslie Corbett, quoted in "Obama Enters New Ring," *Chicago Tribune,* Mendell/Jon Yates, March 2004.

245 "When I first . . . ,": Deborah Landis interview with author, March 2004.

245 "Tonight surely . . . ,": Jesse Jackson Sr., at Obama victory party, Chicago, March 2004.

246 "At its best . . . ,": Obama victory speech, Chicago, March 2004.

CHAPTER 18: A DASH TO THE CENTER

248 "Barack was . . . ,": Cauley discussion with author, April 2004.

249 "markets fail . . . ,": Obama at various press interviews, April–November 2004.

249 "Obama figures out . . . ,": David Wilhelm interview with author, January 2005.

249 "all I have to say . . . ,": Obama discussion with author, March 2004.

250 "I don't think . . . ,": Obama, and *Tribune* story reference, "Obama Has Center in Sights," *Chicago Tribune,* Mendell, April 2004.

251 "I didn't find . . . ,": Obama interview with author, April 2004.

251 "It's impossible . . . ,": Obama aide interview with author, January 2005.

252 "Robert stepped . . . ,": Jarrett interview with author, May 2005.

254 "This is someone . . . ,": Obama discussion with author, April 2005.

254 "Robert is . . . ,": former Obama aide discussion with author, October 2006.

255 "Plain and simple . . . ,": Cauley interview with author, December 2004.

255 Obama-Gibbs conversation: Gibbs recounted to author, November 2006.

256 "I told Barack . . . ,": Gibbs interview with author, July 2005.

256 "You know, it would . . . ,": Obama discussion with author, March 2004.

257 "Absolutely . . . ,": Jarrett interview with author, May 2005.

257 "Nice having you . . . ,": Obama discussion with author, March 2004.

258 "He was fine . . . ,": Peter Coffey interview with author, January 2005.

259 "You really didn't . . . ,": Shomon interview with author, January 2005.

259 "He knows . . . ,": Jarrett interview with author, May 2005.

CHAPTER 19: THE RYAN FILES

261 "left of Mao Tse-tung . . . ,": Steve Rauschenberger, quoted in various media, March 2004.

261 "soft on crime . . . ,": Kirk Dillard interview with author, September 2004.

262 "Meet my stalker . . . ,": Obama speaking to various media, April 2004.

263 Ryan divorce files: various media reports, June 2004.

264 "I've tried to make . . . ,": Obama interview with author, June 2004.

264 "I wanna . . . ,": Obama conversation with aides, June 2004.

265 "I don't think . . . ,": Gibbs discussion with author, June 2004.

265 "I don't take any pleasure . . . ,": Obama at a press conference in Peoria, June 2004.

265 "Do you think . . . ,": Radio reporter in questioning Obama, Peoria, June 2004.

266 "all of us . . . ,": Obama speaking to reporters, Chicago, June 2004.

267 "He's tired . . . ,": Gibbs discussion with author, June 2004.

268 "There's no other . . . ,": Rahul Sangwan, "Jack Ryan a Conservative Idealist," *The Dartmouth Independent,* October 4, 2004.

269 "Barack just killed . . . ,": Nate Tamarin interview with author, December 2004.

269 "We wanted . . . ,": Cauley discussion with author, January 2007.

271 "This was not laborious . . . ,": Obama interview with author, August 2005.

CHAPTER 20: THE SPEECH

273–285 Obama's convention day material: David Mendell, adapted from "Obama Finding Himself Flush with Media Attention," *Chicago Tribune,* July 28, 2004.

274 Tim Russert interview: *Meet the Press,* NBC News, July 2004.

275 Oomph quote: Ryan Lizza, "The Natural," *Atlantic Monthly,* July 2004.

275 "Chicagoans have grown . . . ,": Joe Frolik, "A Newcomer to the Business of Politics Has Seen Enough to Reach Some Conclusions about Restoring Voters' Trust," *Cleveland Plain Dealer,* August 1996.

277 More oomph: Scott Fornek, "Is Oomph an Oops for Obama?" *Chicago Sun-Times,* July 2004.

277 "Is that good . . . ,": Obama conversation with author, Boston, July 2004.

277 "You're the . . . ,": Bob Schieffer conversation with Obama, Boston, July 2004.

278 "Talk to . . . ,": Obama conversation with Schieffer, Boston, July 2004.

278 "We brought her . . . ,": Obama at press conference, Boston, July 2004.

279 "When am I . . . ,": Dick Kay conversation with Green, Boston, July 2004.

279 "I can't . . . ,": Obama speech to League of Conservation Voters, Boston, July 2004.

280 "Even that last try . . . ,": Practice speech observer conversation with author, July 2006.

281–284 various DNC speech quotes: Obama, Boston, July 2004.

283 "Of all . . . ,": Axelrod interview with author, December 2004.

284 "I'm gonna . . . ,": Obama, as told to author by Katz, June 2005.

285 "This is a . . . ,": Jeff Greenfield, as recounted by Axelrod and Gibbs, December 2004.

285 "When I looked . . . ,": Green conversation with author, March 2005.

285 "I was shivering . . . ,": Chris Matthews, *Hardball,* July 28, 2004.

285 "electrified . . . ,": Wolf Blitzer, CNN coverage, July 28, 2004.

285 "I guess . . . ,": Obama, as recounted by Green to author, March 2005.

CHAPTER 21: BACK TO ILLINOIS

286 "I don't want . . . ,": Cauley, Obama quote recounted to author, August 2004.

287 "This was supposed . . . ,": Obama discussion with author, August 2004.

287 David Mendell, material adapted from "Heady Week Yields to Hard Work," *Chicago Tribune,* August 2004.

287 "I don't intend . . . ,": Obama interview with author, August 2004.

288 "In just a few . . . ,": Axelrod interview with author, October 2004.

288 "Apparently . . . ,": Obama to audience in Kewanee, Illinois, August 2004.

288 "This is all . . . ,": Obama discussion with author, August 2004.

289 "I vote for . . . ,": David Bramson interview with author, August 2004.

289 "I've learned . . . ,": Obama discussion with author, August 2004.

290 "I can't go . . . ,": Dick Durbin conversation with Obama, August 2004.

291 "Those little small . . . ,": Obama to audience in Quincy, Illinois, August 2004.

291 "We've lost . . . ,": Mike Daly conversation with campaign aides, August 2004.

292 "When are we . . . ,": Obama conversation with Michelle Obama, August 2004.

292 "What's in . . . ,": Michelle Obama conversation with Obama, August 2004.

292 "We've finally kind of . . . ,": campaign aide conversation with author, August 2004.

293 "This is not . . . ,": Michelle Obama interview with author, August 2004.

293 "They're both . . . ,": campaign aide conversation with author, August 2004.

293 "This is what . . . ,": Michelle Obama conversation with author, August 2004.

294 "It's embarrassing . . . ,": Obama conversation with author, August 2004.

294–295 Smoking anecdote: Tommy Vietor conversation with author, August 2004.

296 "knock the halo . . . ,": Obama, *The Audacity of Hope.*

296–298 Bud Billiken parade material: David Mendell, adapted from "Billiken Crowd Jeers Keyes, Cheers Obama," *Chicago Tribune,* August 2004.

297 "At one point . . . ,": Mike Signator conversation with author, August 2004.

298 "Barack Obama has . . . ,": Liam Ford/David Mendell, "Jesus Wouldn't Vote for Obama, Keyes Says," *Chicago Tribune,* September 2004.

298 "Barack thinks . . . ,": Cauley conversation with author, September 2004.

298 "You know, Barack . . . ,": Axelrod, recounted by Obama in conversation with author, September 2004.

298 "What could I . . . ,": Obama, *The Audacity of Hope.*

299 "We can't lose . . . ,": Axelrod conversation with author, November 2004.

300 "He hates . . . ,": Vietor conversation with author, November 2004.

300 "It was a thing . . . ,": Chicago political consultant conversation with author, June 2005.

301 "I can't believe . . . ,": Obama, recounted to author by campaign aide, August 2004.

301 "He is the smartest . . . ,": Amanda Fuchs conversation with author, November 2004.

302 "I am not looking . . . ,": Obama, recounted by Cauley during interview with author, December 2004.

CHAPTER 22: THE SENATOR

303 "Congratulations" anecdote: conversation between the Obamas, January 2005.

304 "I'm not a . . . ,": Obama conversation with Jesse Jackson Sr., January 2005.

304 "We are bound . . . ,": Obama to supporters in Washington, D.C., January 2005.

306 "If Mike Tackett . . . ,": Gibbs conversation with author, February and March 2005.

307 "house in order . . . ,": Obama interview with author, December 2005.

308 "I'm writing . . . ,": Obama conversation with author, April 2005.

308 "Does he . . . ,": Rahm Emanuel discussion with author, May 2005.

308 "Barack . . . is tightly . . . ,": Shomon conversation with author, April 2006.

310 "There are probably . . . ,": Obama interview with author, December 2006.

311 "We have enough . . . ," and "If you want to position . . . ,": Chris Lu interview with author, December 2006.

312 "My reticence . . . ,": Tom Stoppard quoted in Daphne Merkin, "Playing with Ideas," *New York Times Magazine,* November 2006.

313 "It was an important . . . ,": Gibbs interview with author, May 2007.

314–315 Knox College speech quotes: Obama to audience at college, June 2005.

315 "It's hard for him . . . ,": Jon Favreau interview with author, December 2006.

317 Katrina quotes: Obama on *This Week with George Stephanopoulos,* ABC News, September 2005.

318 "From slave ships": Cornel West quoted in "World: Exiles from a City," *The Observer,* September 2005.

318 "hull of a slave . . . ," and other Jackson quotes: editorial "Katrina's Racial Storm," *Chicago Tribune,* September 2005.

318–319 "Every speaker . . . ," and other Jackson quotes: interview with author, June 2006.

320 "I think there is . . . ,": Obama interview with author, December 2005.

CHAPTER 23: SOUTH AFRICA

324 "humanly possible . . . ," and "Kenya will just be . . . ,": Gibbs discussion with author, March 2005.

325 "I can't hang . . . ,": Obama conversation with official, August 2006.

326 "But the lawyers . . . ,": Axelrod interview with author, December 2006.

329 "It's not an issue . . . ,": Obama at a press conference, South Africa, August 2006.

330 "It sends this message . . . ,": Zackie Achmat interview with reporters, August 2006.

330 "You're going . . . ,": Desmond Tutu to Obama at a press conference, August 2006.

330 "Well, he stayed . . . ,": Gibbs conversation with author, August 2006.

331 "It was very . . . ,": David Wheeler interview with author, August 2006.

332 "He's a total . . . ,": filmmaker conversation with author, August 2006.

333 "It would look . . . ,": U.S. embassy official to reporters, August 2006.

335 "You hear a lot . . . ," and further quotes: Obama at a press conference, August 2006.

336 "That was . . . ,": Pete Souza conversation with author, August 2006.

337 "If it wasn't for . . . ,": Obama statement outside Hector Pieterson Museum, August 2006.

337 "You know . . . ,": Obama conversation with entourage member, August 2006.

CHAPTER 24: NAIROBI

341 "take care of . . . ,": Gibbs conversation with author, Nairobi, August 2006.

341 "You'll be . . . ,": Obama to reporters, August 2006.

342 "How's it . . . ,": Mike Flannery conversation with Gibbs, August 2006.

343 "Oh . . . I'm like . . . ,": Gibbs conversation with Flannery, August 2006.

343 "I'm not sure . . . ,": Gibbs press briefing, August 2006.

343–344 Embassy scene and quotes from press pool report by Laurie Abraham.

344 "Chris Wills . . . ,": Obama to Wills, August 2006.

345 "People have just . . . ,": Dennis Onyango interview with author, August 2006.

346 "This is where . . . ,": Catherine Oganda interview with author, August 2006.

346 "If the Americans . . . ,": John Nyambalo interview with author, August 2006.

347 "May the innocent victims . . . ,": Obama speech at embassy site, August 2006.

348 "Oh my goodness . . . ,": Michelle Obama conversation with author, August 2006.

349 "Be careful . . . ,": Jennifer Barnes conversation with author, August 2006.

350 "This is intense . . . ,": Bob Hercules conversation with author, August 2006.

351 "He is my . . . ,": Gregory Ochieng interview with author, August 2006.

351 "Part of my role . . . ," and subsequent Obama quotes: Obama at a press conference, August 2006.

353 "Where is . . . ,": Obama to reporters, August 2006.

353 "When he is not . . . ,": Abraham, "Mr. Obama Goes to Washington (and Cape Town and Nairobi and . . .)," *Elle* magazine, December 2006.

353 "I feel demeaned . . . ,": filmmaker conversation with author, August 2006.

354 "Kenya is not my . . . ,": Obama at a press conference, August 2006.

CHAPTER 25: SIAYA: A FATHER'S HOME

356 "Obviously . . . ,": Obama to reporters, August 2006.

357 "This is a prelude . . . ,": filmmaker talking under his breath, August 2006.

359 "Stop pushing . . . ,": Obama to a crowd at a hospital, August 2006.

361 "I haven't digested . . . ,": Michelle Obama discussion with reporters, August 2006.

363 "Hopefully, I can . . . ,": Obama in a school dedication ceremony, August 2006.

363 "Does this meet . . . ,": Gibbs conversation with Bill Lambrecht of the *St. Louis Post-Dispatch,* August 2006.

364 "As I was driving . . . ,": Obama speech to Siaya political gathering, August 2006.

365 "It's like . . . ,": Gibbs conversation with author, August 2006.

367 "This is . . . ,": Souza under his breath, August 2006.

367 "Don't trust . . . ,": Obama to reporters outside grandmother's house, August 2006.

367 "It's a . . . ,": military liaison conversation with author, August 2006.

369 "Hello . . . ,": Obama to a crowd in Kibera, August 2006.

370 "Let's walk . . . ,": Gibbs to reporters, August 2006.

371 "Here in Kenya . . . ,": Obama in a speech at University of Nairobi, August 2006.

371 "There are people . . . ,": Onyango interview with author, August 2006.

372 "Barack figured . . . ,": Auma Obama interview with author, August 2006.

373 "It was a little . . . ,": Obama interview with author, December 2006.

373 "Barack and I . . . ,": Michelle Obama interview with author, December 2006.

374 "My father . . . ,": Auma Obama interview with author, August 2006.

CHAPTER 26: LEBRON REVISITED

375 "How bad . . . ,": Tamarin conversation with author, October 2006.

375 "We definitely . . . ,": Gibbs interview with author, May 2007.

376 "The Democrats should . . . ,": Maureen Dowd, "What's Better? His Empty Suit or Her Baggage?" *New York Times,* March 2006.

376 "My attitude is . . . ,": Jacob Weisberg, "Barack Obama: The Path to Power," *Men's Vogue,* September 2006.

376 "I inhaled . . . ,": Obama on NBC's *Tonight Show,* December 2006.

376 "People have always . . . ,": Shomon conversation with author, June 2006.

377 "There are . . . ,": Axelrod interview with author, December 2006.

377 "I saw John Kennedy . . . ,": Newton Minow recounted in interview with author, January 2007.

377–378 Iowa material adapted from "Looking Beyond Obama-mania": David Mendell, "Is He Ready Yet?" *Chicago Tribune,* September 2006.

378 "Oh Oh Oh Obama . . . ,": David Yepsen, "What Triggers the Buzz Around Obama? It's Hope," *Des Moines Register,* September 2006.

379 "committee of scribes . . . ,": Obama, *The Audacity of Hope.*

379 "There was more . . . ,": Obama interview with author, December 2006.

379 "He's not really . . . ,": Penny Reynolds interview with author, December 2006.

380 "The combination . . . ,": Gibbs interview with author, May 2007.

380 "If Barack . . . ,": Butts interview with author, December 2006.

381 "You keep . . . ," and subsequent quotes: Michelle Obama interview with author, December 2006.

382 "I think . . . ,": Obama interview with author, December 2006.

383 Tony Rezko material: David Jackson/Ray Gibson, "Rezko Owns Vacant Lot Next to Obama's Home," *Chicago Tribune,* November 2006.

384 "Look, I came . . . ,": Obama interview with author, December 2006.

385 "He is in fantastic . . . ,": Pete Giangreco interview with author, January 2007.

385 "Freak Show . . . ,": Mark Halperin/John F. Harris, *The Way to Win* (New York: Random House, 2006).

385 "I don't know . . . ,": Axelrod interview with author, December 2006.